de Gruyter Studies in Mathematics 28

de Gruyter Studies in Mathematics

1 Riemannian Geometry, 2nd rev. ed., *Wilhelm P. A. Klingenberg*
2 Semimartingales, *Michel Métivier*
3 Holomorphic Functions of Several Variables, *Ludger Kaup and Burchard Kaup*
4 Spaces of Measures, *Corneliu Constantinescu*
5 Knots, *Gerhard Burde and Heiner Zieschang*
6 Ergodic Theorems, *Ulrich Krengel*
7 Mathematical Theory of Statistics, *Helmut Strasser*
8 Transformation Groups, *Tammo tom Dieck*
9 Gibbs Measures and Phase Transitions, *Hans-Otto Georgii*
10 Analyticity in Infinite Dimensional Spaces, *Michel Hervé*
11 Elementary Geometry in Hyperbolic Space, *Werner Fenchel*
12 Transcendental Numbers, *Andrei B. Shidlovskii*
13 Ordinary Differential Equations, *Herbert Amann*
14 Dirichlet Forms and Analysis on Wiener Space, *Nicolas Bouleau and Francis Hirsch*
15 Nevanlinna Theory and Complex Differential Equations, *Ilpo Laine*
16 Rational Iteration, *Norbert Steinmetz*
17 Korovkin-type Approximation Theory and its Applications, *Francesco Altomare and Michele Campiti*
18 Quantum Invariants of Knots and 3-Manifolds, *Vladimir G. Turaev*
19 Dirichlet Forms and Symmetric Markov Processes, *Masatoshi Fukushima, Yoichi Oshima and Masayoshi Takeda*
20 Harmonic Analysis of Probability Measures on Hypergroups, *Walter R. Bloom and Herbert Heyer*
21 Potential Theory on Infinite-Dimensional Abelian Groups, *Alexander Bendikov*
22 Methods of Noncommutative Analysis, *Vladimir E. Nazaikinskii, Victor E. Shatalov and Boris Yu. Sternin*
23 Probability Theory, *Heinz Bauer*
24 Variational Methods for Potential Operator Equations, *Jan Chabrowski*
25 The Structure of Compact Groups, *Karl H. Hofmann and Sidney A. Morris*
26 Measure and Integration Theory, *Heinz Bauer*
27 Stochastic Finance, *Hans Föllmer and Alexander Schied*

Valerii I. Gromak · Ilpo Laine · Shun Shimomura

Painlevé Differential Equations in the Complex Plane

Walter de Gruyter
Berlin · New York 2002

Authors

Valerii I. Gromak
Department of Mathematics
and Mechanics
Belarussian State University
F. Skorina av. 4
220050 Minsk
Belarus

Ilpo Laine
Department of Mathematics
University of Joensuu
P. O. Box 111
80101 Joensuu
Finland

Shun Shimomura
Department of Mathematics
Keio University
3-14-1 Hiyoshi
Kohoku-ku
Yokohama, 223-8522
Japan

Series Editors

Carlos E. Kenig
Department of Mathematics
University of Chicago
5734 University Ave
Chicago, IL 60637, USA

Andrew Ranicki
Department of Mathematics
University of Edinburgh
Mayfield Road
Edinburgh EH9 3JZ, Scotland

Michael Röckner
Fakultät für Mathematik
Universität Bielefeld
Universitätsstraße 25
33615 Bielefeld, Germany

Mathematics Subject Classification 2000: Primary: 34-02, 34M55; secondary: 34M05, 30D35, 37K35, 39A12

Keywords: Painlevé differential equations, Painlevé transcendents, Bäcklund transformations, Discrete Painlevé equations

♾ Printed on acid-free paper which falls within the guidelines of the ANSI to ensure permanence and durability.

Library of Congress – Cataloging-in-Publication Data

Gromak, V. I. (Valerii Ivanovich)
Painlevé differential equations in the complex plane / Valerii I. Gromak, Ilpo Laine, Shun Shimomura.
p. cm. – (De Gruyter studies in mathematics ; 28)
Includes bibliographical references and index.
ISBN 3 11 017379 4 (cloth : alk. paper)
1. Painlevé equations. 2. Functions of complex variables. I. Laine, Ilpo. II. Shimomura, Shun. III. Title. IV. Series.
QA372 .G775 2002
515'.532–dc21 2002031158

ISBN 3-11-017379-4

Bibliografische Information Der Deutschen Bibliothek

Die Deutsche Bibliothek verzeichnet diese Publikation in der Deutschen Nationalbibliografie; detaillierte bibliografische Daten sind im Internet über <http://dnb.ddb.de> abrufbar.

Printed in Germany.
Cover design: Rudolf Hübler, Berlin.
Typeset using the authors' TEX files: I. Zimmermann, Freiburg.
Printing and binding: Hubert & Co. GmbH & Co. KG, Göttingen.

Preface

This book was originally conceived by the first two authors in 1994, and a preliminary draft has been available for several years. However, recent developments concerning the meromorphic nature of Painlevé transcendents and their value distribution materially affected our plan. At this time, the third author joined us. An important role in the preparation of this book may be attributed to two summer schools at the Mekrijärvi Research Station of the University of Joensuu in August 2000 and 2001. The first of these enabled the entire author team to gather together, and the final book plan took shape during this event. The second summer school in 2001 taught us the important role of discrete Painlevé equations in this field. The second author also acknowledges a visit to the Mittag-Leffler Institute.

Several colleagues offered generous help in this project. In particular, the second author expresses his gratitude to Prof. Aimo Hinkkanen (Urbana) for an intensive collaboration on problems concerning the meromorphic nature of solutions. The third author is grateful to Prof. Kyoichi Takano (Kobe) and Prof. Kazuo Okamoto (Tokyo) for their valuable suggestions on the work of Prof. Masuo Hukuhara, as well as to Prof. Katsuya Ishizaki (Saitama) and Prof. Kazuya Tohge (Kanazawa) for useful discussions about the Nevanlinna theory. Finally, Prof. Jarmo Hietarinta (Turku) has turned our attention to the importance of discrete Painlevé equations.

Several of our friends have been kind enough to comment on parts of the manuscript. In particular, we thank Dr. Janne Heittokangas and Risto Korhonen (Joensuu), who also offered important assistance in checking numerous computations and in assisting us to prepare the figures. We are also grateful for the important assistance of Galina Filipuk (Minsk) in numerical calculations, and for technical assistance in preparing the text. Our special thanks go to Minna Pylkkönen, the present secretary to the second author, who assisted in our seemingly endless communications, and in our continuously changing needs of text-processing with an admirable patience and skill. Of course, the authors retain sole responsibility for any errors and defects which remain in the book.

We would also like to express our gratitude to the publisher for accepting this book in the series Studies in Mathematics and also to its mathematics staff for their expertise, kind assistance and patience.

For financial assistance, we are grateful to the INTAS organization in the first phase of the preparation, to Academy of Finland, to a grant from the Japanese Ministry of Education, Science, Sports and Culture, to the Väisälä Fund of Finnish Academy of Sciences and Letters and to our home universities.

Minsk, Joensuu and Yokohama, September 2002

Valerii I. Gromak
Ilpo Laine
Shun Shimomura

Contents

Introduction

The Painlevé equations were first derived, between 1895–1910, in the investigations by Painlevé and Gambier while studying the following problem originally posed by Picard [1]: Given $R(z, w, w')$ rational in w and w' and analytic in z, what are the second order ordinary differential equations of the form

$$w'' = R(z, w, w') \tag{0.1}$$

with the property that the singularities other than poles of any solution of (0.1) depend on the equation in question only and not on the constants of integration?

Painlevé [1] and Gambier [1] proved that there are fifty canonical equations of the form (0.1) with the property proposed by Picard. This property is known as the Painlevé property, and the differential equations, respectively difference equations, possessing this property are called equations of Painlevé type (P-type). For later presentations of this classical background, see Ince [1] and Bureau [1]. The method introduced by Painlevé to solve the classification problem by Picard was completely different from the classical Fuchs method for solving the similar problem in the case of first order differential equations. The Painlevé method relies on an application of a Poincaré theorem concerning the expansion of solutions in a series of powers of small parameters, called the α-method. By this method, finding necessary conditions for the Painlevé property is relatively easy, while finding sufficient conditions becomes more complicated.

Among the fifty equations obtained, the following six, known as the Painlevé differential equations, appear to be the most interesting ones:

$$w'' = 6w^2 + z, \tag{P_1}$$

$$w'' = 2w^3 + zw + \alpha, \tag{P_2}$$

$$w'' = \frac{(w')^2}{w} - \frac{1}{z}w' + \frac{1}{z}(\alpha w^2 + \beta) + \gamma w^3 + \frac{\delta}{w}, \tag{P_3}$$

$$w'' = \frac{(w')^2}{2w} + \frac{3}{2}w^3 + 4zw^2 + 2(z^2 - \alpha)w + \frac{\beta}{w}, \tag{P_4}$$

$$w'' = \frac{3w-1}{2w(w-1)}(w')^2 - \frac{1}{z}w' + \frac{1}{z^2}(w-1)^2\left(\alpha w + \frac{\beta}{w}\right) + \frac{\gamma w}{z} + \frac{\delta w(w+1)}{w-1}, \tag{P_5}$$

$$\begin{aligned} w'' = {} & \frac{1}{2}\left(\frac{1}{w} + \frac{1}{w-1} + \frac{1}{w-z}\right)(w')^2 - \left(\frac{1}{z} + \frac{1}{z-1} + \frac{1}{w-z}\right)w' \\ & + \frac{w(w-1)(w-z)}{z^2(z-1)^2}\left(\alpha + \frac{\beta z}{w^2} + \frac{\gamma(z-1)}{(w-1)^2} + \frac{\delta z(z-1)}{(w-z)^2}\right) \end{aligned} \tag{P_6}$$

where α, β, γ, δ are arbitrary complex constants. Indeed, the solutions of eleven of other equations of Painlevé type may be expressed in terms of solutions of the six equations above (actually of (P_1), (P_2) or (P_4)), while the 33 remaining equations are solvable in terms of solutions of linear differential equations of second or third order, or in terms of elliptic functions.

In the early history of Painlevé equations, the investigations due to Boutroux [1], [2] and to Garnier [1]–[4] are indispensable, inducing a lot of motivation and influence to modern work in this field. Indeed, the Boutroux' studies on the growth and the asymptotic behavior of Painlevé solutions are in the background of the present corresponding efforts. After Boutroux, it was Garnier who continued studying analytic properties of Painlevé transcendents up to 1960's on a wide scope, including the behavior around fixed singular points, the relation to the Riemann–Hilbert problem and isomonodromic deformations.

Due to the intuitive nature of possible singularities, it seems plausible that all solutions of (P_1), (P_2) and (P_4) would be meromorphic in the complex plane in the sense that all local solutions admit an analytic continuation which is single-valued and meromorphic in $\mathbb{C}$. Similarly, it seems natural that (P_3) and (P_5) could be transformed by $z = \exp t$ to corresponding modified equations $(\widetilde{P}_3)$ and $(\widetilde{P}_5)$ such that all solutions of these modified equations would be meromorphic in $\mathbb{C}$ as well. Finally, since $0, 1, \infty$ are potential fixed critical points for the solutions of (P_6), it is clear that no transformation of type $z = \varphi(t)$, φ entire, would apply to (P_6) to produce meromorphic solutions only. These heuristic ideas arose almost simultaneously as the systematic research on Painlevé equations started, see Painlevé [1], [2], [3]. In fact, Painlevé claimed in [2] that all solutions of (P_1) are meromorphic in the complex plane, proposing a proof. Unfortunately, his original idea remained incomplete in technical details although the idea itself can be saved. This may be seen from a number of later presentations due to Hukuhara [1], Hinkkanen and Laine [1]–[4] and Steinmetz [4], see also Okamoto and Takano [1]. Our Chapter 1 is now devoted to presenting the essential technical contents of these later completions to the original idea of Painlevé to prove the meromorphic nature of solutions of the Painlevé equations.

As all solutions of the Painlevé equations (P_1), (P_2), (P_4), $(\widetilde{P}_3)$ and $(\widetilde{P}_5)$ are meromorphic functions, it is natural to continue by investigating their growth and value distribution. These considerations have been included in Chapter 2 and Chapter 3. Boutroux started, around 1910, to investigate these questions. Credit for more extensive studies of value distribution properties of Painlevé transcendents is due to Schubart and Wittich in 1950's and to Steinmetz about a quarter of century later. However, intensive research in this field really started only very recently. In Chapter 2, we apply the rescaling idea due to Steinmetz to prove that all first Painlevé transcendents are of finite order of growth, while a reasoning due to Shimomura will be applied to prove that these functions are of regular growth. We then make use of the more geometric reasoning due to Shimomura to consider the growth of solutions of (P_2), (P_4), $(\widetilde{P}_3)$ and $(\widetilde{P}_5)$.

In Chapter 3, value distribution of Painlevé transcendents will be studied. While the main part of this chapter is devoted to considering deficiencies and ramification indices as well as to analyzing the contents of the second main theorem with respect to these functions, we also include a section to look at their value distribution with respect to small target functions. This is perhaps the subfield of value distribution of the Painlevé transcendents with the largest number of open problems; we included this section in the book in the hope of prompting new research in this fascinating topic.

Due to the birth history of Painlevé differential equations, much of the fundamental work by Painlevé was made while keeping in mind the relations between the singularity structure and integrability. This connection was first observed by Kowalevskaya [1], [2] in her work on the equations of a spinning top. The essence of these classical investigations was that whenever the movable singularities of a given system of differential equations are nothing but poles, the system turns out to be integrable, in some sense. More recently, the integrability of Painlevé equations has been conformed by the inverse scattering method, originally developed by Gardner et al. [1] in order to solve the Cauchy problem for the Korteweg–de Vries equation. Since there exist several modern presentations about the connection between the singularity structure and integrability, see e.g. Ablowitz, Ramani and Segur [2], [3], including the inverse scattering method, in relation to the Painlevé equations, this aspect will mostly be omitted in this book, while we take another approach in Chapters 4 through 10.

Indeed, it is well-known presently, see e.g. Nishioka [1] and Umemura [1], that the Painlevé differential equations cannot be integrated, in general, in terms of solutions of linear differential equations. Therefore, the Painlevé transcendents define, in a generic sense, new transcendental functions which, in fact, may be considered as some kind of nonlinear special functions. However, for some special values of the equation parameters, the Painlevé equations may be integrated in terms of elementary functions or in terms of some classical transcendental functions such as the Airy, Bessel or hypergeometric functions. In the context of the Painlevé equations, such special solution families are usually constructed by certain recurrence relations, usually called as Bäcklund transformations. Informally, these transformations are understood as systems of equations relating a given solution to another solution of the same equation, possibly with different values of the equation parameters, or perhaps to a solution of another equation. This kind of approach to investigate solutions of Painlevé equations is in the core of Chapters 4 through 9, one chapter devoted to each of the six equations. As a necessary background, we also provide a detailed analysis of the behavior of the Painlevé transcendents in a neighborhood of a singularity. Moreover, for (P_1) and (P_2), we have included an introduction to their higher order analogues, although their investigation is not very developed as of today.

Our final Chapter 10 has been devoted to describe some applications of the Painlevé differential equations. Due to the important role of Painlevé equations in a number of physical applications in the fields of hydrodynamics, plasma physics, nonlinear optics and solid state physics, we first offer a short (and certainly not complete) presentation on connections between partial differential equations and Painlevé equations. In most cases, this connection appears while looking at special solutions of solitary type of the partial differential equations in question. Typical such partial differential equations are the Boussinesq, Korteweg–de Vries and sine-Gordon equations, shortly considered along with some of their modifications. The second part of the last chapter has been devoted to discrete Painlevé equations. Indeed, there has been a lot of interest in integrable mappings and discrete systems recently. Although discrete versions of Painlevé equations seem to be a most natural field of studying, due to the importance of the continuous Painlevé equations, not very have been done yet. In fact, investigations of complex difference equations have not abounded in the last decades. Therefore, this part is relatively short, mostly written as be a reminder of an extremely promising field of research in the borderline of mathematics and its applications.

The two appendices contain (1) basic local existence and uniqueness results for solutions of complex differential equations and (2) elements of the Nevanlinna theory. In both parts, we have restricted ourselves to include such results only that are needed to understand the preceding chapters.

Our list of references is quite extensive, but far from being complete anyway. Indeed, the Mathematical Reviews database today contains about 1900 items with the keyword Painlevé in the item title or the corresponding review text. Clearly, by the same database, the scientific interest in this field of research has been steadily increasing since 1980. Despite of the incompleteness of our bibliography, we have made some effort to include, as far as possible, references known to us which are treating Painlevé differential equations and Painlevé transcendents in the complex domain, while other references have been restricted to the most essential ones only.

Finally, we would like mention here that the major part of the book, Chapters 4 through 10, has been built up on a base draft written by the first author, while the second author, respectively the third author, has contributed the first draft for Chapter 1, Chapter 2 and Appendix A, respectively Chapter 3 and Appendix B. Revising of the numerous subsequent drafts has been made in continuous collaboration between the whole author team. Reading the entire book in all details needs a large number of technical computations. We have mostly omitted such detailed calculations including, however, hints about the line of reasoning to be followed. We have made this decision mostly by space requirements, but also by the standard availability of mathematical software such as Mathematica or Maple.

Chapter 1

Meromorphic nature of solutions

As described in the Introduction, the original idea leading to the six Painlevé equations, namely the lack of movable singularities other than poles for solutions of a second order differential equation in the complex plane, proposes that the solutions might be meromorphic. The first attempt to offer a proof for this was made by Painlevé himself in [2], p. 227–238, for the solutions of (P_1). However, as it is not difficult to observe, the original proof as well as several subsequent proofs in well-known textbooks, see e.g. Bieberbach [1], Golubev [2], Hille [1], Ince [1], appear to be incomplete. For a more detailed analysis, see Hinkkanen and Laine [1], p. 346–348.

The defects in the classical treatments have been observed in Hukuhara [1], see Okamoto and Takano [1]. Unfortunately, these lecture notes have not been widely available outside of Japan. Independently of Hukuhara [1], the meromorphic nature of Painlevé solutions for (P_1), (P_2), (P_4) and for a modified form of (P_3) and (P_5) has been rigorously proved recently, see Hinkkanen and Laine [1], Steinmetz [4], Hinkkanen and Laine [2], [3]. See also Shimomura [11]

The purpose of this chapter is to present the reader an idea of these proofs, detailed enough to enable him/her to work out complete proofs by relying, partially, to the original references mentioned above and to a suitable symbolic software.

This chapter has been organized as follows: For the first Painlevé transcendents, i.e. for solutions of (P_1), we prove their meromorphic nature by giving a modified version of Hukuhara [1], combined with some ideas from Hinkkanen and Laine [1]. For (P_2), we offer here a proof which partially follows the treatment in Hinkkanen and Laine [1], though not given in detail there. The modified forms of (P_3) and (P_5) have been treated in detail in Hinkkanen and Laine [2], [3]. Unfortunately, these proofs are technically complicated. Due to space limitations, they are omitted here, and we only refer to the original papers. To give the reader an idea of the differential inequality technique applied in the case of (P_3), (P_4) and (P_5), we include a complete proof for (P_4), essentially following the pattern of reasoning in Steinmetz [4].

§1 The first Painlevé equation (P_1)

Let $w(z)$ be an arbitrary local solution of (P_1) satisfying the initial conditions $w(z_0) = w_0$, $w'(z_0) = w'_0$, and being defined in a disk centred at the point $z_0 \in \mathbb{C}$. We may assume that $w_0 \neq \infty$, $w'_0 \neq \infty$ for if this is not the case, we may move the point z_0 slightly to achieve this. By the local existence and uniqueness theorem of complex differential equations, see Theorem A.3, $w(z)$ has to be locally single-valued and analytic around z_0. We now proceed to prove

Theorem 1.1. *All local solutions of*

$$w'' = z + 6w^2 \qquad (P_1)$$

can be analytically continued to single-valued meromorphic solutions in the complex plane.

Remark 1. The proof below is essentially based on the lecture notes by Hukuhara [1], while part of the reasoning is that one proposed by Hinkkanen and Laine [1].

Proof of Theorem 1.1, *first part.* Let $w(z)$ be a local solution of (P_1) with $w(z_0) = w_0 \neq \infty$, $w'(z_0) = w_0' \neq \infty$, defined in a disk centred at z_0. Let $B(z_0, R)$ be the largest disk to which $w(z)$ can be continued as a single-valued meromorphic function. By the uniqueness theorem of meromorphic functions, the extended function obviously satisfies (P_1) in $B(z_0, R)$. If $R = \infty$, there remains nothing to prove.

Therefore, assume now that $R < \infty$ in $D := B(z_0, R)$. If, for each $z_1 \in S(z_0, R) := \partial D$, there exists $\delta > 0$ such that $w(z)$ can be continued as a single-valued meromorphic function to the domain $D \cup B(z_1, \delta)$, then we may cover $S(z_0, R)$ by finitely many such disks (by reducing some of the radii, if necessary) so that any disk overlaps exactly two others. The union of D with these disks covers a disk $D_1 = B(z_0, R_1)$ for some $R_1 > R$. If two disks B and B' of the form $B(z_1, \delta)$ overlap, then the extensions of w to $D \cup B$ and $D \cup B'$ agree in $D \cap B \cap B' \neq \emptyset$ so that they agree in $B \cap B'$. We conclude that w may be analytically (more precisely, meromorphically) continued without restriction in D_1. Since D_1 is simply connected, it follows from the monodromy theorem that we have defined w as a single-valued meromorphic function in D_1. This contradicts the definition of R.

Therefore, there exists a point $a \in S(z_0, R)$ for which there is no $\delta > 0$ such that $w(z)$ can be continued as a single-valued meromorphic function from D to $D \cup B(a, \delta)$.

(1) We are now looking for a pair of analytic functions u, v defined in a neighborhood of a possible pole z_1 of w, due towards constructing the necessary auxiliary functions needed along with the subsequent proof. To motivate the selection of u, v, recall that poles are the only singularities which may appear in $B(z_0, R)$. A usual analysis shows that the poles are of double multiplicity. Substituting now the Laurent expansion of $w(z)$ around $z = z_1$ into (P_1), and denoting $\zeta := z - z_1$, we obtain by standard coefficient comparison

$$w(z) = \frac{1}{\zeta^2} - \frac{z_1}{10}\zeta^2 - \frac{1}{6}\zeta^3 + h\zeta^4 + \frac{z_1^2}{300}\zeta^6 + \frac{z_1}{150}\zeta^7 + \sum_{k=8}^{\infty} a_k\zeta^k, \qquad (1.1)$$

where the coefficient h depends on the initial conditions. Moreover,

$$w'(z) = -2\zeta^{-3} - \frac{z_1}{5}\zeta - \frac{1}{2}\zeta^2 + 4h\zeta^3 + \frac{z_1^2}{50}\zeta^5 + \frac{7z_1}{150}\zeta^6 + \sum_{k=8}^{\infty} ka_k\zeta^{k-1}. \qquad (1.2)$$

Taking into account that $\zeta = z - z_1$, we obtain from (1.1) and (1.2),

$$\begin{cases} w(z) = \zeta^{-2} - \frac{z}{10}\zeta^2 - \frac{1}{15}\zeta^3 + h\zeta^4 + \cdots \\ w'(z) = -2\zeta^{-3} - \frac{z}{5}\zeta - \frac{3}{10}\zeta^2 + 4h\zeta^3 + \cdots . \end{cases} \tag{1.3}$$

Since all poles $w(z)$ are double, it is natural to propose such an auxiliary function v that $w(z) = v(z)^{-2}$, at least locally around a pole, say z_1. Taking now a branch of $v(z)$, we obtain from the first equation of (1.3) by the usual formal power series computations that

$$\zeta = v(z)\Big(1 - \frac{z}{20}v(z)^4 - \frac{1}{30}v(z)^5 + \frac{1}{2}hv(z)^6 + \cdots\Big).$$

Substituting this into the second equation of (1.3), we get

$$w'(z) = -2v(z)^{-3} - \frac{1}{2}zv(z) - \frac{1}{2}v(z)^2 + 7hv(z)^3 + \cdots . \tag{1.4}$$

Now, this leads us to propose the auxiliary functions u, v defined by

$$w = v^{-2}, \quad w' = -2v^{-3} - \frac{1}{2}zv - \frac{1}{2}v^2 + uv^3. \tag{1.5}$$

Invoking (P_1), we obtain

$$\begin{cases} u' = \frac{1}{8}z^2v + \frac{3}{8}zv^2 + \left(\frac{1}{4} - zu\right)v^3 - \frac{5}{4}uv^4 + \frac{3}{2}u^2v^5 \\ v' = 1 + \frac{1}{4}zv^4 + \frac{1}{4}v^5 - \frac{1}{2}uv^6. \end{cases} \tag{1.6}$$

It is immediate to see that whenever u, v are meromorphic functions satisfying (1.6), then $w = v^{-2}$ is a solution of (P_1).

(2) In addition to u and v, we need to consider

$$\Phi(z) := (w')^2 + \frac{w'}{w} - 4w^3 - 2zw. \tag{1.7}$$

It is now a straightforward computation to see that

$$\Phi(z) = -\frac{1}{4w^2} + \frac{z^2}{4w} + \frac{u^2}{w^3} - \frac{zu}{w^2} - 4u. \tag{1.8}$$

In fact, it suffices to substitute (1.5) into (1.7) and make use of $w = v^{-2}$. Multiplying both sides of (P_1) by $2w'$, we observe that

$$\frac{d}{dz}\left(\Phi(z) - \frac{w'(z)}{w(z)}\right) = -2w(z).$$

Using (P_1) once more, it is immediate to verify that $\Phi(z)$ satisfies the non-homogenous first order linear differential equation

$$\Phi'(z) + \frac{1}{w(z)^2}\Phi(z) = -\frac{z}{w(z)} + \frac{w'(z)}{w(z)^3}. \tag{1.9}$$

Let now Γ be the line segment joining z_0 and a. We may assume that Γ avoids zeros of $w(z)$ modifying Γ by small half-circles around the zeros of $w(z)$ on Γ. In fact, it suffices to choose the radii of these half-circles so that the modified line segment remains rectifiable. For simplicity, we continue to speak of Γ as a line segment. Let $\Gamma(z_0, t)$ denote the initial part of Γ from z_0 to t. We now solve (1.9) along $\Gamma(z_0, t)$. By partial integration, we get

$$\begin{aligned}\Phi(z) = E(z_0, z)^{-1}\Big[\Phi(z_0) - \frac{E(z_0, z)}{2w(z)^2} + \frac{1}{2w(z_0)^2} \\ - \int_{\Gamma(z_0,z)} \frac{E(z_0, t)}{2w(t)^4}(2tw(t)^3 - 1)\,dt\Big],\end{aligned} \tag{1.10}$$

where

$$E(z_0, t) = \exp\Big(\int_{\Gamma(z_0,t)} w(\tau)^{-2}\,d\tau\Big). \tag{1.11}$$

By (1.10) and (1.11) we immediate conclude

Lemma 1.2. *If $|w(z)|^{-1}$ is bounded on Γ, then $\Phi(z)$ is bounded on Γ.* □

(3) To prove the assertion of Theorem 1.1, denote

$$A := \liminf_{\Gamma \ni z \to a} |w(z)|,$$

and consider three possible cases separately as follows.

Case I. $0 < A < +\infty$.

By Lemma 1.2, $\Phi(z)$ is bounded on Γ near $z = a$, since $|w(z)| \geq \frac{A}{2} > 0$ there. Take now a sequence (z_n) on Γ, $z_n \to a$, such that $w(z_n)$ and $w(z_n)^{-1}$ remain bounded as $n \to \infty$. Regarding (1.7) with $z = z_n$ as a quadratic equation for $w'(z_n)$ we observe that $w'(z_n)$ is bounded as $n \to \infty$. Applying now the standard Cauchy estimates reasoning, see Corollary A.4, to the pair

$$\begin{cases} w' = g \\ g' = 6w^2 + z \end{cases}$$

of differential equations we conclude that $w(z)$ is analytic around $z = a$, which is a contradiction.

Case II. $A = +\infty$.

In this case, $\lim_{\Gamma \ni z \to a} w(z) = \infty$, and so, by Lemma 1.2, $\Phi(z)$ is bounded on Γ near $z = a$. Recalling that $w = v^{-2}$, we may write (1.8) in the form

$$\Phi(z) = -\frac{1}{4}v^4 + \frac{1}{4}z^2v^2 - u(zv^4 + 4 - uv^6).$$

Therefore, $u(zv^4+4-uv^6)$ is bounded on Γ near $z = a$. If $u(z_n)$ is bounded near $z = a$ for a sequence $z_n \to a$, the Cauchy estimates method may be applied to the pair (1.6) to conclude that $w(z)$ has a pole at $z = a$, a contradiction. Hence $\lim_{\Gamma \ni z \to a} u(z) = \infty$, implying that $\lim_{\Gamma \ni z \to a} uv^6 = 4$. Writing $u(zv^4 + 4 - uv^6) = uv^6(\frac{z}{v^2} + \frac{4}{v^6} - u)$, we observe that

$$g := u - \frac{4}{v^6} - \frac{z}{v^2} \tag{1.12}$$

remains bounded on Γ as $z \to a$. Differentiation gives

$$g' = u' + \frac{24v'}{v^7} - \frac{1}{v^2} + \frac{2zv'}{v^3}. \tag{1.13}$$

In the second equation of (1.6), we may express u in terms of g and v by using (1.12). This results in

$$v' = -1 - \frac{1}{4}zv^4 + \frac{1}{4}v^5 - \frac{1}{2}gv^6. \tag{1.14}$$

Substituting (1.14) into (1.13), taking the first equation of (1.6) and using (1.12) again to write u in terms of v and g, we obtain

$$g' = \frac{1}{8}z^2v - \frac{3}{8}zv^2 + \frac{1}{4}v^3 + zgv^3 - \frac{5}{4}gv^4 + \frac{3}{2}g^2v^5. \tag{1.15}$$

As v and g are bounded on Γ near $z \to a$, the Cauchy method applies once more to (1.14) and (1.15) implying that $w(z)$ has a pole at $z = a$, a contradiction. □

Remark 2. Before proceeding to the final case, we remark that the above argument remains valid for any rectifiable arc $\Gamma \subset B(z_0, R)$ from z_0 to a avoiding zeros of $w(z)$.

Lemma 1.3. *Suppose that for a first Painlevé transcendent $w(z)$ and a point $c \in B(z_0, R)$, $|w(c)| \le 1/5$ and $|w'(c)| \ge 3|c|^{1/2} + 12$. Then $w(z)$ is analytic and bounded in $|z - c| < |w'(c)|^{-1}$, and $|w(z)| \ge 1/4$ on the circle $|z - c| = \frac{1}{2}|w'(c)|^{-1}$.*

Proof. We make the following change of variables for (P_1): $\rho := w'(c)$, $\xi := \rho(z-c)$, $\eta(\xi) := w(z)$. Denoting by $\dot{}$ the differentiation with respect to ξ, (P_1) transforms into

$$\ddot{\eta} = 6\rho^{-2}\eta^2 + \rho^{-3}\xi + \rho^{-2}c, \quad \eta(0) = w(c), \quad \dot{\eta}(0) = 1.$$

Integrating twice (from 0 to ξ), we obtain

$$\eta(\xi) = \eta(0) + \dot{\eta}(0)\xi + \xi^2 g(\xi), \tag{1.16}$$

where

$$g(\xi) := 6\rho^{-2}\xi^{-2}\int_0^\xi\Big(\int_0^\tau \eta(t)^2\,dt\Big)d\tau + \frac{1}{2}\rho^{-2}\Big(c+\rho^{-1}\frac{\xi}{3}\Big). \tag{1.17}$$

Define now $\delta_0 := \sup\{\delta \mid |\eta(\xi)| < 2 \text{ for } |\xi| < \delta\}$ and assume, for a while, that $\delta_0 < 1$. Since $|\rho| \geq 3|c|^{1/2} + 12$, we obtain from (1.16) and (1.17)

$$|g(\xi)| \leq 6\cdot 12^{-2}|\xi|^{-2}\cdot 2|\xi|^2 + \frac{|c|+1/3}{2(9|c|+12^2)} < \frac{1}{6}$$

and

$$|\eta(\xi)| < \frac{1}{5} + 1 + \frac{1}{6} < \frac{3}{2}$$

for all $|\xi| < \delta_0$. This is in contradiction to the definition of δ_0, implying that we have $\delta_0 \geq 1$. Therefore, $|\eta(\xi)| < 2$, $|g(\xi)| < \frac{1}{6}$ hold in $|\xi| < 1$. Invoking (1.16) and (1.17) once more, we observe that

$$|\eta(\xi)| \geq |\xi| - |\eta(0)| - |\xi|^2|g(\xi)| \geq \frac{1}{2} - \frac{1}{5} - \Big(\frac{1}{2}\Big)^2\frac{1}{6} > \frac{1}{4},$$

on the circle $|\xi| = \frac{1}{2}$. Expressing these inequalities in terms of the original variables, we obtain the conclusion. □

Proof of Theorem 1.1, *second part.* We now complete the proof of Theorem 1.1. By the notations of the first part of the proof, this amounts to proving

Case III. $A = 0$.

Consider the subset $\Gamma_0 := \{z \in \Gamma \mid |w(z)| \leq 1/5\}$ of Γ. By the condition $A = 0$, Γ_0 extends to $z = a$. We may also assume that $|w'(z)| \geq 3R^{1/2} + 12 \geq 3|z|^{1/2} + 12$ for $z \in \Gamma_0$, at least close enough to $z = a$. In fact, if this is not true, then there exists a sequence (z_n) on Γ_0, $z_n \to a$, such that $\big(w(z_n)\big)$ and $\big(w'(z_n)\big)$ are bounded, and we may apply the final reasoning of Case I to conclude that $w(z)$ is analytic around $z = a$, a contradiction.

Consider now that part of Γ which is close enough to $z = a$ in the above sense, and let $z = c_1$ be the first point in Γ_0. By Lemma 1.3, we find a semicircle Γ_1, defined by $|z - c_1| = \frac{1}{2}|w'(c_1)|^{-1}$, $\operatorname{Im}\big((z - c_1)/(a - c_1)\big) \geq 0$, such that $|w(z)| \geq 1/4$ on Γ_1 and that $w(z)$ is analytic and bounded on $|z - c_1| < |w'(c_1)|^{-1}$. The point $z = a$ cannot be located inside of $|z - c_1| = |w'(c_1)|^{-1}$, since the analytic continuation of $w(z)$ over $z = a$ along Γ would follow. Replace now the "diameter" $[c_1^-, c_1^+] \subset \Gamma$ by the semicircle Γ_1, see Figure 1.1, and call this modified path $\widetilde{\Gamma}_1$. Obviously, on the part of $\widetilde{\Gamma}_1$, from the starting point to c_1^+, $|w(z)| \geq 1/5$. Restart from $z = c_1^+$, and let $z = c_2$ be the first point on Γ, beyond c_1^+ and in Γ_0. Again we find a semicircle Γ_2 : $|z - c_2| = \frac{1}{2}|w'(c_2)|^{-1}$, $\operatorname{Im}\big((z-c_2)/(a-c_2)\big) \geq 0$, with the "diameter" $[c_2^-, c_2^+] \subset \Gamma$. Let $\Gamma_2' \subset \Gamma_2$ be the arc on Γ_2 with end-points $\widetilde{c}_2^-, c_2^+$ on $\widetilde{\Gamma}_1$, see Figure 1.1 again,

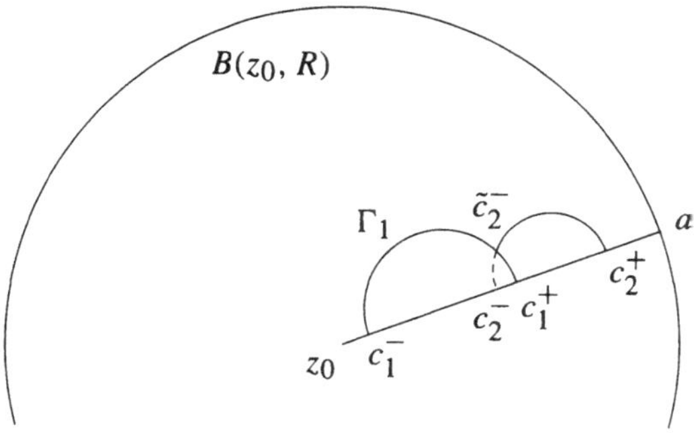

Figure 1.1.

and denote $\widetilde{\Gamma}_2$ to be the path obtained by replacing the part on $\widetilde{\Gamma}_1$ from $\tilde{c}_2^-$ to c_2^+ by Γ_2'. Repeating this process inductively we obtain a modified path $\widetilde{\Gamma}(= \widetilde{\Gamma}_\infty)$, with the corresponding sequence (c_n) contained in Γ_0. Observe that this process cannot terminate in finitely many steps by the same argument as to Γ_1. Moreover, (c_n) converges to $z = a$. In fact, if $c_n \to c_0 \in \Gamma \setminus \{a\}$ then the radius of Γ_n satisfies $\frac{1}{2}|w'(c_n)|^{-1} \leqq |c_{n+1} - c_n| \to 0$ as $n \to \infty$. Hence, $|w(c_0)| \leq 1/5$, $|w'(c_0)| = \infty$, contradicting the fact that $w(z)$ is meromorphic at $z = c_0$. Finally, the modified path $\widetilde{\Gamma}$ is rectifiable, its length being $\sigma(\widetilde{\Gamma}) \leqq (1 + 2\pi)\sigma(\Gamma)$, see Figure 1.1 once more. Hence, by Remark 2 before Lemma 1.3, Case III now reduces back to either Case I or Case II, since $|w(z)| \geq 1/5$ on $\widetilde{\Gamma}$, and we arrive at the final contradiction. □

§2 The second Painlevé equation (P_2)

Let now $w(z)$ be an arbitrary local solution of (P_2) satisfying the initial conditions $w(z_0) = w_0$, $w'(z_0) = w_0'$ defined at least in a disk centred at $z_0 \in \mathbb{C}$. Again, we may assume that $w_0 \neq \infty$, $w_0' \neq \infty$. We now prove

Theorem 2.1. *All local solutions of*

$$w'' = 2w^3 + zw + \alpha, \quad \alpha \in \mathbb{C}, \tag{P_2}$$

can be analytically continued to single-valued meromorphic solutions in the complex plane.

Remark. The proof below is again based on the original idea due to Painlevé [2]. In the details, we now follow the reasoning proposed by Hinkkanen and Laine [1]. However, their coefficient estimate technique for the Case V below, see Hinkkanen and Laine [1], p. 372–375, has been replaced by a reasoning through elementary differential inequalities. This idea has been proposed by Langley [1]. We are grateful for his permission to apply this idea here.

The proof below is indirect, being divided in five subcases. The actual contradiction results by Case I and Case II, while Case III, Case IV and Case V eventually go back to Case II. Observe that Case I, Case III, Case IV and Case V form the logical entity exhausting all possibilities. Therefore, Case II can be understood as a technical lemma.

Proof of Theorem 2.1, *first part.* Similarly as to the proof of Theorem 1.1, let $B(z_0, R)$ be the largest disc to which $w(z)$ can be continued as a single-valued meromorphic function, with $R < \infty$, and let $a \in S(z_0, R)$ and $\delta > 0$ be such that $w(z)$ cannot be continued as a single-valued meromorphic function from $B(z_0, R)$ to $B(z_0, R) \cup B(a, \delta)$.

We are now looking for a pair of analytic functions u, v defined in a neighborhood of a possible pole z_1 of w, and satisfying equations similar to the case of (P_1), see (1.6) above. This is due towards constructing the necessary auxiliary functions needed along with the subsequent proof. To motivate the selection of these auxiliary functions, recall that poles are the only singularities which may appear in $B(z_0, R)$. A usual analysis shows that the poles are simple with residue $\mu = \pm 1$, hence $\mu^2 = 1$. Substituting now the Laurent expansion of $w(z)$ around $z = z_1$ into (P_2), we obtain by standard coefficient comparison

$$w(z) = \frac{\mu}{\zeta} - \frac{\mu z_1}{6}\zeta - \frac{\alpha + \mu}{4}\zeta^2 + A\zeta^3 + \frac{\mu + 3\alpha}{72} z_1 \zeta^4 + O(\zeta^5) \tag{2.1}$$

and

$$w'(z) = -\frac{\mu}{\zeta^2} - \frac{\mu z_1}{6} - \frac{\alpha + \mu}{2}\zeta + O(\zeta^2), \tag{2.2}$$

where $\zeta := z - z_1$ and A is an arbitrary complex parameter. Defining now $v := 1/w$, a simple computation results in

$$v = \mu\zeta + \frac{\mu z_1}{6}\zeta^3 + \frac{\alpha + \mu}{4}\zeta^4 + \left(\frac{\mu z_1^2}{36} - A\right)\zeta^5 + \frac{3\alpha + 5\mu}{72} z_1 \zeta^6 + O(\zeta^7). \tag{2.3}$$

Inverting the series (2.3) by the usual formulas for formal power series, we obtain

$$\begin{aligned} \zeta &= \mu v - \frac{\mu z_1}{6} v^3 - \frac{1 + \alpha\mu}{4} v^4 + O(v^5) \\ &= \mu v \left(1 - \frac{z_1}{6} v^2 - \frac{\alpha + \mu}{4} v^3 + O(v^4)\right). \end{aligned} \tag{2.4}$$

Therefore

$$\frac{1}{\zeta} = \frac{\mu}{v}\left(1 + \frac{z_1}{6} v^2 + \frac{\alpha + \mu}{4} v^3 + O(v^4)\right)$$

and

$$\frac{1}{\zeta^2} = \frac{1}{v^2}\left(1 + \frac{z_1}{3}v^2 + \frac{\alpha+\mu}{2}v^3 + O(v^4)\right). \tag{2.5}$$

Inserting (2.4) and (2.5) into (2.2), we obtain

$$w' = -\frac{\mu}{v^2}\left(1 + \frac{z_1}{3}v^2 + \frac{\alpha+\mu}{2}v^3 + O(v^4)\right) - \frac{\mu z_1}{6} - \frac{\alpha+\mu}{2}\mu v + O(v^2).$$

Writing $z_1 = z - \zeta$, and expressing ζ in terms of v again in the constant term, we get

$$w' = -\frac{\mu}{v^2} - \frac{\mu z}{2} + \frac{v}{2} - (1+\alpha\mu)v + O(v^2).$$

This motivates us to define the function u by

$$w' = -\frac{\mu}{v^2} - \frac{\mu z}{2} - \left(\frac{1}{2} + \alpha\mu\right)v + uv^2, \tag{2.6}$$

hence

$$v' = -v^2 w' = \mu + \frac{\mu z}{2}v^2 + \left(\frac{1}{2} + \alpha\mu\right)v^3 - uv^4. \tag{2.7}$$

Differentiating (2.6) we obtain

$$w'' = (w')' = \frac{2\mu v'}{v^3} - \frac{\mu}{2} - \left(\frac{1}{2} + \alpha\mu\right)v' + u'v^2 + 2uvv'. \tag{2.8}$$

Writing (P_2) in the form $w'' = 2v^{-3} + zv^{-1} + \alpha$, equating with (2.8), substituting the right hand side of (2.7) for v', and recalling (2.7), a formal computation results in

$$\begin{cases} u' = \dfrac{\mu z}{2}\left(\dfrac{1}{2} + \alpha\mu\right) - \mu z u v + \left(\dfrac{1}{2} + \alpha\mu\right)^2 v - 3\left(\dfrac{1}{2} + \alpha\mu\right)uv^2 + 2u^2v^2 \\ v' = \mu + \dfrac{\mu z}{2}v^2 + \left(\dfrac{1}{2} + \alpha\mu\right)v^3 - uv^4, \end{cases} \tag{2.9}$$

corresponding to (1.5) in the case of (P_1). To construct the auxiliary functions, we first observe by (P_2) that

$$\left((w')^2 - w^4 - 2\alpha w - zw^2\right)' = -w^2.$$

Since (2.1) implies that

$$w^2 = \frac{1}{\zeta^2} - \frac{z_1}{3} + O(\zeta),$$

we readily obtain

$$(w')^2 - w^4 - 2\alpha w - zw^2 = \frac{1}{\zeta} + \gamma + \frac{z_1}{3}\zeta + O(\zeta^2),$$

where γ is a constant. Therefore, the auxiliary function

$$V := (w')^2 - w^4 - 2\alpha w - zw^2 + \frac{w'}{w} \tag{2.10}$$

is holomorphic in a neighborhood of any pole z_1 of w. In addition, we shall make use of

$$W := \frac{V'}{V} = \frac{w^4 - (w')^2 + zw^2 + \alpha w}{w\big(w(w')^2 - w^5 + w' - 2\alpha w^2 - zw^3\big)}. \tag{2.11}$$

To obtain the form (2.11) of W, the expression of w'' from the equation (P_2) has to be substituted.

We now continue the proof in the absence of singularities other than poles. Let z_0, R, and a be as before. We join z_0 and a by an arc Γ of finite length contained in $B(z_0, R)$, such that Γ avoids the zeros and poles of $w(z)$. This implies that $w(z)$, $u(z)$, $v(z)$ and $V(z)$ are holomorphic along $\Gamma \setminus \{a\}$. We initially join z_0 and a by a line segment Γ. The assumption that Γ avoids the zeros and poles of $w(z)$ can be satisfied by modifying Γ, similarly as to (P_1) above, by small half-circles around the zeros and poles of $w(z)$ on Γ. For simplicity, however, we first speak of Γ as a line segment. Later on, Γ will have to be modified for another reason. The proof now divides in five subcases as follows:

Case I. We first show that if at least one of $|w(z)|$, $|w'(z)|$ is bounded along Γ, then $w(z)$ is analytic at $z = a$, leading to a contradiction. In fact, from

$$w(z) = w(z_0) + \int_{z_0}^{z} w'(t)\,dt,$$
$$w'(z) = w'(z_0) + \int_{z_0}^{z} \big(2w(t)^3 + tw(t) + \alpha\big)\,dt,$$

where the integrals are taken along Γ, we observe that $w(z)$, $w'(z)$ are simultaneously bounded along Γ. Writing (P_2) as the pair of differential equations

$$\begin{cases} w' = g \\ g' = 2w^3 + zw + \alpha \end{cases} \tag{2.12}$$

for $w(z)$ and $g(z)$, the Cauchy estimates imply that w, g must be analytic around $z = a$.

Before proceeding, we again recall that the above argument applies, if $|w(z_n)|$, $|w'(z_n)|$ are bounded for a sequence (z_n) on Γ, tending to $z = a$.

Case II. We next prove that if $|w(z)|$ is unbounded on Γ while $|u(z)|$ stays bounded, then $z = a$ is a pole of $w(z)$. This assumption means that for some fixed $A < \infty$ and for any pre-assigned $\varepsilon > 0$, there are points z_1 on Γ in any neighborhood of $z = a$ small enough where $|u(z_1)| < A$ and $|v(z_1)| < \varepsilon$. At such a point z_1, the pair (2.9) has a solution $u(z)$, $v(z)$ which takes on pre-assigned values $u(z_1)$, $v(z_1)$ at $z = z_1$

such that the functions $u(z)$, $v(z)$ are holomorphic in a disk $|z - z_1| < r$, where r has a positive lower bound. We may take r to be a fixed constant independent of z_1 (under the conditions $z_1 \in \Gamma$, $|u(z_1)| < A$, $|v(z_1)| < \varepsilon$). This follows from Corollary A.4. If $|z_1 - a| < r$, then $u(z)$ and $v(z)$ are holomorphic at $z = a$. Hence $w = 1/v$ is well defined in a neighborhood of a. Since $\liminf |v(z)| = 0$ as $z \to a$ along Γ and $\lim_{z\to a} v(z)$ exists, it follows that $v(a) = 0$. Thus $w(z)$ has a pole at $z = a$, and a contradiction again follows.

Case III. We now assume that there is a sequence of points $\{z_n\}$ on Γ such that $z_n \to a$, $|w(z_n)| \to \infty$, and $|V(z_n)| < A$ for some finite A. Note that $|v(z_n)| \to 0$. From (2.10) and the second equation in (2.9) we obtain

$$\begin{aligned} V &= \frac{(v')^2 - 1 - 2\alpha v^3 - zv^2 - v'v^3}{v^4} \\ &= \frac{z^2}{4} - 2\mu u + \alpha z v - \mu z u v^2 - 2\alpha\mu u v^3 + \left(\alpha^2 - \frac{1}{4}\right) v^2 + u^2 v^4 \\ &= \frac{z^2}{4} + \alpha z v + \left(\alpha^2 - \frac{1}{4}\right) v^2 + \frac{uv^4}{v^4}(uv^4 - 2\mu - \mu z v^2 - 2\alpha\mu v^3) \qquad (2.13) \\ &= \frac{z^2}{4} + \alpha z v + \left(\alpha^2 - \frac{1}{4}\right) v^2 + u(uv^4 - 2\mu - \mu z v^2 - 2\alpha\mu v^3) \\ &= \frac{z^2}{4} + \alpha z v + \left(\alpha^2 - \frac{1}{4}\right) v^2 + u(-2\mu - \mu z v^2 - 2\alpha\mu v^3) + u^2 v^4. \end{aligned}$$

The last expression in (2.13) offers a quadratic equation for u. Solving this equation, we get

$$2uv^4 = 2\mu + \mu z v^2 + 2\alpha\mu v^3 \pm \sqrt{\Phi},$$

where

$$\Phi = (2\mu + \mu z v^2 + 2\alpha\mu v^3)^2 + \left[4V - z^2 - 4\alpha z v - 4\left(\alpha^2 - \frac{1}{4}\right) v^2\right] v^4.$$

Therefore, uv^4 remains bounded on $\{z_n\}$ as $n \to \infty$. Since $|v(z_n)| \to 0$, the third last expression of (2.13) implies that either $uv^4 \to 0$ or $uv^4 \to 2\mu \neq 0$ on $\{z_n\}$ as $n \to \infty$.

If $uv^4 \to 0$, then $u^2v^4 = o(|u|)$ and (2.13) gives $u\big(-2\mu + o(1)\big) = O(1)$. Hence u is bounded on z_n, and this situation reduces back to Case II. Therefore, it suffices to consider the subcase $uv^4 \to 2\mu$ in more detail. We now conclude from the second last expression of (2.13) that

$$u(uv^4 - 2\mu - \mu z v^2 - 2\alpha\mu v^3)$$

remains bounded $\{z_n\}$, and so

$$h = v^{-4}(uv^4 - 2\mu - \mu z v^2 - 2\alpha\mu v^3)$$

as well. Writing

$$h = u - \frac{2\mu}{v^4} - \frac{\mu z}{v^2} - \frac{2\alpha\mu}{v}, \tag{2.14}$$

we may differentiate (2.14) to obtain

$$h' = u' - \frac{\mu}{v^2} + \left(\frac{2\alpha\mu}{v^2} + \frac{2\mu z}{v^3} + \frac{8\mu}{v^5}\right) v'. \tag{2.15}$$

Expressing u in terms of h and v from (2.14), we rewrite the second equation of (2.9) in the form

$$v' = -\mu - \frac{1}{2}\mu z v^2 + \left(\frac{1}{2} - \alpha\mu\right) v^3 - hv^4. \tag{2.16}$$

Substituting (2.16) into (2.15) and the first equation of (2.9) for u', expressing u in terms of h and v from (2.14), and noting that $\mu^2 = 1$, we finally obtain

$$h' = \left(\frac{\alpha}{2} - \frac{\mu}{4}\right) z + \left(\frac{1}{4} - \alpha\mu + \alpha^2 + h\mu z\right) v + \left(3\alpha h\mu - \frac{3}{2}h\right) v^2 + 2h^2v^3. \tag{2.17}$$

Now, (2.16) and (2.17) is a pair of differential equations for $v(z)$ and $h(z)$. Since h and v remain bounded on $\{z_n\}$, the same reasoning as in Case II above shows that $w(z) = 1/v(z)$ has a simple pole at $z = a$.

Case IV. We now assume that $|w(z)|$ and $V(z)$ are unbounded along Γ as $z \to a$, while $|w(z)| \geq \varepsilon > 0$ on Γ. The idea is to construct a sequence $\{z_n\}$ on Γ such that $z_n \to a$, $|w(z_n)| \to \infty$ and $|V(z_n)| \to 0$. This means that Case IV reduces back to Case III.

Now if $|V(z)|$ is unbounded on Γ, the function $W(z)$ of (2.11) must also be unbounded since

$$V(z) = V(z_0) \exp\left[\int_{z_0}^{z} W(t)\,dt\right].$$

Hence there is a sequence $\{z_n\}$ on Γ with $z_n \to a$ and $|W(z_n)| \to \infty$. At least one of the two sequences $\{w(z_n)\}$ and $\{w'(z_n)\}$ has to be unbounded, since otherwise the argument under Case I would apply, and $w(z)$ would be holomorphic at $z = a$, contradicting the assumption that $w(z)$ is unbounded on Γ. If now $\{w(z_n)\}$ were bounded, $\{w'(z_n)\}$ must be unbounded and we may assume that $\lim_{n\to\infty} |w'(z_n)| = \infty$. Writing now (2.11) in the form

$$W = \frac{-1 + \frac{1}{(w')^2}(w^4 + zw^2 + \alpha w)}{w^2 + \frac{w}{w'} + \frac{1}{(w')^2}(-w^6 - 2\alpha w^3 - zw^4)},$$

we immediately conclude

$$\limsup_{n\to\infty} |W(z_n)| \leqq \left[\liminf_{n\to\infty} |w(z_n)|\right]^{-2}.$$

This is finite by the assumption that $|w(z)| \geq \varepsilon > 0$ on Γ, contradicting the assumption that $|W(z_n)| \to \infty$. It follows that $|w(z_n)| \to \infty$ as $z_n \to a$. From (2.10) and (2.11), we obtain

$$\begin{aligned} w^2 V W &= w^4 - (w')^2 + zw^2 + \alpha w = -(V - w'/w) - \alpha w \\ &= (w'/w) - V - \alpha w, \end{aligned}$$

so that $w' = w^3 V W + w V + \alpha w^2$ and $w'/w = w^2 V W + V + \alpha w$. Substituting these into (2.10), we obtain

$$\begin{aligned} (w^6 W^2 + 2w^4 W + w^2)V^2 &+ (2\alpha w^5 W + 2\alpha w^3 + w^2 W)V \\ &+ (\alpha^2 - 1)w^4 - zw^2 - \alpha w = 0. \end{aligned}$$

This may be written as a quadratic equation

$$h_1 V^2 + h_2 V + h_3 = 0$$

for $V(z)$, where $h_1 = w^2(w^2 W + 1)^2$ and

$$\frac{h_2}{h_1} = \frac{2\alpha w^5 W + 2\alpha w^3 + w^2 W}{w^2(w^2 W + 1)^2}, \qquad \frac{h_3}{h_1} = \frac{(\alpha^2 - 1)w^4 - zw^2 - \alpha w}{w^2(w^2 W + 1)^2}.$$

It is immediate to see that h_3/h_1 and h_2/h_1 tend to zero on the sequence (z_n) as $n \to \infty$. Therefore, we observe that $V(z_n) \to 0$, and this subcase now reduces back to Case III.

Case V. In this final case, $w(z)$ and $V(z)$ are unbounded on Γ as $z \to a$, while

$$\liminf_{\Gamma \ni z \to a} |w(z)| = 0.$$

Now, the only properties of Γ that matter in Case IV are that Γ has a finite length and avoids the zeros and poles of $w(z)$. Hence, we are free to modify Γ to another arc $\widetilde{\Gamma}$ as long as these properties are preserved. Therefore, the idea in this case is to construct a modified arc $\widetilde{\Gamma}$ such that

$$\liminf_{\widetilde{\Gamma} \ni z \to a} |w(z)| > 0.$$

Let $\varepsilon > 0$ be given. By assumption, there is a sequence of points $\{z_n\}$ on Γ such that $|w(z)| < \varepsilon$ for $z \in \Gamma_k$, where Γ_k is the subarc of Γ joining z_{2k-1} to z_{2k}, for $k \geq 1$, while $|w(z)| \geq \varepsilon$ on the remaining subarcs of Γ. Consider now the sequence $\{|w'(z_n)|\}$. None of its infinite subsequences can be bounded, for, if there were such a bounded subsequence, then the method of Case I would apply (since also $|w(z_n)| \leq \varepsilon$ for all n) and $w(z)$ would be holomorphic at $z = a$, contradicting the assumption that $|w(z)|$ is unbounded on Γ. Hence, given any small $\kappa > 0$, we may assume that

$$|w'(z_n)| > 1/\kappa \quad \text{for all } n > N.$$

We set $w(z_n) = w_n$, $w'(z_n) = 1/\zeta_n$ and consider the differential equation for the local inverse $z = f(w)$,

$$-f''(w) = (2w^3 + wf(w) + \alpha)\big(f'(w)\big)^3 \tag{2.18}$$

at $w = w_{2k-1}$ with the initial conditions

$$f(w_{2k-1}) = z_{2k-1}, \quad f'(w_{2k-1}) = \zeta_{2k-1}.$$

In fact, the inverse function $f(w)$ of $w(z)$ exists in a neighborhood of w_n, since $w'(z_n) \neq 0$. □

To continue, we now apply a differential inequality technique due to Langley [1]. This presents a considerable improvement to the complicated reasoning in Hinkkanen and Laine [1].

Lemma 2.2 (Langley). *Given $M \geq \max(16, |\alpha|)$, let f be a solution of*

$$-f''(w) = (2w^3 + wf(w) + \alpha)f'(w)^3$$

in a neighborhood of ξ, where

$$|\xi| \leq 1, \quad |f(\xi)| \leq M, \quad |f'(\xi)| \leq \frac{1}{2}.$$

Then, f has a continuation to a single-valued analytic function on the disc $B\left(\xi, \frac{1}{12M}\right)$. Moreover, if $0 < \sigma < \frac{1}{12M}$, then

$$\left| \frac{f'(s)}{f'(t)} - 1 \right| < 3Me\sigma \tag{2.19}$$

for all $s, t \in B(\xi, \sigma)$.

Proof. Let first ρ be the supremum for all $t > 0$ such that f is analytic with $|f'(w)| \leq 1$ on $B(\xi, t)$. Clearly, $\rho > 0$ and we may assume $\rho \leqq 1$, for otherwise there is nothing to prove concerning the first assertion. Next we find $w^* \in B(\xi, \rho)$ such that $|f'(w^*)| \geq \frac{3}{4}$. In fact, if this is not the case, then $|f'(w)| \leq \frac{3}{4}$ and

$$|f(w)| = \left| f(\xi) + \int_\xi^w f'(s)\, ds \right| \leqq M + \tfrac{3}{4}\rho < M + 1$$

hold on $B(\xi, \rho)$. By the argument based on Cauchy estimates as in Case I, f could now be extended to an analytic function with $|f'(w)| \leq 1$ on a disc $B(\xi, \rho')$ with $\rho' > \rho$, a contradiction.

Now, on $B(\xi, \rho)$ we have

$$|f'(w)| \leq 1, \quad |f(w)| \leqq M + \rho \leqq M + 1.$$

By (2.18), this implies

$$|f''(w)| \leqq 2|w|^3 + |w||f(w)| + |\alpha| \leq 2 + M + 1 + M \leq 3M.$$

Therefore,

$$\frac{1}{4} \leqq |f'(w^*) - f'(\xi)| = \left| \int_\xi^{w^*} f''(s)\,ds \right| \leqq 3M\rho$$

and so $\rho \geq \frac{1}{12M}$, proving the first assertion.

To prove (2.19), we first observe that, for $|z| \leqq 1$,

$$|e^z - 1| \leqq e|z|. \tag{2.20}$$

In fact,

$$|e^z - 1| = \left| \sum_{j=1}^{\infty} \frac{1}{j!} z^j \right| = |z| \left| \sum_{j=1}^{\infty} \frac{1}{j!} z^{j-1} \right|$$
$$\leqq |z| \sum_{j=0}^{\infty} \frac{1}{(j+1)!} |z|^j \leqq |z| \sum_{j=0}^{\infty} \frac{1}{(j+1)!} \leqq e|z|.$$

Take now $0 < \sigma < \frac{1}{12M} \leqq \rho$ and $w, s, t \in B(\xi, \sigma)$. By (2.18), we now obtain

$$\left| \frac{f''(w)}{f'(w)} \right| \leqq \big(2|w|^3 + |w||f(w)| + |\alpha|\big)|f'(w)|^2 \leqq 3M.$$

Therefore,

$$\left| \log \frac{f'(s)}{f'(t)} \right| = |\log f'(s) - \log f'(t)| = \left| \int_t^s \frac{f''(u)}{f'(u)}\,du \right|$$
$$\leqq \int_t^s \left| \frac{f''(u)}{f'(u)} \right| du \leqq 3M\sigma \big(< \tfrac{1}{4} \big).$$

By (2.20), we conclude that

$$\left| \frac{f'(s)}{f'(t)} - 1 \right| = \left| e^{\log \frac{f'(s)}{f'(t)}} - 1 \right| \leqq e \left| \log \frac{f'(s)}{f'(t)} \right| \leqq 3Me\sigma,$$

proving Lemma 2.2. □

Proof of Theorem 2.1, *second part.* We are now prepared to complete the proof of Theorem 2.1. Following the reasoning below might be easier, if the reader makes use of Figure 2.1 while reading the proof. Recall that we have a sequence (z_n) on Γ such that $z_n \to a$, $|w(z_n)| \to \infty$ and $|W(z_n)| \to \infty$, while $\liminf_{\Gamma \ni z \to a} |w(z)| = 0$, and we would like to modify Γ into a rectifiable arc $\widetilde{\Gamma}$ from z_0 to a such that

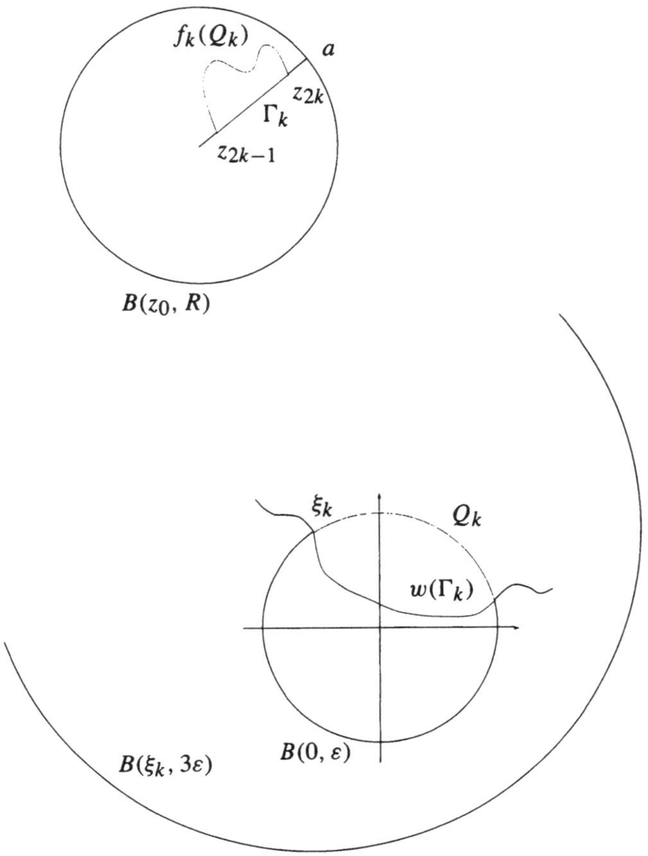

Figure 2.1.

$\liminf_{\widetilde{\Gamma}\ni z\to a} |w(z)| \geq \varepsilon > 0$. Fix now $M := \max(16, |\alpha|, |z_0| + R)$ and $\varepsilon > 0$ such that $3\varepsilon < \frac{1}{12M}$. Let f_k denote the branch of the inverse function of w, mapping $\xi_k := w(z_{2k-1})$ to z_{2k-1}. Since $\lim_{n\to\infty} |w'(z_n)| = \infty$, we may assume $|f_k'(\xi_k)| \leqq \frac{1}{2}$. Obviously, $|f_k(\xi_k)| \leq M$, and $|\xi_k| = \varepsilon \leq 1$. By Lemma 2.2, f_k extends to be analytic on $B(\xi_k, \frac{1}{12M})$, and so on $B(\xi_k, 3\varepsilon) =: B_k$ in particular. By (2.19),

$$\left| \frac{f_k'(w_1)}{f_k'(w_2)} - 1 \right| \leq 3Me\sigma < \frac{e}{4} < 0,75$$

on B_k. Hence, f_k is univalent on B_k, mapping B_k onto a simply connected domain D_k

containing z_{2k-1}. Moreover,

$$\left|\frac{f_k'(w_1)}{f_k'(w_2)}\right| < 2$$

on B_k, and therefore

$$\left|\frac{w'(z_1)}{w'(z_2)}\right| < 2$$

for all $z_1, z_2 \in D_k$. We also observe that $B(0, 2\varepsilon) \subset B_k$.

Now $w(\Gamma_k)$ is a path in B_k, starting at ξ_k, and $f_k(w(z)) = z$ close to z_{2k-1}. Therefore, f_k maps $w(\Gamma_k)$ to Γ_k. So the arc Γ_k lies is D_k, as well as the arc $f_k(Q_k)$, where Q_k is the shorter arc of the circle $|w| = \varepsilon$ joining $w_1 := w(z_{2k-1}) = \xi_k$ to $w_2 := w(z_{2k})$, and $f_k(w_2) = z_{2k}$. Since $|w'(z)| \leqq 2|w'(z_{2k-1})|$ for all $z \in D_k$, the length L_k of Γ_k satisfies

$$L_k \geq |w_2 - w_1|/(2|w'(z_{2k})|) = \frac{1}{2}|w_2 - w_1||f_k'(\xi_k)|.$$

On the other hand, $|f'(w)| \leqq 2|f'(\xi_k)|$ for all $w \in B_k$, and so the length $\widetilde{L}_k$ of $f_k(Q_k)$ is at most

$$\widetilde{L}_k \leqq \frac{\pi}{2}|w_2 - w_1| \cdot 2|f_k'(\xi_k)| \leq 2\pi L_k.$$

Hence, replacing each Γ_k by $f_k(Q_k)$, we obtain a modified path $\widetilde{\Gamma}$ whose length is at most 2π times the length of Γ, hence finite. By construction, $\liminf_{z\to a, z\in\widetilde{\Gamma}} w(z) \geq \varepsilon > 0$.

Applying now the argument of Case IV to $\widetilde{\Gamma}$, considering $\widetilde{\Gamma}$ close enough to a, we conclude that w extends meromorphically to a neighborhood of a, and we have the final contradiction. However, we have to take into account the possible need of constructing a Riemann surface around the modified path $\widetilde{\Gamma}$ before applying the final argument, see Hinkkanen and Laine [1], p. 367, for details. □

§3 The third Painlevé equation (P_3)

The third Painlevé equation

$$w'' = \frac{(w')^2}{w} - \frac{w'}{z} + \frac{\alpha w^2 + \beta}{z} + \gamma w^3 + \frac{\delta}{w}, \qquad (P_3)$$

where $\alpha, \beta, \gamma, \delta \in \mathbb{C}$, has a behavior different of (P_1) and (P_2) with respect to the meromorphic nature of solutions. In fact, if $\beta = \gamma = 0$, and $\alpha\delta \neq 0$, then it is immediate to see that

$$w(z) = Cz^{1/3}, \quad \text{where } C^3 = -\delta/\alpha,$$

is a non-meromorphic solution of (P_3). However, by the transformation

$$W(\zeta) = zw(z) = e^{\zeta/2}w(e^{\zeta/2}), \quad z = e^{\zeta/2}, \tag{3.1}$$

the meromorphic nature of all solutions can be achieved. In fact, replacing $(\alpha, \beta, \gamma, \delta)$ with $(4\alpha, 4\beta, 4\gamma, 4\delta)$ in the transformed equation, and writing again w, z instead of W, ζ, we obtain

Theorem 3.1. *All local solutions of the modified third Painlevé equation*

$$w'' = \frac{(w')^2}{w} + \alpha w^2 + \gamma w^3 + \beta e^z + \frac{\delta e^{2z}}{w} \tag{$\widetilde{P}_3$}$$

can be analytically continued to single-valued meromorphic solutions in the complex plane.

For a detailed proof of Theorem 3.1, we only refer to Hinkkanen and Laine [2]. For what follows later on in Chapter 7, it is useful to observe here as well that the pairs of coefficients (α, γ) and$(-\beta, -\delta)$ are in a symmetric role. In fact, multiplying $(\widetilde{P}_3)$ by $1/w$ we get

$$\left(\frac{w'}{w}\right)' = \alpha w + \gamma w^2 + \frac{\beta e^z}{w} + \frac{\delta e^{2z}}{w^2}. \tag{3.2}$$

By an additional transformation $\frac{1}{t} = e^{-z}w$, we obtain

$$\left(\frac{t'}{t}\right)' = -\beta t - \delta t^2 - \frac{\alpha e^z}{t} - \frac{\gamma e^{2z}}{t^2}, \tag{3.3}$$

showing the anticipated symmetry. Moreover, defining

$$V := \frac{1}{2}\left(\frac{w'}{w}\right)^2 - \alpha w - \frac{1}{2}\gamma w^2 + \frac{\beta e^z}{w} + \frac{1}{2}\frac{\delta e^{2z}}{w^2}, \tag{3.4}$$

and combining (3.2) and (3.4), we immediately obtain

$$V' = \left(\frac{w'}{w}\right)' - \alpha w - \gamma w^2 = \frac{\beta e^z}{w} + \frac{\delta e^{2z}}{w^2}. \tag{3.5}$$

Therefore, by (3.5), the auxiliary function $V(z)$ reduces to a constant as soon as $\beta = \delta = 0$ (and similarly, if $\alpha = \gamma = 0$). Actually, this case reduces back to a Riccati equation, see Hinkkanen and Laine [2], p. 326–327, for details.

§4 The fourth Painlevé equation (P_4)

The fourth Painlevé equation

$$w'' = \frac{1}{2}\frac{(w')^2}{w} + \frac{3}{2}w^3 + 4zw^2 + 2(z^2 - \alpha)w + \frac{\beta}{w}, \tag{P_4}$$

where $\alpha, \beta, \in \mathbb{C}$, again behaves similarly as to (P_1) and (P_2) with respect to the meromorphic nature of solutions. In fact, we apply the idea of reasoning in Steinmetz [4] to prove

Theorem 4.1. *All local solutions of* (P_4) *can be analytically continued to single-valued meromorphic solutions in the complex plane.*

Proof of Theorem 4.1, *first part.* In this proof, the division in subcases is not completely the same as in Hinkkanen and Laine [1], [2], see §2. However, the basic idea of the proof is the same, by letting $B(z_0, R)$ be the largest disk to which a local solution $w(z)$ around z_0 can be analytically continued as a single-valued meromorphic function, and deducing a contradiction from $R < \infty$. To this end, let $a \in S(z_0, R)$ be a point for which there exists no $\delta > 0$ such that $w(z)$ can be continued as a single-valued meromorphic function from $B(z_0, R)$ to $B(z_0, R) \cup B(a, \delta)$, and let Γ be the line segment $[z_0, a)$, possibly slightly modified to avoid zeros, poles and one-points of $w(z)$.

Case I. In this case, assume that there exists $M > 1$ and a sequence (z_n) on Γ with $z_n \to a$ such that

$$\frac{1}{M} \le |w(z_n)| \le M \quad \text{and} \quad |w'(z_n)| \le M. \tag{4.1}$$

Then we may write (P_4) as a pair of first order differential equations

$$\begin{cases} w' = g \\ g' = \dfrac{1}{2}\dfrac{g^2}{w} + \dfrac{3}{2}w^3 + 4zw^2 + 2(z^2 - \alpha)w + \dfrac{\beta}{w}, \end{cases} \tag{4.2}$$

with bounded right hand sides on the sequence (z_n). Therefore, by the standard Cauchy estimates reasoning, w must be analytic around $z = a$, a contradiction. □

Auxiliary functions and path modification. In $B(z_0, R)$, we may now define a meromorphic function $V(z)$ by

$$V := \frac{1}{4w}\big((w')^2 - w^4 - 4zw^3 - 4(z^2 - \alpha)w^2 + 2\beta\big). \tag{4.3}$$

Differentiating and making use of (P_4), we immediately see that

$$V' = -w^2 - 2zw. \tag{4.4}$$

Moreover, we define another meromorphic function $U(z)$ in $B(z_0, R)$ by

$$U := V + \frac{w'}{w - 1}. \tag{4.5}$$

As a preparation for the path modification, we first prove two lemmas.

Lemma 4.2. *If* $\liminf_{\Gamma \ni z \to a} |w(z) - 1| > 0$, *then* U *is bounded on* Γ.

Proof. We first proceed to find four polynomials, Q_1, Q_2, Q_3, Q_4 in the variables z and $t := (w-1)^{-1}$ such that

$$\big(U' - (Q_1 + Q_2 U)\big)^2 = Q_3 + Q_4 U. \tag{4.6}$$

To this end, observe that

$$U' = V' + \Big(\frac{w'}{w-1}\Big)' = -w^2 - 2zw + \frac{w''}{w-1} - \Big(\frac{w'}{w-1}\Big)^2. \tag{4.7}$$

By (4.3) and (4.5),

$$(w')^2 = w^4 + 4zw^3 + 4(z^2 - \alpha)w^2 - 2\beta + 4w\Big(U - \frac{w'}{w-1}\Big), \tag{4.8}$$

which may be used to solve w' as

$$w' = -\frac{2w}{w-1} \pm \bigg\{\Big(\frac{4w^2}{(w-1)^2} + w^4 + 4zw^3 + 4(z^2-\alpha)w^2 - 2\beta\Big) + 4wU\bigg\}^{1/2}. \tag{4.9}$$

Substituting now into (4.7) w'' from (P_4), $(w')^2$ from (4.8) and w' from (4.9), and simplifying the result by a suitable mathematical software, we obtain (4.6) with $t := w - 1$ and

$$\begin{cases} Q_1 := \frac{1}{t^4}\big((-1-2z)t^4 + (-2+4\alpha-6z-4z^2)t^3 \\ \qquad + (-5+4\alpha+2\beta-4z-4z^2)t^2 - 12t - 8\big), \\ Q_2 := -\frac{2}{t^2}(t+2), \\ Q_3 := \frac{4(t+2)^2}{t^8}\big(t^6 + 4(1+z)t^5 + (6-4\alpha+12z+4z^2)t^4 \\ \qquad + 4(1-2\alpha+3z+2z^2)t^3 + (5-4\alpha-2\beta+4z+4z^2)t^2 \\ \qquad + 8t + 4\big), \\ Q_4 := \frac{16}{t^6}(4 + 8t + 5t^2 + t^3). \end{cases} \tag{4.10}$$

Recalling that $Q_j(z)$, $j = 1, \dots, 4$, remain bounded on Γ, we may apply (4.6) and (4.10) to conclude that

$$|U'(z)| \le K_1|U(z)| + K_2$$

on Γ for some constants $K_1 \ge 0$, $K_2 \ge 0$. Therefore,

$$\frac{d|U|}{K_1|U| + K_2} \leqq \frac{|dU|}{K_1|U| + K_2} \le dz,$$

and so $|U|$ is bounded on Γ due to the rectifiability of Γ. □

Lemma 4.3. *Let $\Phi(w,z)$ be a polynomial in both variables and $z = f(w)$ be the local solution of the initial value problem*

$$f''(w) = -\frac{f'(w) + \Phi\big(w, f(w)\big)f'(w)^3}{2w}, \quad f(w_0) = z_0, \quad f'(w_0) = \eta, \tag{4.11}$$

where $w_0 \neq 0$ and $0 < |\eta| < 1$. If $|z_0| \leq K$ and $\frac{1}{K} \leqq |w_0| \leq K$, then there exists $\delta = \delta(K, \Phi) > 0$ such that $z = f(w)$ exists in $|w - w_0| \leq \frac{1}{2}|w_0|$ and that $\frac{1}{2}|\eta| \leq |f'(w)| \leq 2|\eta|$, provided $0 < |\eta| < 1$.

Proof. Define $\delta > 0$ to be the largest radius to satisfy

$$\delta \leq \frac{1}{2}|w_0|, \quad f(w) \text{ exists in } |w - w_0| < \delta \text{ and } |f(w) - z_0| < 1. \tag{4.12}$$

Moreover, define $M = M(K)$ as

$$M := \sup\left\{|\Phi(w,z)| \;\Big|\; |w - w_0| \leq \frac{1}{2}|w_0|,\ |z - z_0| \leq 1,\ |z_0| \leq K,\ \frac{1}{K} \leq |w_0| \leq K\right\}.$$

Since $|w| \geq |w_0| - |w - w_0| \geq \frac{1}{2}|w_0|$ in $|w - w_0| < \delta$, we observe that

$$|f''(w)| \leq \frac{|f'(w)| + M|f'(w)|^3}{|w_0|} \tag{4.13}$$

in $|w - w_0| < \delta$. Consider now an arbitrary, but fixed radius $\{w = w_0 + re^{i\alpha} \mid \alpha \in \mathbb{R}\}$ and define

$$v(r) := |f'(w_0 + re^{i\alpha})|.$$

Since

$$|v'(r)| = \left|\frac{d}{dr}|f'(w_0 + re^{i\alpha})|\right| \leq |f''(w_0 + re^{i\alpha})| \tag{4.14}$$

we may combine (4.13) and (4.14) to obtain

$$-\frac{v + Mv^3}{|w_0|} \leq v' \leq \frac{v + Mv^3}{|w_0|}, \quad v(0) = |\eta|. \tag{4.15}$$

Observe now that the initial value problems

$$y' = \pm\frac{y + My^3}{|w_0|}, \quad y(0) = |\eta|$$

are solved (uniquely) by

$$y_{\pm}(r) = \pm|\eta|\frac{e^{\frac{r}{|w_0|}}}{\left\{1 + M|\eta|^2 - M|\eta|^2 e^{2\frac{r}{|w_0|}}\right\}^{1/2}}. \tag{4.16}$$

By standard differential inequality techniques, see Walter [1], Theorem XI, p. 69, we conclude that

$$y_-(r) \le v(r) \le y_+(r). \tag{4.17}$$

By (4.16) it readily follows that

$$y_+(r) \le |\eta| e^{\frac{r}{|w_0|}} (1 - Me|\eta|^2)^{-1/2}$$

in $0 \le r \le \frac{1}{2}|w_0|$, provided $0 < |\eta| < (4eM)^{-1/2}$. Therefore, by (4.17), we infer that

$$|f'(w)| \le \frac{|\eta| e^{1/2}}{\sqrt{3/4}} < 2|\eta|, \quad w = w_0 + re^{i\alpha},$$

whenever $0 \le r < \delta \le \frac{1}{2}|w_0|$. Moreover, by radial integration,

$$|f(w) - z_0| \le 2|\eta|\delta \le |\eta||w_0| \le |\eta| K$$

for $|w - w_0| < \eta$. Assuming now η small enough to satisfy

$$|\eta| < \min\left((4eM)^{-1/2}, 1/K\right),$$

we have obtained

$$|f'(w)| < 2|\eta|, \quad |f(w) - z_0| < 1 \tag{4.18}$$

in $|w - w_0| < \eta$. Writing now (4.11) as the pair of differential equations

$$\begin{cases} f' = g \\ g' = -\dfrac{g + \Phi(w, f)g^3}{2w} \end{cases}$$

we may apply the standard Cauchy estimates technique, taking into account (4.11) and (4.18), that $f(w)$ continues analytically over the circle boundary $|w - w_0| = \delta$. Therefore, $\delta = \frac{1}{2}|w_0|$. The asserted lower estimate for $|f'(w)|$ similarly follows from

$$\begin{aligned} |f'(w_0 + re^{i\alpha})| \ge y_-(r) &> \frac{y_-(r)}{(1 + My_-^2(r))^{1/2}} \\ &\ge \frac{\eta e^{-r/w_0}}{(1 + M|\eta|^2)^{1/2}} \ge \frac{|\eta|}{(e + 1/4)^{1/2}} > \frac{1}{2}|\eta|. \end{aligned}$$

□

We are now able to proceed to the required path modification, i.e. to modify Γ into a rectifiable curve $\widetilde{\Gamma}$ from z_0 to a such that $\liminf_{\widetilde{\Gamma} \ni z \to a} |w(z) - 1| > 0$ on $\widetilde{\Gamma}$. Of course, this is only needed, if $\liminf_{\Gamma \ni z \to a} |w(z) - 1| = 0$. If $\limsup_{\Gamma \ni z \to a} |w(z) - 1| = 0$, then V' is bounded on Γ by (4.4). Consequently, V is bounded on Γ and, by (4.3), w' as well. Hence, this situation reduces back to Case I. Therefore, we may assume that $\liminf_{\Gamma \ni z \to a} |w(z) - 1| = 0$ and $\limsup_{\Gamma \ni z \to a} |w(z) - 1| > 0$. We now consider

$$\Gamma_\varepsilon := \left\{ z \in \Gamma \mid |w(z) - 1| < \varepsilon \right\},$$

assuming that $\varepsilon \le 1/5$. The set Γ_ε clearly consists of countably many arcs $\Gamma_{\varepsilon,j}$, to be replaced by arcs $\widetilde{\Gamma}_{\varepsilon,j}$ of length $l(\widetilde{\Gamma}_{\varepsilon,j}) \le 4\pi l(\Gamma_{\varepsilon,j})$ such that $|w(z)-1| = \varepsilon$ on $\widetilde{\Gamma}_{\varepsilon,j}$. Therefore, the rectifiability of $\widetilde{\Gamma}$ is obvious.

We may assume that $\lim_{\Gamma_\varepsilon \ni z \to a} |w'(z)| = +\infty$, since otherwise the situation again reduces back to Case I by taking a sequence (z_n) from Γ_ε converging to a. Take now a subarc $\Gamma_{\varepsilon,j}$ close enough to a so that $|w'|$ is sufficiently large on $\Gamma_{\varepsilon,j}$. Given $z_0 \in \Gamma_{\varepsilon,j}$ with $w_0 = w(z_0)$, $w'(z_0) = 1/\eta$, $w(z)$ admits a local inverse $z = f(w)$ satisfying

$$f''(w) = -\frac{1}{2w}\big(f'(w) + (3w^4 + 8zw^3 + 4(z^2-\alpha)w^2 + 2\beta)(f'(w))^3\big),$$

with initial values $f(w_0) = z_0$, $f'(w_0) = \eta$. By Lemma 4.3, $f(w)$ exists in $|w-w_0| \le \frac{1}{2}|w_0|$. Since $|w_0 - 1| \le \varepsilon \le 1/5$, $\frac{1}{2}|w_0| \ge \frac{2}{5}$, hence $f(w)$ exists in $|w-1| \le \varepsilon$ contained in $|w - w_0| \le \frac{1}{2}|w_0|$. Replace now the arc $w(\Gamma_{\varepsilon,j})$ by the shorter arc μ_j on $|w-1| = \varepsilon$ joining the end-points of $w(\Gamma_{\varepsilon,j})$. Then

$$l\big(w(\Gamma_{\varepsilon,j})\big) \le \pi l(\mu_j).$$

By Lemma 4.3 again, if $|w_j - 1| \le \varepsilon$ for $j = 1, 2$, $|f'(w_1)/f'(w_2)| \le 4$. Replace now the original subarc $\Gamma_{\varepsilon,j}$ by $\widetilde{\Gamma}_{\varepsilon,j} := f(\mu_j)$, and we are done with the path modification.

Proof (of Theorem 4.1, second part). We may now proceed under the assumption that Γ has been modified as described above to ensure that $(w-1)^{-1}$ and $U = V + \frac{w'}{w-1}$ are bounded on Γ. Moreover, (4.8) may be rewritten in the form

$$\left(w' + \frac{2w}{w-1}\right)^2 = w^4 + 4zw^3 + 4(z^2-\alpha)w^2 - 2\beta + 4Uw + \frac{4w^2}{(w-1)^2} \tag{4.19}$$

to continue into

Case II (i). Assume that for a sequence (z_n) on Γ, converging to a, $w(z_n) \to 0$ as $n \to \infty$. Moreover, we assume that $\beta \ne 0$ and we fix γ to satisfy $-2\beta = \gamma^2$. By (4.19), $w'(z_n)^2 \to \gamma^2$ as $n \to \infty$. Taking a subsequence, if needed, we may assume that $w'(z_n) \to \gamma$ as $n \to \infty$. We now define u by

$$w' = \gamma(1 + uw). \tag{4.20}$$

Therefore, $u(z_n)w(z_n) \to 0$ as $n \to \infty$. By (4.5), $V(z_n)$ remains bounded on (z_n). Combining (4.3) and (4.20) we obtain

$$-2\beta u(2 + uw) = w^3 + 4zw^2 + 4(z^2-\alpha)w + 4V,$$

and so $u(z_n)$ remains bounded on (z_n). Differentiating now (4.20) and applying (P_4), resp. (4.20), to eliminate w'', resp. w', we get

$$u' = -\frac{\gamma}{2}u^2 + \frac{1}{2\gamma}\big(3w^2 + 8zw + 4(z^2-\alpha)\big). \tag{4.21}$$

Now, the Cauchy estimates may be applied to (4.20) and (4.21) to see that w may be continued analytically over $z = a$, a contradiction.

Case II (ii). Assume now that $w(z_n) \to 0$ for a sequence (z_n) on Γ converging to a while $\beta = 0$. We now consider v satisfying $v^2 = w$. By (4.8), we deduce

$$\begin{aligned}\left(v' + \frac{v}{v^2-1}\right)^2 &= \frac{(w')^2}{4v^2} + \frac{2vv'}{v^2-1} + \frac{v^2}{(v^2-1)^2} \\ &= \frac{1}{4}w^3 + zw^2 + (z^2-\alpha)w + U + \frac{w}{(w-1)^2},\end{aligned}$$

implying that $v'(z_n)$ remains bounded on (z_n). Substituting (P_4) with $\beta = 0$ into $2vv'' = w'' - 2(v')^2$ we immediately get

$$\begin{cases} v' = g \\ g' = \dfrac{3}{4}v^5 + 2zv^3 + (z^2-\alpha)v, \end{cases}$$

resulting in a contradiction by the Cauchy estimates technique again.

Case III. We now assume that $w(z_n) \to \infty$ for a sequence (z_n) on Γ converging to a. Substituting now $w = \frac{1}{v}$ into (4.19), we infer that

$$\left(v' - \frac{2v^2}{1-v}\right)^2 = 1 + 4zv + 4(z^2-\alpha)v^2 - 2\beta v^4 + 4Uv^3 + \frac{4v^4}{(1-v)^2}. \tag{4.22}$$

Looking at (4.22), we note that $v(z_n) \to 0$ as $n \to \infty$ implies $v'(z_n)^2 \to 1$ as $n \to \infty$. Taking a subsequence, if needed, we may assume that $v'(z_n) \to \delta$, where $\delta = \pm 1$, as $n \to \infty$. By (4.22),

$$\frac{(v')^2 - 1}{v} = 4z + 4(z^2-\alpha)v - 2\beta v^3 + 4Uv^2 + 4\frac{vv'}{1-v}. \tag{4.23}$$

Differentiating $vw = 1$ twice, and substituting (P_4) and (4.23), we get

$$\begin{aligned} v'' &= 2\frac{(v')^2}{v} - v^2\left(\frac{1}{2}\frac{(w')^2}{w} + \frac{3}{2}w^3 + 4zw^2 + 2(z^2-\alpha)w + \frac{\beta}{w}\right) \\ &= \frac{3}{2}\frac{(v')^2 - 1}{v} - 4z - 2(z^2-\alpha)v - \beta v^3 \\ &= 2z + 4(z^2-\alpha)v - 4\beta v^3 + 6Uv^2 + 6\frac{vv'}{1-v}. \end{aligned}$$

Since $v(z_n) \to 0$ and $v'(z_n) \to \delta \neq 0$, we may apply the Cauchy estimates reasoning to

$$\begin{cases} v' = g \\ g' = 2z + 4(z^2-\alpha)v - 4\beta v^3 + 6Uv^2 + 6\dfrac{vg}{1-v}. \end{cases}$$

Therefore, v has an analytic continuation over $z = a$ and $v(a) = 0$. Hence, w continues analytically across $z = a$ with a simple pole at $z = a$, a contradiction again. □

Remark. As in the case of (P_2), the Riemann surface construction around $\widetilde{\Gamma}$ may have to be applied.

§5 The fifth Painlevé equation (P_5)

The fifth Painlevé equation

$$w'' = \left(\frac{1}{2w} + \frac{1}{w-1}\right)(w')^2 - \frac{w'}{z} + \frac{(w-1)^2}{z^2 w}(\alpha w^2 + \beta) + \frac{\gamma w}{z} + \frac{\delta w(w+1)}{w-1}, \quad (P_5)$$

where again $\alpha, \beta, \gamma, \delta \in \mathbb{C}$, behaves similarly as to (P_3) with respect to the meromorphic nature of solutions. As to examples of non-meromorphic solutions, it is noted at once that

$$w(z) = Cz^{1/2} + 1, \quad \text{where } C \neq 0,$$

is a solution of (P_5), provided $\alpha = -\beta = \frac{1}{8}$, $\delta = 0$ and $\gamma = -\frac{1}{8}C^2$. Moreover, in the case of $\alpha = \beta = \gamma = \delta = 0$, i.e. in the reduced case of

$$w'' = \left(\frac{1}{2w} + \frac{1}{w-1}\right)(w')^2 - \frac{w'}{z},$$

non-meromorphic solutions may be found in the form

$$w(z) = \left(\frac{1 + Cz^a}{1 - Cz^a}\right)^2,$$

provided $C \neq 0$ and $a \notin \mathbb{Z}$.

Parallel to the case of (P_3), we apply the transformation $z = e^t$, and revert to write z instead of t, after the transformation. We then obtain

Theorem 5.1. *All local solutions of the modified fifth Painlevé equation*

$$w'' = \left(\frac{1}{2w} + \frac{1}{w-1}\right)(w')^2 + (w-1)^2\left(\alpha w + \frac{\beta}{w}\right) + \gamma e^z w + \frac{\delta e^{2z} w(w+1)}{w-1} \quad (\widetilde{P}_5)$$

can be analytically continued to single-valued meromorphic solutions in the complex plane.

For a detailed proof of this theorem, we refer to Hinkkanen and Laine [3].

§6 The sixth Painlevé equation (P_6)

For the sixth Painlevé equation

$$\begin{aligned} w'' &= \frac{1}{2}\left(\frac{1}{w}+\frac{1}{w-1}+\frac{1}{w-z}\right)(w')^2 - \left(\frac{1}{z}+\frac{1}{z-1}+\frac{1}{w-z}\right)w' \\ &+ \frac{w(w-1)(w-z)}{z^2(z-1)^2}\left(\alpha+\frac{\beta z}{w^2}+\gamma\frac{z-1}{(w-1)^2}+\delta\frac{z(z-1)}{(w-z)^2}\right), \end{aligned} \tag{P_6}$$

perhaps the most detailed presentation for meromorphic nature of its solutions may be found in Hukuhara [2], see also Okamoto and Takano [1]. Another program to this end has been sketched in Steinmetz [4], p. 374–376. For a further treatment in the original spirit of Painlevé [4], see Hinkkanen and Laine [4].

The presentation offered by Hukuhara may be summarized as follows, using the notations $(\alpha_0, \alpha_1, \alpha_\infty, \alpha_z)$ in place of $(-\beta, \gamma, \alpha, \frac{1}{2}-\delta)$ for the coefficients of (P_6):

(1) For an arbitrary solution $w(z)$, consider a line segment Γ terminating in $z = a \in \mathbb{C} \setminus \{0, 1\}$. Supposing that $w(z)$ is continued meromorphically along $\Gamma \setminus \{a\}$, we would like to show that $z = a$ is at most a pole of $w(z)$. In the right-hand side of (P_6), the singular values of $w(z)$ around $z = a$ are $w = 0, 1, a, \infty$. Note that they have a symmetric property implied by the following three transformations

$$z = 1 - Z,\ w = 1 - W; \quad z = 1/Z,\ w = 1/W; \quad z = Z,\ w = Z/W.$$

(2) Let β be a complex number such that $\beta \neq 0, 1, a, \infty$. Take the auxiliary function

$$\begin{aligned} U := &\frac{z(z-1)(w')^2}{2w(w-1)(w-z)} - \frac{\alpha_0}{(z-1)w} + \frac{\alpha_1}{z(w-1)} \\ &- \frac{\alpha_\infty w}{z(z-1)} - \frac{\alpha_z - 1/2}{w-z} - \frac{(z-\beta)w'}{(w-z)(w-\beta)}. \end{aligned}$$

Under the supposition $\liminf_{\Gamma \ni z \to a} |w(z) - \beta| > 0$, the function U is bounded on Γ, and then, by the use of this property, it is shown that $z = a$ is at most a pole of $w(z)$. In this argument, we treat the case where $w(a_k)$ tends to one of the singular values as $a_k \to a$ along Γ, and by the symmetry of the singular values, it is sufficient to examine the case where $w(a_k) \to 0$ as $a_k \to a$.

(3) Finally, the case where $\liminf_{\Gamma \ni z \to a} |w(z) - \beta| = 0$ is treated. By a suitable modification of the path Γ, it is reduced to the case mentioned above. In this process, we employ essentially the same method as in §1, Case III.

(4) The proof is described under the supposition $\alpha_0 \neq 0$. In case $\alpha_0 = 0$, we make the change of variables $w = v^2$, $v' = u$. Then $(w')^2/(2w) = 2u^2$; and (P_6) is equivalent to the pair

$$\begin{cases} v' = u, \\ u' = \dfrac{Q(z, u, v)}{z^2(z-1)^2(v^2-1)^2(v^2-z)^2}, \end{cases}$$

of differential equations, where Q is a polynomial. A similar reasoning as for the case $\alpha_0 \neq 0$ may now be applied to complete the proof.

Chapter 2

Growth of Painlevé transcendents

In this chapter, we proceed to consider the growth of solutions of (P_1), (P_2), (P_4), $(\widetilde{P}_3)$ $(\widetilde{P}_5)$, known to be meromorphic by the preceding Chapter 1. Examining the growth of Painlevé transcendents, Nevanlinna theory plays a background role in our considerations. For all necessary notations and basic results needed, see Appendix B. A twofold approach has been adopted in this chapter. Namely, we make use of the Steinmetz' rescaling method for (P_1), while the more geometric approach due to the third author will be applied for (P_2) and (P_4). Finally, for $(\widetilde{P}_3)$ and $(\widetilde{P}_5)$, we just add a couple of remarks, in addition to a reference to Shimomura [9].

§7 Growth of first Painlevé transcendents

In this section, we prove that all first Painlevé transcendents $w(z)$ are of regular growth of order $\rho(w) = 5/2$. This fact actually goes back to Boutroux [1], [2], although no definition for the order of a meromorphic function was available at that time. See Steinmetz [5] for a more detailed background. The classical paper by Boutroux remained in the category of mathematical folklore until N. Steinmetz [5] and Shimomura [8], [10] recently offered the first rigorous proofs, independently.

Let now $w(z)$ be a fixed solution of (P_1), meromorphic in $\mathbb{C}$ by Theorem 1.1. By Laine [1], p. 178, $w(z)$ is transcendental, with infinitely many poles by the Clunie argument, see Lemma B.11. For each point $z_0 \neq 0$ regular for $w(z)$, we define a local scaling unit by

$$r(z_0) := \min\left\{|w(z_0)|^{-1/2},\ |w'(z_0)|^{-1/3},\ |z_0|^{-1/4}\right\}, \tag{7.1}$$

with the proviso that $0^{-\alpha} = +\infty$ for $\alpha > 0$. The local scaling unit is the key notion in this section.

Lemma 7.1. *Let (z_n) be a sequence of points regular for $w(z)$ with $z_n \to \infty$, and let $r_n := r(z_n)$ be the corresponding local scaling units. Moreover, we assume that the three limits*

$$y_0 := \lim_{n\to\infty} r_n^2 w(z_n), \quad y_0' := \lim_{n\to\infty} r_n^3 w'(z_n), \quad a := \lim_{n\to\infty} r_n^4 z_n \tag{7.2}$$

exist. If y now denotes the unique solution of the initial value problem

$$y'' = 6y^2 + a, \quad y(0) = y_0, \quad y'(0) = y_0', \tag{7.3}$$

then

$$r_n^2 w(z_n + r_n z) \to y(z) \tag{7.4}$$

locally uniformly in $\mathbb{C}$ *as* $n \to \infty$.

Proof. We first define

$$y_n(z) := r_n^2 w(z_n + r_n z). \tag{7.5}$$

Therefore, by (P_1),

$$y_n'' = 6y_n^2 + z_n r_n^4 + r_n^5 z. \tag{7.6}$$

By analytic dependence on initial values and parameters, see e.g. Horn and Wittich [1], p. 141,

$$y_n(z) \to y(z)$$

in a neighborhood of $z = 0$. By a classical result due to Poincaré, see Bieberbach [1], p. 14, the convergence holds locally uniformly in $\mathbb{C}$. □

Before proceeding, we remark that the limit function $y(z)$ does not vanish identically. In fact, if so, then $\max(|y(0)|, |y'(0)|, |a|) = 0$ by (7.3). But (7.2) now implies that

$$\begin{aligned} \max(|y(0)|, |y'(0)|, |a|) &= \max(|y_0|, |y_0'|, |a|) \\ &= \max\left(|\lim_{n\to\infty} r_n^2 w(z_n)|, |\lim_{n\to\infty} r_n^3 w'(z_n)|, |\lim_{n\to\infty} r_n^4 z_n|\right) = 1, \end{aligned}$$

a contradiction.

Lemma 7.2. *Let* (q_n) *be a sequence of zeros of* $w(z)$ *and let* (z_n) *be any other sequence such that* $|z_n - q_n| = o(r(z_n))$. *Then* $r(z_n) \ll r(q_n)$.

Proof. By definition we have $\rho_n := r(q_n) = \min\{|w'(q_n)|^{-1/3}, |q_n|^{-1/4}\}$ since $|w(q_n)|^{-1/2} = \infty$. Note that $r_n^2 w(z_n + r_n\zeta) \to y(\zeta)$ as $n \to \infty$ uniformly for $|\zeta| < \delta_0$, where δ_0 is some positive constant. Hence, by $|z_n - q_n| = o(r_n)$, we may suppose that

$$\sup\left\{r_n^2 |w(z)| \mid z \in [z_n, q_n]\right\} \le M_0 = \sup\{|y(\zeta)| \mid |\zeta| < \delta_0\} + 1,$$

where $[z_n, q_n]$ denotes the line segment joining z_n to q_n. Integrating both sides of $w''(z) = 6w(z)^2 + z$, and using this estimate, we obtain

$$\begin{aligned} |w'(q_n) - w'(z_n)| &\le \left|\int_{z_n}^{q_n} (6w(t)^2 + t)dt\right| \le o(r_n) \cdot \max_{s\in[z_n, q_n]} (|w(s)|^2 + |s|) \\ &\le o(r_n^{-3})\left[\max_{t\in[z_n, q_n]} (r_n^2|w(t)|)^2 + r_n^4(|z_n| + o(r_n))\right] \\ &= o(r_n^{-3}). \end{aligned}$$

Therefore,

$$w'(q_n) = w'(z_n) + o(r_n^{-3}),$$

and, by assumption,

$$q_n = z_n + o(r_n).$$

Observing that $|r_n^3 w'(z_n)| \le 1$, and that $|r_n^4 z_n| \le 1$, we obtain

$$|w'(q_n)| \le |w'(z_n)| + o(r_n^{-3}) \le r_n^{-3}(1 + o(1)),$$

and

$$|q_n| \le |z_n|(1 + o(1)) \le r_n^{-4}(1 + o(1)).$$

Hence,

$$|w'(q_n)|^{-1/3} \ge r_n(1 + o(1)), \quad |q_n|^{-1/4} \ge r_n(1 + o(1)),$$

which implies $r_n \ll q_n$. □

Letting now (q_n) be the sequence of zeros of w, and $\varepsilon > 0$ be given, we denote by

$$Q(\varepsilon) := \bigcup_{n=1}^{\infty} \{z \mid |z - q_n| < \varepsilon r(q_n)\}$$

a neighborhood of the sequence (q_n).

Lemma 7.3. *Given $\varepsilon > 0$, $|w'(z)|/|w(z)|^{3/2}$ is bounded outside of $Q(\varepsilon)$. Moreover, the spherical derivative $w^{\#}(z) := |w'(z)|/(1 + |w(z)|^2)$ is bounded there as well.*

Proof. The last assertion is trivial. In fact, if $|w(z)| \ge 1$, then $1+|w(z)|^2 > |w(z)|^2 \ge |w(z)|^{3/2}$ and if $|w(z)| < 1$, then $1 + |w(z)|^2 \ge 1 > |w(z)|^{3/2}$.

Suppose now, contrary to the assertion, that

$$|w'(z_n)|/|w(z_n)|^{3/2} \to \infty \tag{7.7}$$

for a sequence (z_n) tending to ∞. We may assume that $w(z_n)$ is finite for all z_n. Moreover, taking a subsequence, if needed, we may assume that the limits in (7.2) exist. Therefore, Lemma 7.1 may be applied. Since

$$r_n^2 w(z_n) = \left(r_n^3 |w(z_n)|^{3/2}\right)^{2/3} = o(|w'(z_n)| r_n^3)^{2/3} = o(1)$$

by (7.2) and (7.7), we observe, by (7.2) again, that $y_0 = 0$. Hence, from (7.4) and the classical Hurwitz' theorem, we find a sequence $z_n' \to 0$ with $w(z_n + r_n z_n') = 0$. Denoting now $\tilde{q}_n := z_n + r_n z_n'$ we infer that

$$|z_n - \tilde{q}_n| = o(r_n) = o(r(\tilde{q}_n))$$

by Lemma 7.2. Hence $z_n \in Q(\varepsilon)$ for n large enough, a contradiction. □

A similar reasoning results in

Lemma 7.4. *Given* $\varepsilon > 0$,

$$|z| \ll |w(z)|^2 \quad \text{and} \quad |w'(z)|/|w(z)|^3 \ll |z|^{-3/4}$$

as $z \to \infty$ *outside of* $Q(\varepsilon)$.

Proof. To prove the first assertion, suppose there is a sequence tending to infinity such that

$$|w(z_n)| = o(|z_n|^{1/2}). \tag{7.8}$$

Assuming, as we may, that the limits in (7.2) again exist, we observe that

$$|w(z_n)| = o(r(z_n)^{-2})$$

by (7.8) and (7.2). Therefore, $y_0 = \lim_{n\to\infty} r(z_n)^2 w(z_n) = 0$. A contradiction now follows exactly as in the proof of Lemma 7.3. □

To proceed, we first observe by routine computation that

$$F(z) := -zw(z) - 2w(z)^3 + \frac{1}{2}(w'(z))^2 \tag{7.9}$$

is a primitive function of $-w(z)$. Recalling the Laurent expansion of $w(z)$ at a pole z_0, see (1.1),

$$w(z) = \frac{1}{(z-z_0)^2} - \frac{z_0}{10}(z-z_0)^2 - \frac{1}{6}(z-z_0)^3 + h(z-z_0)^4 + \cdots, \tag{7.10}$$

and substituting (7.10) into (7.9), we get

$$F(z) = \frac{1}{z-z_0} - 14h + \frac{z_0}{30}(z-z_0)^3 + \frac{1}{24}(z-z_0)^4 - \frac{h}{5}(z-z_0)^5 + \cdots. \tag{7.11}$$

Defining now

$$V(z) := -2F(z) - w'(z)/w(z), \tag{7.12}$$

then (7.12) and (7.11) readily imply that

$$V(z_0) = 28h. \tag{7.13}$$

Lemma 7.5. *Given* $\sigma > 0$ *and* $\varepsilon > 0$, *there exists* $K > 0$ *such that for any* $z \in \mathbb{C}\backslash Q(\varepsilon)$, *either* $|V(z)|/|w(z)|^2 \le K|z|^{1/2}$ *or* $|w(z)|^{-2} \le \sigma|z|^{-1}$ *holds. Therefore,*

$$\frac{|V(z)|}{|w(z)|^2} \le \sigma\frac{|V(z)|}{|z|} + K|z|^{1/2}$$

for all $z \in \mathbb{C}\backslash Q(\varepsilon)$.

Proof. Suppose that for a sequence (z_n) in $\mathbb{C}\backslash Q(\varepsilon)$, tending to infinity,

$$0 < \frac{|V(z_n)|}{|w(z_n)|^2|z_n|^{1/2}} \to +\infty$$

as $n \to \infty$. Again, we may assume that the points in (z_n) are no poles of $w(z)$. By Lemma 7.4, $|z_n| \ll |w(z_n)|^2$. By (7.9) and (7.12),

$$(w'(z))^2 = 2zw(z) + 4w(z)^3 - V(z) - w'(z)/w(z). \tag{7.14}$$

Therefore,

$$\frac{|V(z_n)|}{|w(z_n)|^2|z_n|^{1/2}} \le \frac{|w'(z_n)|^2}{|w(z_n)|^2|z_n|^{1/2}} + \frac{4|w(z_n)|}{|z_n|^{1/2}} + \frac{2|z_n|^{1/2}}{|w(z_n)|} + \frac{|w'(z_n)|}{|w(z_n)|^3|z_n|^{1/2}}.$$

Since the last two terms remain bounded as $n \to \infty$, by Lemma 7.4, we must have

$$\frac{|w'(z_n)|^2}{|w(z_n)|^2|z_n|^{1/2}} + \frac{4|w(z_n)|}{|z_n|^{1/2}} \to \infty \tag{7.15}$$

as $n \to \infty$. If now $4|w(z_n)|/|z_n|^{1/2}$ remains bounded as $n \to \infty$, at least in a subsequence, then by Lemma 7.4,

$$\frac{|w'(z_n)|^2}{|w(z_n)|^2|z_n|^{1/2}} = \frac{|w'(z_n)|^2|w(z_n)|}{|w(z_n)|^3|z_n|^{1/2}} \ll \frac{|w(z_n)|}{|z_n|^{1/2}},$$

and so $|w'(z_n)|^2/(|w(z_n)|^2|z_n|^{1/2})$ also remains bounded, contradicting (7.15). Therefore, $|w(z_n)|/|z_n|^{1/2} \to +\infty$ as $n \to \infty$ and so

$$\frac{|V(z_n)|}{|w(z_n)|^2} = \frac{|V(z_n)|}{|z_n|}\frac{|z_n|}{|w(z_n)|^2} = o\left(\frac{|V(z_n)|}{|z_n|}\right)$$

holds and the assertion follows. □

Lemma 7.6. *Given $\varepsilon > 0$ sufficiently small, consider the corresponding neighborhood $Q(\varepsilon)$ of the zero-sequence of $w(z)$. Then there exist $R_0 > 1$ and a sequence of mutually disjoint disks*

$$\Delta_n(\varepsilon) = \{z \mid |z - z_n| < 3\varepsilon r(z_n)\}$$

such that

$$Q(\varepsilon) \cap \{z \mid |z| \ge R_0\} \subset \Delta(\varepsilon) := \bigcup_{n=1}^{\infty} \Delta_n(\varepsilon).$$

Proof. Suppose first that there exists a sequence of disks

$$D_n := \{z \mid |z - z_n| < \eta_n r(z_n)\},$$

where $0 < \eta_n \to 0$, such that D_n contains at least three zeros of w, including z_n. By Lemma 7.1 and the Hurwitz' theorem, the limit function $y(z)$ has at least a triple zero at $z = 0$. By (7.3), this implies that $a = 0$, contradicting the fact that $|a| = 1$, see the remark after the proof of Lemma 7.1. Therefore, for $\varepsilon > 0$ sufficiently small and n sufficiently large, the disk

$$E_n := \{z \mid |z - q_n| < \varepsilon r(q_n)\},$$

has a non-empty intersection with at most one disk $E_m, m \neq n$. The assertion follows by taking $\Delta_n(\varepsilon) = E_n$ in the case of an empty intersection with all $E_m, m \neq n$, and by taking $\Delta_n(\varepsilon)$ to be the larger of $\{z \mid |z - q_n| < 3\varepsilon r(q_n)\}$ and $\{z \mid |z - q_n| < 3\varepsilon r(q_m)\}$ otherwise. □

Lemma 7.7. *Given $\varepsilon > 0$ sufficiently small,*

$$V(z) \ll |z|^{3/2} \tag{7.16}$$

outside of $\Delta(\varepsilon)$. In particular,

$$V(p_n) = O(|p_n|^{3/2}) \tag{7.17}$$

as $n \to \infty$, where (p_n) denotes the sequence of distinct poles of $w(z)$.

Proof. Fix $R_0 > 1$ and let $\Delta(\varepsilon)$ be as in Lemma 7.6. Denote further, for $R > R_0$,

$$M(R) := \max\{|V(z)||z|^{-3/2} \mid R_0 \le |z| \le R,\ z \notin \Delta(\varepsilon)\}.$$

Suppose this maximum will be attained at $z_1 = \rho\exp(i\alpha)$, $R_0 < \rho \le R$. Let L be the radial line segment from $z_0 = R_0\exp(i\alpha)$ to z_1, modified to a path Γ as follows, if needed: If $L \cap \Delta_n(\varepsilon) \neq \emptyset$, we replace that intersection by the shorter part of the boundary $\partial\Delta_n(\varepsilon)$, and if $z_0 \in \Delta_m(\varepsilon)$ for some m, we replace z_0 by a point $L \cap \partial\Delta_n(\varepsilon)$. Obviously, $\int_\Gamma |t|^{1/2}|dt| \le \pi|z_1|^{3/2}$. Clearly, $V(z_0)$ is finite, say $|V(z_0)| \le \gamma$, where γ is independent of R. Moreover,

$$\frac{|V(t)|}{|t|} = \frac{|V(t)|}{|t|^{3/2}}|t|^{1/2} \le M(R)|t|^{1/2} \tag{7.18}$$

on the path of integration Γ.

To obtain a differential equation for $V(z)$, differentiate (7.13), and invoke (P_1) and (7.15) to obtain

$$\begin{aligned} V' &= 2w - \left(\frac{w'}{w}\right)' = 2w - \frac{w''}{w} + \left(\frac{w'}{w}\right)^2 = -\frac{z}{w} - 4w + \left(\frac{w'}{w}\right)^2 \\ &= -\frac{z}{w} - 4w + \frac{1}{w^2}\left(2zw + 4w^3 - V - \frac{w'}{w}\right) = \frac{z}{w} - \frac{w'}{w^3} - \frac{1}{w^2}V, \end{aligned}$$

from which we obtain by integration along Γ

$$V(z_1) = V(z_0) + \int_\Gamma \left(\frac{t}{w(t)} - \frac{w'(t)}{w(t)^3} - \frac{1}{w(t)^2} V(t) \right) dt.$$

Therefore, by Lemma 7.4 and 7.5,

$$|V(z_1)| \leqq |V(z_0)| + \int_\Gamma \left(\sigma \frac{|V(t)|}{|t|} + K|t|^{1/2} \right) |dt|;$$

here we have used the observation that $|t|/|w(t)| = |t|^{1/2} \left(\frac{|t|}{|w(t)|^2} \right)^{1/2} \ll |t|^{1/2}$ remains bounded on Γ by Lemma 7.4 and that $|w'(t)|/|w(t)|^3 = O(|t|^{-3/4})$, hence being bounded as well. Making now use of (7.18), we obtain

$$\begin{aligned} |V(z_1)| &\leqq \gamma + \sigma M(R) \int_\Gamma |t|^{1/2} |dt| + K \int_\Gamma |t|^{1/2} |dt| \\ &\leqq \gamma + \pi \sigma M(R) |z_1|^{3/2} + \pi K |z_1|^{3/2}. \end{aligned}$$

Therefore,

$$\frac{|V(z_1)|}{|z_1|^{3/2}} = M(R) \leqq \frac{\gamma}{|z_1|^{3/2}} + \pi\sigma M(R) + \pi K \leqq \frac{\gamma}{R_0^{3/2}} + \pi\sigma M(R) + \pi K.$$

Hence, selecting σ to satisfy $\pi\sigma < 1/2$, we obtain

$$M(R) \leqq 2\left(\frac{\gamma}{R_0^{3/2}} + \pi K \right) =: B$$

and so

$$\frac{|V(z)|}{|z|^{3/2}} \leqq M(R) \leqq B$$

whenever $z \notin \Delta(\varepsilon)$ and $R > R_0$.

Finally, to see that (7.16) implies (7.17), let z_n now stand for the centres of the disks $\Delta_n(\varepsilon)$. Lemma 7.1 again applies, resulting in the limit function $y(z) = y_0 + y_0' z + \cdots$. Since $y(0) = y_0$ is finite, there are no poles of $y(z)$ in a certain disk around the origin. Therefore, if n is large enough and ε small enough, $\Delta_n(\varepsilon)$ contains no poles of $w(z)$, hence (7.17) is valid by (7.16). □

We now proceed to prove the final lemma to prepare our preliminary growth estimate for first Painlevé transcendents. To this end, recall again the Laurent expansion (1.1) of $w(z)$ at a pole $z = p$, with the notation $\xi = z - p$:

$$w(z) = \sum_{k=0}^{\infty} a_k \xi^{k-2} = \frac{1}{\xi^2} - \frac{p}{10}\xi^2 - \frac{1}{6}\xi^3 + h\xi^4 + \frac{p^2}{300}\xi^6 + \sum_{k=9}^{\infty} a_k \xi^{k-2}. \quad (7.19)$$

Lemma 7.8. *Suppose that the coefficients in* (7.19) *satisfy, for some* $K > 0$,

$$|a_k| \leqq (K|p|^{1/4})^k \tag{7.20}$$

for $k = 0, \dots, 8$. *Then* (7.20) *holds for all* $k \in \mathbb{N}$.

Proof. Substituting (7.19) into (P_1), we obtain by a standard coefficient comparison that

$$\big((k-1)(k-2) - 12\big)a_{k+1} = 6\sum_{j=1}^{k} a_j a_{k+1-j} \tag{7.21}$$

for all $k \geq 8$. Taking moduli and observing that the right-hand side in (7.21) contains at most $k-3$ non-zero terms, we get, inductively,

$$\big((k+2)(k-5)\big)|a_{k+1}| \leqq 6(k-3) \cdot (K|p|^{1/4})^{k+1}.$$

Since $6(k-3)/\big((k+2)(k-5)\big) \leqq 1$ for all $k \geq 8$, the assertion follows. □

Proposition 7.9. *For any first Painlevé transcendent* $w(z)$ *with the distinct pole sequence* (p_n),

$$\sum_{0<|p_n|\leq r} |p_n|^{-1/2} = O(r^2) \tag{7.22}$$

as $r \to \infty$. *Moreover,* $N(r, w) = O(r^{5/2})$ *and* $T(r, w) = O(r^{5/2})$.

Proof. Restricting to poles of sufficiently large modulus, recall that $28h = V(p_n)$ by (7.13). Therefore, by Lemma 7.7,

$$|h| = O(|p_n|^{3/2}).$$

It is now immediate to check that (7.20) holds for $k = 0, \dots, 8$ in (7.19) for some $K > 0$, hence for all $k \in \mathbb{N}$ by Lemma 7.8. By the standard formula to compute the radius of convergence for a power series, we observe that (7.19) converges at least in $0 < |z - p_n| < c|p_n|^{-1/4}$, c being a constant independent of p_n. Therefore, the polar neighborhoods $|z - p_n| < \frac{c}{2}|p_n|^{-1/4}$ must be mutually disjoint (for n sufficiently large, say $|p_n| \geq r_0$). Comparing the total area of these polar neighborhoods with the area of the disk of radius r, centred at the origin, we obtain

$$\sum_{r_0 \leq |p_n| \leq r} \pi \frac{c^2}{4} |p_n|^{-1/2} \leqq \pi(2r)^2,$$

proving (7.22). Moreover, since $|p_n|^{-1/2} \geq r^{-1/2}$ for all p_n in the disk, we get

$$4\pi r^2 \geq \pi \frac{c^2}{4} |p_n|^{-1/2}\big(\bar{n}(r, w) - \bar{n}(r_0, w)\big) \geq \pi \frac{c^2}{4} r^{-1/2}\big(\bar{n}(r, w) - \bar{n}(r_0, w)\big),$$

hence

$$\bar{n}(r, w) - \bar{n}(r_0, w) \leqq 16c^{-2}r^{5/2}$$

and so

$$n(r, w) = 2\bar{n}(r, w) = O(r^{5/2}). \tag{7.23}$$

Therefore,

$$N(r, w) = O(r^{5/2}),$$

and since $m(r, w) = S(r, w)$ by an immediate application of the Clunie lemma, see Lemma B.11, we infer that

$$T(r, w) = O(r^{5/2}). \qquad \square$$

Theorem 7.10. *The first Painlevé transcendents $w(z)$ are of order $\rho(w) = 5/2$. Moreover, they are of regular growth, and therefore the lower order $\mu(w) = 5/2$ as well.*

Remark. By Proposition 7.9, it suffices to prove the reversed inequality $\rho(w) \geq 5/2$ to get the first assertion. Actually, this has been already proved by Mues and Redheffer [1], p. 419–420. However, we prefer another approach, due to Shimomura [10], as this enables to obtain the regularity of growth at the same time. Similarly as to Proposition 7.9, the proof of Theorem 7.10 follows by a series of lemmas. To this end, let $\gamma > 1$ (in fact > 2, as shown along with the proof) denote a real number, and assume that (p_n), the distinct pole sequence of $w(z)$, has been arranged according increasing moduli. Moreover, we shall use the shortened notation $\bar{n}(\rho) := \bar{n}(\rho, w)$.

Lemma 7.11. *For each ρ sufficiently large, there exists a point z_ρ such that $\rho/2 \leq |z_\rho| < \rho$ and that*

$$\sum_{|p_n|<\gamma\rho} |z_\rho - p_n|^{-2} \leq 6\bar{n}(\gamma\rho)\rho^{-2}\log\rho.$$

Proof. Given a suitable constant $\delta > 0$, we first define polar neighborhoods $|z - p_n| < \delta\omega(|p_n|)$, where $\omega(t) := \min\{1, t^{-1/4}\}$ for $t \geq 0$. Moreover, denote $H_\delta := \bigcup_{n=1}^{\infty}\{z \mid |z - p_n| < \delta\omega(|p_n|)\}$ and $\Delta_\delta(\rho) = B(\rho) \setminus H_\delta$, where $B(\rho) := B(0, \rho)$. By (7.23),

$$\begin{aligned}\sum_{|p_n|\leq\rho+1} \omega(|p_n|)^2 &= \int_1^{\rho+1} t^{-1/2}d\bar{n}(t) + O(1)\\ &\ll (\rho+1)^{-1/2}\bar{n}(\rho+1) + \int_1^{\rho+1} t^{-3/2}\bar{n}(t)dt + O(1)\\ &\ll (\rho+1)^2 \ll \rho^2.\end{aligned}$$

Therefore, choosing $\delta > 0$ small enough, we may assume that

$$A(H_\delta \cap B(\rho)) \leq \pi\rho^2/4, \tag{7.24}$$

where $A(\cdot)$ denotes the area of the set. For $|p_n| < \gamma\rho$, we obtain

$$\omega(|p_n|) = \min(1, |p_n|^{-1/4}) \geq \min(1, (\gamma\delta)^{-1/4}) \geq \min(1, (\gamma + 1)^{-1/4}\delta^{-1/4})$$

by the definition of $\omega(t)$. Hence, integrating over the annulus $R_n := \{z \mid \delta\omega(|p_n|) \leq |z - p_n| \leq (\gamma + 1)\rho\}$, we get

$$\begin{aligned}\iint_{\delta\omega(|p_n|)\leq|z-p_n|\leq(\gamma+1)\rho} |z - p_n|^{-2}\, dxdy &= \int_0^{2\pi} \int_{\delta\omega(|p_n|)}^{(\gamma+1)\rho} r^{-1}\, drd\varphi \\ &= 2\pi \log\left((\gamma + 1)\rho\delta^{-1}\omega(|p_n|)^{-1}\right) \leq 2\pi \log\left(\delta^{-1}(\gamma + 1)^{5/4}\rho^{5/4}\right) \\ &\leq 3\pi \log \rho\end{aligned}$$

as $\rho \to \infty$. Summing them up, we conclude that

$$I(\rho) = \iint_{\Delta_\delta(\rho)} S_\rho(z)\, dxdy \leq 3\pi\bar{n}(\gamma\rho) \log \rho, \tag{7.25}$$

where

$$S_\rho(z) := \sum_{|p_n|<\gamma\rho} |z - p_n|^{-2},$$

provided that ρ is large enough. On the other hand, by (7.24),

$$I(\rho) \geq F(\rho)A\big(\Delta_\delta(\rho) \setminus B(\rho/2)\big) \geq (\pi\rho^2/2)F(\rho),$$

where

$$F(\rho) = \inf\left\{S_\rho(z) \mid z \in \Delta_\delta(\rho) \setminus B(\rho/2)\right\}.$$

Combining this with (7.25), we get $F(\rho) \leq 6\bar{n}(\gamma\rho)\rho^{-2} \log \rho$. Thus we obtain the conclusion. □

Lemma 7.12. *Given $\rho > 1$, $\gamma > 2$, there exists a constant $C_1 > 0$ independent of ρ and γ such that*

$$\sum_{|p_n|\geq\gamma\rho} \left|\frac{1}{(z - p_n)^2} - \frac{1}{p_n^2}\right| \leq C_1 \left(\frac{\rho}{\gamma}\right)^{1/2}, \quad \sum_{|p_n|\geq\gamma\rho} |z - p_n|^{-4} \leqq C_1$$

for every $z \in B(\rho)$.

Proof. Since $|z - p_n| = |p_n||1 - \frac{z}{p_n}| \geq \frac{1}{2}|p_n|$ by $|z/p_n| \leq 1/\gamma < 1/2$, we immediately observe that

$$\left|\frac{1}{(z - p_n)^2} - \frac{1}{p_n^2}\right| = \left|\frac{z(2p_n - z)}{p_n^2(z - p_n)^2}\right| \leqq \frac{10|z||p_n|}{|p_n|^4} \leq 10\rho|p_n|^{-3}.$$

Therefore, recalling that $\bar{n}(r, w) = O(r^{5/2})$, see (7.23), it follows that

$$\sum_{|p_n| \geq \gamma\rho} \left| \frac{1}{(z - p_n)^2} - \frac{1}{p_n^2} \right| \leq 10\rho \sum_{|p_n| \geq \gamma\rho} |p_n|^{-3}$$
$$= 10\rho \int_{\gamma\rho}^{\infty} t^{-3}\, d\bar{n}(t) \leq 30\rho \int_{\gamma\rho}^{\infty} t^{-4}\bar{n}(t)\, dt = O(\rho/\gamma)^{1/2}.$$

Moreover,

$$\sum_{|p_n| \geq \gamma\rho} |z - p_n|^{-4} \leqq 16 \sum_{|p_n| \geq \gamma\rho} |p_n|^{-4} = 16 \int_{\gamma\rho}^{\infty} t^{-4}\, d\bar{n}(t)$$
$$\leq 64 \int_{2\rho}^{\infty} t^{-5}\bar{n}(t)\, dt = O(1).$$

□

Lemma 7.13. *There exists a set $E^* \subset (0, +\infty)$ of finite linear measure such that*

$$\sum_{n=1}^{\infty} \left| \frac{1}{(z - p_n)^2} - \frac{1}{p_n^2} \right| = O(r^9)$$

for all $r = |z| \in (0, +\infty) \backslash E^*$.

Proof. It suffices to define

$$E^* := (0, |p_1| + 1) \cup \Big(\bigcup_{n=2}^{\infty} [|p_n| - |p_n|^{-3}, |p_n| + |p_n|^{-3}] \Big).$$

Making use of $\bar{n}(r, w) = O(r^{5/2})$ again, we conclude that

$$\sigma(E^*) \leq |p_1| + 1 + 2\sum_{n=2}^{\infty} |p_n|^{-3} = |p_1| + 1 + 2\int_{|p_2|}^{\infty} t^{-3}\, d\bar{n}(t) < \infty.$$

By Lemma 7.12, we now see that, for $\gamma > 2$ and $|z| = r \notin E^*$,

$$\sum_{n=1}^{\infty} \left| \frac{1}{(z - p_n)^2} - \frac{1}{p_n^2} \right| \leq \sum_{1 \leq |p_n| \leq \gamma r} (|p_n|^6 + 1) + \sum_{|p_n| \geq \gamma r} \left| \frac{1}{(z - p_n)^2} - \frac{1}{p_n^2} \right|$$
$$\leq (1 + \gamma^6 r^6)\bar{n}(\gamma r, w) + C_1 r^{1/2} = O(r^9).$$

□

Lemma 7.14. *For any η, $0 < \eta < 1$, and some $L > 0$ independent of γ and η,*

$$\sum_{|p_n| \leq \gamma\rho} |p_n|^{-2} \leqq \eta^{-4}\bar{n}(\gamma\rho)\rho^{-2} + \eta L\rho^{1/2} + O(1).$$

Proof. Since $\bar{n}(t) \leq Lt^{5/2}$ for some $L > 0$, we get

$$\begin{aligned}
\sum_{|p_n|\leq\gamma\rho} |p_n|^{-2} &= \int_{\rho_0}^{\gamma\rho} t^{-2}\, d\bar{n}(t) + O(1) \\
&= (\gamma\rho)^{-2}\bar{n}(\gamma\rho) + 2\int_{\rho_0}^{\gamma\rho} t^{-3}\bar{n}(t)\, dt + O(1) \\
&= (\gamma\rho)^{-2}\bar{n}(\gamma\rho) + 2\int_{\rho_0}^{\eta^2\rho} t^{-3}\bar{n}(t)\, dt + 2\int_{\eta^2\rho}^{\gamma\rho} t^{-3}\bar{n}(t)\, dt + O(1) \\
&\leqq (\gamma\rho)^{-2}\bar{n}(\gamma\rho) + 2L\int_{\rho_0}^{\eta^2\rho} t^{-1/2}\, dt + 2\bar{n}(\gamma\rho)\int_{\eta^2\rho}^{\gamma\rho} t^{-3}\, dt + O(1) \\
&= 4\eta L\rho^{1/2} + \eta^{-4}\rho^{-2}\bar{n}(\gamma\rho) + O(1).
\end{aligned}$$

□

Proof of Theorem 7.10. Recalling that all poles of $w(z)$ are double, we may apply the Mittag–Leffler representation for $w(z)$ to obtain

$$w(z) = \varphi(z) + \Phi(z) = \varphi(z) + \sum_{n=1}^{\infty}\left(\frac{1}{(z-p_n)^2} - \frac{1}{p_n^2}\right),$$

where $\varphi(z)$ is an entire function. In case $p_1 = 0$, the corresponding term in $\Phi(z)$ should be replaced by $1/z^2$. In what follows, we may assume that $r > 1$ is sufficiently large.

We split the summation of $\Phi(z)$ in two parts by

$$\Phi(z) = \sum_{|p_n|<\gamma r}\left(\frac{1}{(z-p_n)^2} - \frac{1}{p_n^2}\right) + \sum_{|p_n|\geq\gamma r}\left(\frac{1}{(z-p_n)^2} - \frac{1}{p_n^2}\right).$$

Choosing now $\gamma \geq \gamma_0 := \max(2, 400C_1^2)$, we may apply Lemma 7.12 to conclude that

$$\sum_{|p_n|\geq\gamma_0 r}\left|\frac{1}{(z-p_n)^2} - \frac{1}{p_n^2}\right| \leqq \frac{1}{20}r^{1/2}, \tag{7.26}$$

whenever $z \in B(r)$. By Lemma 7.11, and the estimate (7.23), we find a point $z_0 \in B(r)\backslash B(r/2)$ such that

$$\sum_{|p_n|<\gamma_0 r} \frac{1}{|z_0 - p_n|^2} \leq K'\bar{n}(\gamma_0 r)r^{-2}\log r \tag{7.27}$$

for some constant $K' > 0$. Choosing now $\eta = \eta_0 := L^{-1}/20$ in Lemma 7.14, we obtain

$$\sum_{|p_n|<\gamma_0 r} |p_n|^{-2} \leq \eta_0^{-4}\bar{n}(\gamma_0 r)r^{-2} + \frac{1}{20}r^{1/2} + O(1). \tag{7.28}$$

Hence, combining (7.26), (7.27) and (7.28), we obtain

$$|\Phi(z_0)| \leq K^* \bar{n}(\gamma_0 r) r^{-2} \log r + \frac{1}{10} r^{1/2} + O(1) = O(r^{1/2} \log r) \tag{7.29}$$

for some $z_0 \in B(r) \setminus B(r/2)$; here $K^* = K' + (\eta_0)^{-4}$. Moreover, we may estimate $\Phi''(z_0)$ by observing first that

$$\begin{aligned}
|\Phi''(z_0)| &\leqq 6 \sum_{n=1}^{\infty} \frac{1}{|z_0 - p_n|^4} \\
&\leqq 6 \sum_{|p_n| < \gamma_0 r} \frac{1}{|z_0 - p_n|^4} + 6 \sum_{|p_n| \geq \gamma_0 r} \frac{1}{|z_0 - p_n|^4} \\
&\leqq 6 \Big(\sum_{|p_n| < \gamma_0 r} \frac{1}{|z_0 - p_n|^2} \Big)^2 + 6 \sum_{|p_n| \geq \gamma_0 r} \frac{1}{|z_0 - p_n|^4}
\end{aligned}$$

and then applying (7.27) and Lemma 7.12 to obtain

$$|\Phi''(z_0)| \leq 6 (K^*)^2 \bar{n}(\gamma_0 r)^2 r^{-4} (\log r)^2 + 6 C_1 = O(r (\log r)^2). \tag{7.30}$$

We next observe that $\varphi(z)$ in the decomposition $w(z) = \varphi(z) + \Phi(z)$ is a polynomial and, even more, a constant. In fact, by Lemma 7.13,

$$T(\rho, \varphi) = m(\rho, \varphi) \leqq m(\rho, w) + m(\rho, \Phi) = O(\log \rho),$$

provided $\rho \notin E^*$. Here $m(\rho, w) = O(\log \rho)$ follows by the Clunie argument and the fact that $w(z)$ is of finite order. As E^* is of finite linear measure, a standard reasoning, see Laine [1], Lemma 1.1.1, may be applied to remove the exceptional set, hence $\varphi(z)$ is a polynomial.

Since $|w(z_0)|^2 = \frac{1}{6}(w''(z_0) - z_0)$, we may first write

$$\begin{aligned}
|\varphi(z_0)| &\leq |\Phi(z_0)| + |w(z_0)| \leqq |\Phi(z_0)| + \frac{1}{\sqrt{6}} |w''(z_0) - z_0|^{1/2} \\
&\leq |\Phi(z_0)| + \big(|\varphi''(z_0)| + |\Phi''(z_0)| + |z_0| \big)^{1/2}.
\end{aligned}$$

Invoking now (7.29) and (7.30), we obtain a contradiction, unless $\varphi(z)$ is a constant, say $\varphi(z) \equiv C_0 \in \mathbb{C}$. This follows from the fact that we may find a sequence of points z_0 tending to ∞.

To arrive at the final conclusion, we now write $z_0 = w''(z_0) - 6 w(z_0)^2$. Then for all r sufficiently large,

$$\begin{aligned}
(r/2)^{1/2} \leqq |z_0|^{1/2} &\leqq \big(|\Phi''(z_0)| + 6 (|\Phi(z_0)| + |C_0|)^2 \big)^{1/2} \\
&\leqq |\Phi''(z_0)|^{1/2} + \sqrt{6} \big(|\Phi(z_0)| + |C_0| \big) \\
&\leqq 2\sqrt{6} K^* \bar{n}(\gamma_0 r) r^{-2} \log r + \frac{\sqrt{6}}{10} r^{1/2} + O(1).
\end{aligned}$$

This readily results in

$$\bar{n}(\gamma_0 r) \geq \frac{1}{12}\frac{1}{K^*} r^{5/2}(\log r)^{-1},$$

and so

$$T(r,w) \geq N(r,w) = \frac{1}{2}\bar{N}(r,w) \gg r^{5/2}(\log r)^{-1}, \tag{7.31}$$

which proves the assertion. □

§8 Growth of second and fourth Painlevé transcendents

For the second and fourth Painlevé transcendents, we may also show the finiteness of their order of growth.

Theorem 8.1. *For an arbitrary solution $w(z)$ of (P_2), resp. (P_4), we have $T(r,w) = O(r^C)$, where C is a positive number independent of $w(z)$ and of the coefficients α, β.*

Remark. More precise results are known, see Steinmetz [5] and Shimomura [8]. In fact, for every solution of (P_2), resp. (P_4), $T(r,w) = O(r^3)$, resp. $T(r,w) = O(r^4)$. It is well-known that (P_2), resp. (P_4), may admit, for suitable values of parameters, rational solutions and transcendental solutions of order $3/2$, resp. 2. See Chapters 5 and 6 for more details. It seems likely that these are the only exceptional possibilities with respect to the order of growth for solutions of (P_2), resp. (P_4). However, it remains open to find a rigorous proof.

In Section 8.1 below, using the method of Shimomura [6], we prove Theorem 8.1 for solutions of (P_2). The same argument applies for (P_4); the proof will be sketched in Section 8.2.

8.1. Proof of Theorem 8.1 for (P_2). Let $w(z)$ be an arbitrary solution of (P_2). The main part of the proof is devoted to constructing a polar neighborhood around each pole p. In this reasoning, we treat $w(z)$ in the sector $|\arg z - \arg p| < \pi/2$, $0 \leq \arg p < 2\pi$, containing the pole p. In what follows, we assume that the branch of $z^{1/2}$ together with the value $a^{1/2}$ needed below is determined by $z^{1/2} = |z|^{1/2}\exp(i\phi/2)$, $|\phi - \arg p| < \pi/2$. Similarly as to the case of (P_1), we organize the proof through a sequence of lemmas. To this end, define

$$\theta := 2^{-5}, \quad R_0(\alpha) := 1200(|\alpha|+1). \tag{8.1}$$

We begin with the following lemma, which is proved by a modification of Hukuhara's argument, see Okamoto and Takano [1].

Lemma 8.2. *Let a be a point satisfying $|a| > R_0(\alpha)$. If $|w(a) - \theta a^{1/2}| \leq \theta^3|a|^{1/2}/6$, then*

(i) *$w(z)$ is analytic and bounded for $|z-a| < \delta_a$,*

(ii) $|w(z)-\theta a^{1/2}| \ge \theta^3|a|^{1/2}/5$ *for* $(5/6)\delta_a \le |z-a| \le \delta_a$,

where

$$\theta|a|^{-1/2}\min\{1, \theta^2|a|/|w'(a)|\} < \delta_a \le 3\theta|a|^{-1/2}. \tag{8.2}$$

Proof. We put $z = a + \rho t$, $\rho = a^{-1/2}$, $w(z) = \theta a^{1/2}(1+\theta v(t))$ in (P_2). Then (P_2) becomes

$$\ddot{v}(t) = \theta^{-1} + v(t) + \rho^3 t(\theta^{-1} + v(t)) + \theta^{-2}\rho^3\alpha + 2\theta(1+\theta v(t))^3,$$

where $\dot{}$ again stands for the differentiation with respect to t. Integrating both sides twice, we obtain

$$v(t) = v(0) + \dot{v}(0)t + \theta^{-1}t^2/2 + g(t), \tag{8.3}$$

where

$$g(t) = \frac{\theta^{-2}\rho^3}{6}(\theta t^3 + 3\alpha t^2) + \int_0^t\int_0^\tau \left(v(s) + \rho^3 s v(s) + 2\theta(1+\theta v(s))^3\right) ds\, d\tau. \tag{8.4}$$

Here

$$v(0) = \theta^{-2}a^{-1/2}(w(a) - \theta a^{1/2}), \quad \dot{v}(0) = \theta^{-2}a^{-1}w'(a).$$

By assumption,

$$|v(0)| \le \theta^{-2}|a|^{-1/2}|w(a) - \theta a^{1/2}| \le \theta/6. \tag{8.5}$$

We now divide the proof in two subcases.

(1) *Case* $|\dot{v}(0)| \le 1$. We put

$$M(\eta) := \max\left\{|v(t)| \mid |t| \le \eta\right\}, \quad \eta_0 := \sup\left\{\eta \mid M(\eta) \le 8\theta\right\}.$$

By (8.5), $\eta_0 > 0$. Suppose that $\eta_0 < 3\theta$. Since $|a| > R_0(\alpha)$, we may use (8.1) and (8.4) to conclude that, for $|t| \le \eta_0$,

$$\begin{aligned} |g(t)| &\le \frac{\theta^{-2}|a|^{-3/2}}{6}(3\theta^2 + 3|\alpha|)(3\theta)^2 \\ &\quad + \frac{(3\theta)^2}{2}\left(8\theta + |a|^{-3/2}\cdot 24\theta^2 + 2\theta(1+8\theta^2)^3\right) \\ &\le \frac{9}{2}\theta^2 R_0(\alpha)^{-3/2}(1 + 24\theta^2 + \theta^{-2}|\alpha|) + 9\theta^3\left(4 + (1+8\theta^2)^3\right) \qquad (8.6) \\ &\le \frac{9}{2}\theta^2\left(\frac{2+2^{10}|\alpha|}{1200(1+|\alpha|)} + 2^{-4}(4 + (1+10^{-7})^3)\right) \\ &< \frac{9}{2}\theta^2\cdot 1.4 < \frac{\theta}{4}. \end{aligned}$$

Hence, from (8.3) and (8.5) it follows that, for $|t| \le \eta_0$,

$$\begin{aligned}|v(t)| &\le |v(0)| + |t| + \theta^{-1}|t|^2/2 + \theta/4 \\ &\le (1/6 + 3 + 9/2 + 1/4)\theta < 7.92\theta,\end{aligned} \tag{8.7}$$

which contradicts the definition of η_0. This implies that $\eta_0 \ge 3\theta$, and that (8.6) is valid for all t such that $|t| \le 3\theta$. Moreover, by (8.3), if $2.5\theta \le |t| \le 3\theta$, then

$$\begin{aligned}|v(t)| &\ge \theta^{-1}|t|^2/2 - |v(0)| - |t| - |g(t)| \\ &\ge (2.5^2/2 - 1/6 - 2.5 - 1/4)\theta > \theta/5.\end{aligned}$$

Therefore, $|w(z) - \theta a^{1/2}| \ge \theta^3|a|^{1/2}/5$ for $(5/6)\delta_a \le |z-a| \le \delta_a$ with $\delta_a = 3\theta|a|^{-1/2}$.

(2) *Case* $|\dot{v}(0)| =: \kappa > 1$. We now define

$$\eta_1 := \sup\{\eta \mid M(\eta) \le 3\theta\},$$

and suppose that $\eta_1 < (1.2/\kappa)\theta$. Then, by (8.1), we get for $|t| \le \eta_1$,

$$\begin{aligned}|g(t)| &\le \frac{\theta^{-2}|a|^{-3/2}}{6}(1.2\theta^2 + 3|\alpha|)(1.2\theta)^2 \\ &\quad + \frac{(1.2\theta)^2}{2}\left(3\theta + |a|^{-3/2}\cdot 3.6\theta^2 + 2\theta(1+3\theta^2)^3\right) \\ &\le \frac{(1.2\theta)^2}{2}R_0(\alpha)^{-3/2}(1 + 4\theta^2 + \theta^{-2}|\alpha|) \\ &\quad + (1.2\theta)^2\theta(1.5 + (1+3\theta^2)^3) \\ &\le \frac{(1.2\theta)^2}{2}\Big(\frac{2 + 2^{10}|\alpha|}{1200^{3/2}(1+|\alpha|)} + 2^{-4}(1.5 + (1+3\theta^2)^3)\Big) \\ &\le \frac{(1.2\theta)^2}{2}(0.029 + 0.16) < \frac{1.2^2}{2}\cdot\frac{\theta^2}{5} < \frac{\theta}{100}.\end{aligned} \tag{8.8}$$

By (8.3), (8.5) and this inequality (8.8), again for $|t| \le \eta_1$,

$$\begin{aligned}|v(t)| &\le |v(0)| + \kappa|t| + \theta^{-1}|t|^2/2 + \theta/100 \\ &\le (1/6 + 1.2 + 1.2^2/2 + 1/100)\theta < 2.1\theta,\end{aligned} \tag{8.9}$$

which contradicts the definition of η_1. This implies $\eta_1 \ge (1.2/\kappa)\theta$, and hence (8.9) is valid for all t such that $|t| \le (1.2/\kappa)\theta$. For $(0.8/\kappa)\theta \le |t| \le (1.2/\kappa)\theta$, we have

$$\begin{aligned}|v(t)| &\ge \kappa|t| - \theta^{-1}|t|^2/2 - |v(0)| - |g(t)| \\ &\ge (0.8 - 0.8^2/2 - 1/6 - 1/100)\theta > \theta/5.\end{aligned}$$

Therefore $|w(z) - \theta a^{1/2}| \ge \theta^3|a|^{1/2}/5$ in $(5/6)\delta_a \le |z-a| \le \delta_a$ with $\delta_a = (1.2/\kappa)\theta|a|^{-1/2} = 1.2\theta|a|^{-1/2}(\theta^2|a|/|w'(a)|)$, which completes the proof. □

Lemma 8.3. *Under the same supposition as in Lemma 8.2, if* $|w(a) - \theta a^{1/2}| \le \theta^3 |a|^{1/2}/6$, *then*

(ii′) $|w(z) - \theta z^{1/2}| > \theta^3 |z|^{1/2}/5.5$ *for* $(5/6)\delta_a \le |z - a| \le \delta_a$.

Proof. By (8.1), (8.2) and the supposition $|a| > R_0(\alpha)$, we have $|z| \ge |a| - \delta_a > \theta^{-2}$, $|a| > \theta^{-2} = 2^{10}$ and $||z|^{1/2} - |a|^{1/2}| = |z - a|/(|z|^{1/2} + |a|^{1/2}) \le \delta_a/60 \le \theta^2/20$. Hence, by Lemma 8.2, (ii), for $(5/6)\delta_a \le |z - a| \le \delta_a$,

$$\begin{aligned}
&|w(z) - \theta z^{1/2}| - \theta^3 |z|^{1/2}/5.5 \\
&\quad \ge |w(z) - \theta a^{1/2}| - \theta^3 |z|^{1/2}/5.5 - \theta \big| |z|^{1/2} - |a|^{1/2} \big| \\
&\quad \ge \theta^3 (1/5 - 1/5.5) |a|^{1/2} - (\theta^3/5.5 + \theta) \big| |z|^{1/2} - |a|^{1/2} \big| \\
&\quad \ge (2^5/55 - 2/20)\theta^3 > 0,
\end{aligned}$$

which completes the proof. □

We next proceed to a path construction, related to an arbitrary pole, sufficiently large, of $w(z)$, and based on Lemma 8.2 and Lemma 8.3.

Lemma 8.4. *Let* p *be an arbitrary pole of* $w(z)$ *satisfying* $|p| > 2R_0(\alpha)$, *and let* R_* *be a complex number satisfying* $R_0(\alpha) < R_* < R_0(\alpha) + 1$. *Then there exists a curve* $\Gamma(p)$, *defined by* $\phi : [0, x_p] \to \mathbb{C}$ *as follows:*

(1) $|\phi(0)| = R_*$, $\phi(x_p) = p$;

(2) x *is the length of* $\Gamma(p)$ *from* $\phi(0)$ *to* $\phi(x)$;

(3) $|\phi(x)|$ *is monotone increasing on* $[0, x_p]$;

(4) $|dz| \le (6/\sqrt{11}) d|z|$ *along* $\Gamma(p)$;

(5) $|w(z) - \theta z^{1/2}| \ge 2^{-18} |z|^{1/2}$ *along* $\Gamma(p)$.

Proof. For the simplicity of the description, we treat the case where $\arg p = 0$. For p in the generic position, a completely similar argument applies. Consider the segment $S_0 := [R_*, p] \subset \mathbb{R}$. We start from $z = R_*$, and proceed along S_0. If $|w(z) - \theta z^{1/2}| > \theta^3 |z|^{1/2}/6$ on S_0, then we put $\Gamma(p) = S_0$. Suppose now that a point $a \in S_0$ satisfies $|w(a) - \theta a^{1/2}| \le \theta^3 |a|^{1/2}/6$ and $|w(z) - \theta z^{1/2}| > \theta^3 |z|^{1/2}/6$ for $R_* \le z < a$. Draw the semi-circle $C_a : |z - a| = \delta_a$, $\operatorname{Re} z \ge 0$, which intersects $\mathbb{R}$ at a_- and a_+, $a_- < a_+$. Here δ_a is as in Lemma 8.2. Note that $a_+ \in S_0$, because the pole p does not belong to the interior of C_a. Let a_-^*, a_+^* ($\operatorname{Re} a_-^* < \operatorname{Re} a_+^*$) be the points on the semi-circle C_a^* : $|z - a| = (5/6)\delta_a$, $\operatorname{Re} z \ge 0$ such that the segments $[a_-, a_-^*]$ and $[a_+^*, a_+]$ come in contact with the semi-circle C_a^*. Replace the segment $[a_-, a_+]$ by the curve $\gamma(a)$ which consists of the segments $[a_-, a_-^*]$, $[a_+^*, a_+]$ and the shorter arc $(a_-^*, a_+^*)^{\frown} \subset C_a^*$. Then we get a new curve $\Gamma_1 = \big((S_0 \setminus [a_-, a_+]) \cup \gamma(a)\big) \cap \{z \mid |z| \geq R_*\}$. By Lemmas 8.2 and 8.3, we see that

$$|w(z) - \theta z^{1/2}| \geq \theta^3 |z|^{1/2}/6 > 2^{-18}|z|^{1/2} \tag{8.10}$$

holds on Γ_1. By a geometric consideration about the gradient of tangent lines to $\gamma(a)$, we observe that

$$|dz| \leq (6/\sqrt{11})\, d|z|. \tag{8.11}$$

Restart now from $z = a_+$. Suppose that we first meet a point $b \in \Gamma_1$, $b > a_+$ such that $|w(b) - \theta b^{1/2}| = \theta^3 |b|^{1/2}/6$. (If such a point does not exist, then we put $\Gamma(p) = \Gamma_1$.) By the same argument as above, we obtain the curve $\gamma(b)$, which crosses Γ_1 at b'_-, b_+ (Im $b'_- \geq 0$, $b_+ \in S_0$, Re $b'_- < b_+$). Replacing the part Γ_1 from b'_- to b_+ by that of $\gamma(b)$ from b'_- to b_+, we get a curve Γ_2. On Γ_2, (8.10) and (8.11) obviously remain valid. Starting from $z = b_+$, we continue this procedure. As will be shown at once, after repeating this procedure finitely many times, we arrive at the pole p. Thus we get the path $\Gamma(p)$ with the properties (1) through (5). To show the finiteness of the number of steps in the process, suppose, in the contrary, that there exists a sequence $\{a(\nu)\}_{\nu=0}^{\infty} \subset S_0$ satisfying $\sum_{\nu=0}^{\infty} \delta_{a(\nu)} \leq 1$ and $|w(a(\nu)) - \theta a(\nu)^{1/2}| \leq \theta^3 |p|^{1/2}/6$. Hence, by (8.2), we may choose a subsequence $\{a(\nu_j)\}_{j=0}^{\infty}$ satisfying $a(\nu_j) \to a_* \in S_0$, $w(a(\nu_j)) \to w_* \neq \infty$, $w'(a(\nu_j)) \to \infty$ as $j \to \infty$, which implies $w(a_*) = w_* \neq \infty$, $w'(a_*) = \infty$. This is a contradiction. Thus the lemma is proved. □

For the two final lemmas, we need to define an auxiliary function as follows:

$$\Phi(z) = w'(z)^2 + \frac{w'(z)}{w(z) - \theta z^{1/2}} - w(z)^4 - z w(z)^2 - 2\alpha w(z). \tag{8.12}$$

By a suitable mathematical software we may verify that

$$\Phi'(z) + \frac{\Phi(z)}{W(z)^2} = G(z) := G_1(z) + G_2(z), \tag{8.13}$$

where

$$\begin{aligned}
G_1(z) &:= \frac{w'(z)}{W(z)^3}\big(1 - \theta^2/2 + (\theta/2) z^{-1/2} w(z)\big),\\
G_2(z) &:= \frac{-1}{W(z)^2}\big(\theta^2 z w(z)^2 + \theta z^{3/2} w(z) + \alpha w(z) + \theta\alpha z^{1/2}\big),\\
W(z) &:= w(z) - \theta z^{1/2}.
\end{aligned}$$

Since

$$\begin{aligned}
G_1(z) &= -\left(\frac{1/2}{W(z)^2} + \frac{(\theta/2) z^{-1/2}}{W(z)}\right)' - \frac{(\theta/4) z^{-3/2}}{W(z)} + \frac{(\theta/2)^2 z^{-1}}{W(z)^2} + \frac{(\theta/2) z^{-1/2}}{W(z)^3},\\
G_2(z) &= -\theta^2 z - \frac{1}{W(z)}(2\theta^3 z^{3/2} + \theta z^{3/2} + \alpha) - \frac{1}{W(z)^2}(\theta^4 z^2 + \theta^2 z^2 + 2\alpha\theta z^{1/2}),
\end{aligned}$$

the right-hand side of (8.13) may be written in the form

$$\begin{aligned}
G(z) &= -H_1'(z) - H_2(z),\\
H_1(z) &= \frac{(\theta/2)z^{-1/2}}{W(z)} + \frac{1/2}{W(z)^2},\\
H_2(z) &= \theta^2 z + \frac{h_1(z)}{W(z)} + \frac{h_2(z)}{W(z)^2} + \frac{h_3(z)}{W(z)^3},\\
h_1(z) &= O(z^{3/2}), \quad h_2(z) = O(z^2), \quad h_3(z) = O(z^{-3/2}).
\end{aligned}$$

To solve (8.13), we have the following basically elementary

Lemma 8.5. *Let $\gamma(z_0, z)$ be an arbitrary path starting from z_0 and ending at z such that $W(t) \neq 0$ on $\gamma(z_0, z)$. Then*

$$\begin{aligned}
\Phi(z) &= E(z_0, z)^{-1}\Phi(z_0) - H_1(z) + E(z_0, z)^{-1} H_1(z_0)\\
&\quad + E(z_0, z)^{-1} \int_{\gamma(z_0,z)} E(z_0, t)\left(\frac{H_1(t)}{W(t)^2} - H_2(t)\right) dt.
\end{aligned} \tag{8.14}$$

Here $E(z_0, t) := \exp\left(\int_{\gamma(z_0,t)} W(\tau)^{-2} d\tau\right)$ for $t \in \gamma(z_0, z)$, and $\gamma(z_0, t) \subset \gamma(z_0, z)$ is the part of $\gamma(z_0, z)$ from z_0 to t.

Consider now a circle $|z| = R_*$ such that $R_0(\alpha) < R_* < R_0(\alpha) + 1$. Moreover, assume that there are no poles of $\Phi(z)$ on this circle. Let p be an arbitrary pole of $w(z)$ such that $|p| > 2R_0(\alpha)$, and $U(p) = \{z \mid |z - p| < \eta(p)\}$ be a domain defined by

$$\eta(p) := \sup\{\eta \leq 1 \mid |W(z)| > 2|z|^{1/2} \text{ in } |z - p| < \eta\}.$$

Then we have

Lemma 8.6. *In $U(p)$, $|\Phi(z)| \leq K_0|z|^{\Delta}$, where K_0 is a positive number independent of p, and $\Delta \geq 2$ a number independent of $w(z)$ and p.*

Proof. We shall now make use of the path $\Gamma(p)$, constructed in Lemma 8.4, and starting from some point $z_0(p)$ such that $|z_0(p)| = R_*$. Then, $|W(t)| \geq 2^{-18}|t|^{1/2}$, $|dt| \leq (6/\sqrt{11})d|t|$ along $\Gamma(p)$. From these facts, for $t \in \Gamma(p)$, it follows that

$$\begin{aligned}
|E(z_0(p), t)^{\pm 1}| &\leq \exp\left(\int_{\Gamma(p,t)} \frac{|d\tau|}{|W(\tau)|^2}\right)\\
&\leq \exp\left(\frac{2^{37}\cdot 3}{\sqrt{11}} \int_{R_*}^{|t|} \frac{d|\tau|}{|\tau|}\right) = O(t^{\Delta'}).
\end{aligned}$$

Here $\Delta' = 2^{37} \cdot 3/\sqrt{11}$ and $\Gamma(p, t) \subset \Gamma(p)$ denotes the part of $\Gamma(p)$ from $z_0(p)$ to t. Moreover, $H_1(t) = O(t)$, $H_2(t) = O(t)$ along $\Gamma(p)$. Using (8.14) and these estimates, and observing that $|\Phi(z_0(p))| \leq M_0$, we have $\Phi(p) = O(p^{2\Delta'+2})$, where $M_0 = \max\{|\Phi(z)| \mid |z| = R_*\}$. Observing that $H_1(z) = O(1)$ and $H_2(z) = O(z)$ in

$U(p)$, and applying Lemma 8.5 with $z_0 = p$, $\gamma(z_0, z) = [p, z] \subset U(p)$, we obtain $\Phi(z) = O(z^{2\Delta'+2})$ in $U(p)$. This completes the proof. □

Now we are ready to complete the proof of Theorem 8.1. Put $w(z) = 1/u(z)$, $z = p + p^{-\Delta/4}s$ in (8.12). Then (8.12) takes the form

$$u'(z)^2 - \frac{u(z)^3 u'(z)}{1 - \theta z^{1/2} u(z)} - 1 - zu(z)^2 - 2\alpha u(z)^3 - u(z)^4 \Phi(z) = 0.$$

Hence $v(s) = u(p + p^{-\Delta/4}s)$ satisfies

$$(dv/ds)(s) = \pm p^{-\Delta/4}\big(1 + h(s, v(s))\big), \tag{8.15}$$

where

$$|h(s, v(s))| < 1/2, \quad v(0) = 0,$$

as long as

$$\big|z^{\Delta/4}u(z)\big| = \big|(p + p^{-\Delta/4}s)^{\Delta/4}\big||v(s)| < \varepsilon_0, \tag{8.16}$$

and $z \in U(p)$, see Lemma 8.6. Here $\varepsilon_0 = \varepsilon_0(K_0)$ is a sufficiently small positive constant independent of p. If z satisfies (8.16), then $|w(z)| > \varepsilon_0^{-1}|z|^{\Delta/4} \geq \varepsilon_0^{-1}|z|^{1/2}$, and hence $z \in U(p)$. Put now

$$\eta_* = \sup\big\{\eta \mid \text{(8.16) is valid for } |s| < \eta\big\}. \tag{8.17}$$

Suppose that $\eta_* < \varepsilon_0/4$. Then, integrating (8.15), we get

$$|s|/2 \leq \big|p^{\Delta/4}\big||v(s)| \leq 3|s|/2 \leq 3\varepsilon_0/8 \tag{8.18}$$

for $|s| \leq \eta_* < \varepsilon_0/4$. This implies

$$\big|(p + p^{-\Delta/4}s)^{\Delta/4}\big||v(s)| \leq \big|p^{\Delta/4}\big||v(s)|(1 + 1/L)^{\Delta/4} \leq \varepsilon_0/2$$

for $|s| \leq \eta_*$ and for $|p| \geq L$, where L is sufficiently large. For $|p| \geq L$, this contradicts (8.17). Hence $\eta_* \geq \varepsilon_0/4$. Therefore, for $|p| \geq L$, (8.18) is valid for $|s| < \varepsilon_0/4$, and $w(z)$ is analytic for $0 < |z - p| < (\varepsilon_0/4)|p|^{-\Delta/4}$. Thus we have proved

Lemma 8.7. *For every pole p of $w(z)$ satisfying $|p| > L_0 := \max\{2R_0(\alpha), L\}$, $w(z)$ is analytic in the domain $0 < |z - p| < (\varepsilon_0/4)|p|^{-\Delta/4}$.*

Proof of Theorem 8.1. For each pole p, $|p| > L_0$, we allocate a polar neighborhood $U_*(p) := \{z \mid |z - p| < (\varepsilon_0/8)|p|^{-\Delta/4}\}$. Then, for arbitrary distinct poles p_1, p_2, with moduli $> L_0$, these neighborhoods are mutually disjoint. Hence the counting function of poles in the disk $|z| < r$ does not exceed $O(r^{2+\Delta/2})$. Since $m(r, w) = S(r, w)$, using the Clunie lemma, Lemma B.11, again, we have $T(r, w) = O(N(2r, w)) = O(r^{2+\Delta/2})$, which completes the proof for (P_2). □

8.2. Proof of Theorem 8.1 for (P_4). The proof of Theorem 8.1 for (P_4) is parallel to the preceding proof for (P_2). Due to its rather technical nature, we don't repeat the proof in detail here. For the convenience, a short sketch will be offered. Concerning more details, see Shimomura [6]. Let now $w(z)$ be an arbitrary solution of (P_4). In this case, we define $\theta = 2^{-7}$. The counterpart to Lemma 8.2 now reads as follows:

Lemma 8.8. *Let a be a point satisfying $|a| > R_0(\alpha, \beta)$, where $R_0(\alpha, \beta)$ is a sufficiently large positive number. If $|w(a) - \theta a| \le \theta^3 |a|/6$, then*

(i) *$|w(z)|$ is analytic and bounded for $|z - a| < \delta_a$,*

(ii) *$|w(z) - \theta a| \ge \theta^3 |a|/5$ for $(5/6)\delta_a \le |z - a| \le \delta_a$,*

where

$$\theta |a|^{-1} \min\{1, \theta^2 |a|^2 / |w'(a)|\} < \delta_a \le 3\theta |a|^{-1}.$$

Similarly, corresponding to Lemma 8.3, we obtain

Lemma 8.9. *Under the same conditions as in Lemma 8.8, if $|w(a) - \theta a| \le \theta^3 |a|/6$, then*

(ii′) *$|w(z) - \theta z| > \theta^3 |z|/5.5$ for $(5/6)\delta_a \le |z - a| \le \delta_a$.*

Proof of Lemma 8.8. Substituting now $z = a + \rho t$, $\rho = a^{-1}$, $w(z) = \theta a (1 + \theta v(t))^2$ into (P_4), we obtain

$$\ddot{v}(t) = \theta^{-1} + f_0(t, v(t)) + f_1(t, v(t)),$$

where

$$f_0(t, v) := v + 2(1 + \theta v)^3 + \frac{3}{4}\theta (1 + \theta v)^5,$$

$$f_1(t, v) := \rho^2(-\alpha + 2t + \rho^2 t^2)(\theta^{-1} + v) + 2\rho^2 t (1 + \theta v)^3 + \frac{\rho^4 \beta \theta^{-2}}{8(1 + \theta v)^3}.$$

Integrating twice, we derive that

$$v(t) = v(0) + \dot{v}(0) t + \theta^{-1} t^2 / 2 + g(t),$$

$$g(t) = \int_0^t \int_0^\tau f_0(s, v(s))\, ds d\tau + \int_0^t \int_0^\tau f_1(s, v(s))\, ds d\tau.$$

Using this relation, a reasoning parallel to the proof of Lemma 8.2 applies. In estimating $g(t)$, we choose $R_0(\alpha, \beta)$ so large that $\left| \int_0^t \int_0^\tau f_1(s, v(s))\, ds d\tau \right|$ is sufficiently small for all a such that $|a| > R_0(\alpha, \beta)$ and $|t| \le 3\theta$. □

Using Lemmas 8.8 and 8.9, we may repeat the idea of Lemma 8.4 for an arbitrary pole p, $|p| > 2R_0(\alpha, \beta)$, to construct a path $\Gamma(p)$ ending at $z = p$ with the same properties (1) through (4) of Lemma 8.4, while the counterpart to (5) now reads

(5′) $|w(z) - \theta z| \geq 2^{-24}|z|$ *along* $\Gamma(p)$.

The auxiliary function $\Phi(z)$ now takes the form

$$\begin{aligned}\Phi(z) = {} & \frac{w'(z)^2}{w(z)} + \frac{4w'(z)}{w(z) - \theta z} - w(z)^3 - 4zw(z)^2 \\ & - 4(z^2 - \alpha)w(z) + \frac{2\beta}{w(z)},\end{aligned} \tag{8.19}$$

which satisfies the linear differential equation

$$\Phi'(z) + \left(\frac{2}{W(z)} + \frac{4\theta z}{W(z)^2}\right)\Phi(z) = G(z) := G_1(z) + G_2(z), \tag{8.20}$$

where

$$\begin{aligned} W(z) &:= w(z) - \theta z, \\ G_1(z) &:= \frac{4w'(z)}{W(z)^3}\big((2+\theta)w(z) + (2-\theta)\theta z\big), \\ G_2(z) &:= -\frac{4}{W(z)^2}\big(\theta^2 z^2 w(z)^2 + 2\theta z^2 w(z)(w(z) + \theta z) \\ &\qquad + 4\theta(z^2 - \alpha)zw(z) - 2\beta\big). \end{aligned}$$

Since

$$\begin{aligned} G_1(z) &= -\left(\frac{4(\theta+2)}{W(z)} + \frac{8\theta z}{W(z)^2}\right)' + \frac{4\theta(\theta+4)}{W(z)^2} + \frac{16\theta^2 z}{W(z)^3}, \\ G_2(z) &= \frac{h_2^0(z)}{W(z)^2} + \frac{h_1^0(z)}{W(z)} + h_0^0(z) \\ h_0^0(z) &= O(z^2), \quad h_1^0(z) = O(z^3), \quad h_2^0(z) = O(z^4), \end{aligned}$$

the right-hand side of (8.20) may be written in the form

$$\begin{aligned} G(z) &= -\left(\frac{4(\theta+2)}{W(z)} + \frac{8\theta z}{W(z)^2}\right)' + h_0(z) + \frac{h_1(z)}{W(z)} + \frac{h_2(z)}{W(z)^2} + \frac{h_3(z)}{W(z)^3}, \\ h_0(z) &= O(z^2), \quad h_1(z) = O(z^3), \quad h_2(z) = O(z^4), \quad h_3(z) = O(z). \end{aligned}$$

Let now p be an arbitrary pole of $w(z)$ such that $|p| > 2R_0(\alpha, \beta)$. Solving (8.19) with the aid of the path $\Gamma(p)$, we obtain an estimate $|\Phi(z)| \leq K_0|z|^{\Delta}$ in the domain $U(p) = \{z \mid |z - p| < \eta(p)\}$, where $\eta(p) = \sup\{\eta \leq 1 \mid |W(z)| > 2|z| \text{ in } |z - p| < \eta\}$. Substitute now $w(z) = 1/u(z)$, $z = p + p^{-\Delta/3}s$ into (8.19). By the same argument as in Section 8.1, we get distinct polar neighborhoods $|z - p| < (\varepsilon_0/4)|p|^{-\Delta/3}$. Similarly as to the case of (P_2), we obtain $N(r, w) = O(r^{2+2\Delta/3})$, and hence $T(r, w) = O(r^C)$.

§9 Growth of third and fifth Painlevé transcendents

All solutions of $(\widetilde{P}_3)$ and $(\widetilde{P}_5)$ are meromorphic in $\mathbb{C}$, see Chapter 1 and Hinkkanen and Laine [2], [3]. The method used to find growth estimates for solutions of (P_2) and (P_4) in the proceeding section is applicable in the case of $(\widetilde{P}_3)$ and $(\widetilde{P}_5)$ as well. As a result, we obtain estimates of exponential for the growth of solutions of $(\widetilde{P}_3)$ and $(\widetilde{P}_5)$:

Theorem 9.1. *Let $w(z)$ be an arbitrary solution of $(\widetilde{P}_5)$, resp. $(\widetilde{P}_3)$, with coefficients $(\alpha, \beta, \gamma, \delta)$. Then $T(r, w) = O(\exp(\Lambda_5 r))$, resp. $O(\exp(\Lambda_3 r))$, where $\Lambda_5 = \Lambda_5(\alpha, \beta, \gamma, \delta)$, resp. $\Lambda_3 = \Lambda_3(\alpha, \beta, \gamma, \delta)$, is a positive number independent of $w(z)$.*

We omit the proof of Theorem 9.1. The interested reader may consult Shimomura [9]. Concerning the sharpness of these estimates, we offer the following remarks.

Remark. In the case of $(\alpha, \beta, \gamma, \delta) = (1, -1, -1, 1)$, $(\widetilde{P}_3)$ possesses a solution $\chi(z) = e^{z/2}\phi(z)$ such that

$$\phi(z) = \exp\left[R_0(1 + o(1))e^{-z/4}\cos(4e^{z/2} - (1 - R_0)z/2 + \theta_0 + o(1))\right]$$

as $z \to +\infty$, $z > 0$, where R_0 and θ_0 are positive constants, see Shimomura [2]. Since $T(r, \phi) \geq N(r, 1/(\phi - 1)) + O(1) \gg e^{r/3}$, we have $T(r, \chi) \gg e^{r/3}$. Similarly of the case of $(\widetilde{P}_5)$, assume that the coefficients of $(\widetilde{P}_5)$ satisfy $\beta = 0$, $\delta > 0$, $\alpha, \gamma \in \mathbb{R}$. Then, an arbitrary solution $\psi(z)$ of $(\widetilde{P}_5)$ such that $0 < \psi(0) < 1$ and that $\psi'(0) \in \mathbb{R}$ admits the asymptotic representation

$$\psi(z) = R_0(1 + o(1))e^{-z}\cos^2\left(\sqrt{\delta/2}e^z - C(R_0)z + \theta_0 + o(1)\right),$$

where $C(R_0) = (\gamma/4)\sqrt{2/\delta} - \sqrt{\delta/2}R_0$, as $z \to \infty$ along the positive real axis. Here $R_0 > 0$ and $\theta_0 \in \mathbb{R}$ are constants depending on the initial values $\psi(0)$, $\psi'(0)$, see Shimomura [1], [3]. By this expression it is easy to see that $T(r, \psi) \geq N(r, 1/\psi) + O(1) \gg e^{r/2}$. Therefore, the growth estimates in Theorem 9.1 are qualitatively sharp. However, under certain conditions, (P_5) and (P_3) possesses rational or algebraic solutions. Then the corresponding solutions $w(z)$ of $(\widetilde{P}_5)$ or $(\widetilde{P}_3)$ satisfy $T(r, w) = O(r^{\Lambda_0})$ for some $\Lambda_0 > 0$.

Chapter 3
Value distribution of Painlevé transcendents

This chapter is devoted to considering value distribution properties of Painlevé transcendents. Similarly as to the preceding chapter, Nevanlinna theory will be applied, see Appendix B again for notations and necessary background results. Moreover, several of the error term estimates in this chapter are based on the finiteness of order of growth, as established in Chapter 2. In §10, we evaluate basic properties of deficiencies and ramification indices for Painlevé transcendents, except for solutions of (P_6). The next §11 is devoted to treating the second main theorem for Painlevé transcendents. More precisely, we proceed to show, for (P_1), (P_2), (P_4), $(\widetilde{P}_3)$ and $(\widetilde{P}_5)$, how the second main theorem reduces to an asymptotic equality. The final section in this chapter, §12, offers an introduction to deficiencies with respect to small target functions for the first and second Painlevé transcendents. It seems to us that the topics considered in §12 remains open to interesting further developments.

§10 Deficiencies and ramification indices

10.1. Solutions of (P_1). Let $w(z)$ be an arbitrary solution of

$$w'' = z + 6w^2. \qquad (P_1)$$

By Chapter 1, $w(z)$ is transcendental meromorphic. Moreover, by §7, $w(z)$ is of finite order $\rho(w) = \frac{5}{2}$ and of regular growth. We proceed to show the following

Theorem 10.1. *For every $a \in \mathbb{C}$, we have*

$$m(r, 1/(w-a)) = O(\log r) \quad \textit{and} \quad \delta(a, w) = 0.$$

Moreover,

$$m(r, w) = O(\log r) \quad \textit{and} \quad \delta(\infty, w) = 0.$$

Proof. Since, for every $a \in \mathbb{C}$, $w \equiv a$ is not a solution of (P_1), we may apply the Mohon'ko–Mohon'ko lemma, see Lemma B.12, to $w(z)$, recalling that $w(z)$ is of finite order. Hence, we have $m(r, 1/(w-a)) = O(\log r)$ from which $\delta(a, w) = 0$ immediately follows. By the Clunie argument, see Lemma B.11, we also have $m(r, w) = O(\log r)$ and $\delta(\infty, w) = 0$. □

Theorem 10.2. *For every $a \in \mathbb{C}$, we have*

$$N_1(r, 1/(w-a)) \leq \frac{1}{6}T(r, w) + O(\log r) \quad \text{and} \quad \vartheta(a, w) \leq \frac{1}{6}.$$

Moreover,

$$N_1(r, w) = \frac{1}{2}T(r, w) + O(\log r) \quad \text{and} \quad \vartheta(\infty, w) = \frac{1}{2}.$$

Remark 1. In the results above, the estimates $\delta(a, w) = \delta(\infty, w) = 0$, $\vartheta(a, w) \leq 1/6$ have been obtained, originally, by Schubart and Wittich [1]. It remains open whether some of the ramification results in this section could be improved.

Proof. To prove Theorem 10.2 we recall that all poles of $w(z)$ are double, with the Laurent expansion (1.1). Moreover, by the finiteness of order, $m(r, w') = O(\log r)$. Defining now

$$\Psi(z) := w'(z)^2 - 4w(z)^3 - 2zw(z), \tag{10.1}$$

we immediately see that

$$\Psi'(z) = -2w(z). \tag{10.2}$$

For $a \in \mathbb{C}$, consider the set

$$A := \{z \mid w(z) = a,\ w'(z) = 0\}.$$

We may suppose that A is an infinite set, since otherwise, trivially, $\vartheta(a, w) = 0$. Now choose a point $z \in A \setminus \{-6a^2\}$. Then, from (10.1), we obtain

$$\begin{aligned} G(z) &:= w'(z)^2 - 4(w(z)^3 - a^3) - 2z(w(z) - a) \\ &= \Psi(z) + 4a^3 + 2az. \end{aligned} \tag{10.3}$$

Furthermore, by (10.2),

$$\begin{aligned} G'(z) &= 2(a - w(z)), \\ G''(z) &= -2w'(z), \\ G^{(3)}(z) &= -2w''(z) = -2(6w(z)^2 + z). \end{aligned} \tag{10.4}$$

Hence, for every $\sigma \in A \setminus \{-6a^2\}$,

$$G(\sigma) = G'(\sigma) = G''(\sigma) = 0, \quad G^{(3)}(\sigma) = -2(6a^2 + \sigma) \neq 0,$$

and so σ is a triple zero of $G(z)$. This fact means that

$$N_1(r, 1/(w-a)) \leq \frac{1}{3}N(r, 1/G) + O(\log r) \leq \frac{1}{3}T(r, G) + O(\log r). \tag{10.5}$$

By Theorem 10.1, Proposition B.5 and (10.3), $m(r, G) \ll m(r, w') + m(r, w) \ll \log r$. Since all poles of $G'(z)$ are double by (10.4), all poles of $G(z)$ are simple, and therefore $N(r, G) = (1/2)N(r, w) = (1/2)T(r, w) + O(\log r)$. Hence $T(r, G) = (1/2)T(r, w) + O(\log r)$. Combining this with (10.5), we obtain

$$N_1(r, 1/(w-a)) \leq \frac{1}{6}T(r, w) + O(\log r),$$

from which $\vartheta(a, w) \leq 1/6$ immediately follows. From the fact that every pole of $w(z)$ is double, it follows that

$$N_1(r, w) = \frac{1}{2}N(r, w) = \frac{1}{2}T(r, w) + O(\log r).$$

Hence $\vartheta(\infty, w) = 1/2$, which completes the proof. □

10.2. Solutions of (P_2). Recall again that all solutions of

$$w'' = 2w^3 + zw + \alpha, \quad \alpha \in \mathbb{C}, \tag{P_2}$$

are meromorphic functions of finite order $\rho(w) \leqq 3$. Also recall that all poles solutions of (P_2) are simple, see (2.1). Since (P_2) may admit rational solutions, we now restrict ourselves to considering transcendental solutions only. So let $w(z)$ denote an arbitrary transcendental solution of (P_2). Concerning the estimates for deficiencies in Theorem 10.3, see Schubart [1] and Schubart and Wittich [1], and for the ramification indices in Theorem 10.4, see Kießling [1].

Theorem 10.3. (1) *We have*

$$m(r, w) = O(\log r) \quad \textit{and} \quad \delta(\infty, w) = 0.$$

(2) *If* $\alpha \neq 0$, *then, for every* $a \in \mathbb{C}$,

$$m(r, 1/(w-a)) = O(\log r) \quad \textit{and} \quad \delta(a, w) = 0.$$

In the case of $\alpha = 0$, *we have, for every* $a \in \mathbb{C}\backslash\{0\}$,

$$m(r, 1/(w-a)) = O(\log r) \quad \textit{and} \quad \delta(a, w) = 0,$$

and for $a = 0$,

$$m(r, 1/w) \leq \frac{1}{2}T(r, w) + O(\log r) \quad \textit{and} \quad \delta(0, w) \leq \frac{1}{2}.$$

Proof. It is sufficient to show $\delta(0, w) \leq 1/2$ under the condition $\alpha = 0$, because the other assertions of Theorem 10.3 immediately follow from Lemma B.11 and Lemma B.12, and the finiteness of the order of $w(z)$. Assuming that $\alpha = 0$, we define

$$\Psi(z) = w'(z)^2 - w(z)^4 - zw(z)^2, \tag{10.6}$$

hence by differentiating (10.6),

$$\Psi'(z) = -w(z)^2. \tag{10.7}$$

Moreover, we define

$$\Xi(z) = \Psi(z)/w(z). \tag{10.8}$$

Obviously, all poles of $\Psi'(z)$ are double by (10.7), hence Ξ has to be analytic at the poles z_0 of $w(z)$. Differentiating $\Psi = w\Xi$, we obtain $-w = \Xi' + \frac{w'}{w}\Xi$, and making use of the Laurent expansion (2.1), we conclude that $\Xi(z_0) = \pm 1$. Now we suppose that

$$\Xi(z) \not\equiv \pm 1. \tag{10.9}$$

Then

$$\begin{aligned} N(r, w) &\le N(r, 1/(\Xi^2 - 1)) \le 2T(r, \Xi) + O(1) \\ &= 2(N(r, \Xi) + m(r, \Xi)) + O(1), \end{aligned} \tag{10.10}$$

where

$$m(r, \Xi) \ll m(r, w'/w) + m(r, w) + \log r \ll \log r. \tag{10.11}$$

Since every pole of $\Xi(z)$ is necessarily a zero of $w(z)$,

$$N(r, \Xi) \le N(r, 1/w) = T(r, w) - m(r, 1/w) + O(1). \tag{10.12}$$

Substituting (10.11), (10.12) and $N(r, w) = T(r, w) + O(\log r)$ into (10.10), we obtain

$$m(r, 1/w) \le \frac{1}{2}T(r, w) + O(\log r)$$

under the condition (10.9). To derive the desired estimate, it remains to show that (10.9) is fulfilled. To do so, suppose the contrary $\Xi(z) \equiv \pm 1$. Then, by (10.7) and (10.8),

$$\Psi(z) = w'(z)^2 - w(z)^4 - zw(z)^2 \equiv \pm w(z),$$

and so

$$\Psi'(z) = -w(z)^2 = \pm w'(z).$$

Solving this elementary differential equation, we obtain that either $w(z) = 0$ or $w(z) = 1/(\pm z + C)$, $C \in \mathbb{C}$, contradicting the transcendency of $w(z)$. □

Theorem 10.4. (1) *For every* $a \in \mathbb{C}\backslash\{0\}$, *we have*

$$N_1(r, 1/(w-a)) \le \frac{1}{4}T(r, w) + O(\log r) \quad \text{and} \quad \vartheta(a, w) \le \frac{1}{4}.$$

(2) *If* $\alpha \ne 0$, *then*

$$N_1(r, 1/w) \le \frac{1}{5}T(r, w) + O(\log r) \quad \text{and} \quad \vartheta(0, w) \le \frac{1}{5},$$

and if $\alpha = 0$, *then*

$$N_1(r, 1/w) = 0 \quad \textit{and} \quad \vartheta(0, w) = 0.$$

(3) $N_1(r, w) = 0$ *and* $\vartheta(\infty, w) = 0$.

Proof. (1). For $a \in \mathbb{C} \setminus \{0\}$, we consider the set

$$A := \{z \mid w(z) = a, \ w'(z) = 0\}.$$

By the same argument as in the proof of Theorem 10.2, we may suppose that A is an infinite set, and we see that every a-point in $A \setminus \{z^*(a)\}$, $z^*(a) := -2a^2 - \alpha/a$ is double. Denoting

$$\Psi_\alpha(z) := w'(z)^2 - w(z)^4 - zw(z)^2 - 2\alpha w(z),$$

note that $\Psi'_\alpha(z) = -w(z)^2$. Define now

$$\begin{aligned} G(z) &:= w'(z)^2 - (w(z)^4 - a^4) - z(w(z)^2 - a^2) - 2\alpha(w(z) - a) \\ &= \Psi_\alpha(z) + a^4 + 2\alpha a + a^2 z \end{aligned} \tag{10.13}$$

and

$$\Theta(z) := G(z)/(w(z) - a). \tag{10.14}$$

By differentiation, we obtain $G'(z) = a^2 - w(z)^2$, $G''(z) = -2w(z)w'(z)$. Invoking (10.13), we see that $G(\sigma) = G'(\sigma) = G''(\sigma) = 0$ for every $\sigma \in A \setminus \{z^*(a)\}$. Hence each $\sigma \in A \setminus \{z^*(a)\}$ is a zero of $\Theta(z)$. Furthermore, differentiating $G(z) = (w(z) - a)\Theta(z)$ we get $-w(z)^2 + a^2 = G'(z) = w'(z)\Theta(z) + (w(z) - a)\Theta'(z)$. Making use of the Laurent expansion (2.1) we conclude that $\Theta(z)$ is analytic around each pole z_0 of $w(z)$, and $\Theta(z_0) = \pm 1$. These facts imply that

$$\begin{aligned} N(r, \Theta) &\le N(r, 1/(w - a)) - 2N_1(r, 1/(w - a)) + O(\log r) \\ &\le T(r, w) - 2N_1(r, 1/(w - a)) + O(\log r). \end{aligned} \tag{10.15}$$

By (10.13) and (10.14),

$$m(r, \Theta) \ll m(r, w') + m(r, w) + m(r, 1/(w - a)) + \log r \ll \log r.$$

This estimate and (10.15) yield

$$T(r, \Theta) \le T(r, w) - 2N_1(r, 1/(w - a)) + O(\log r). \tag{10.16}$$

For each pole z_0 of $w(z)$, $\Theta(z_0) = \pm 1$, as shown above. On the other hand, $\Theta(\sigma) = 0$ for every $\sigma \in A \setminus \{z^*(a)\} \neq \emptyset$, which implies that $\Theta(z) \not\equiv \pm 1$. Hence

$$N(r, w) \le N(r, 1/(\Theta^2 - 1)) \le 2T(r, \Theta) + O(1). \tag{10.17}$$

Observing that $N(r, w) = T(r, w) + O(\log r)$, and combining (10.17) with (10.16), we obtain

$$N_1(r, 1/(w-a)) \le \frac{1}{4}T(r, w) + O(\log r),$$

proving the first assertion.

(2) and (3). Since every pole of $w(z)$ is simple, the assertion (3) is trivial. If $\alpha = 0$, and $w(z_1) = w'(z_1) = 0$, then $w(z) \equiv 0$ by Theorem A.3. Therefore, for each transcendental solution of (P_2) with $\alpha = 0$, every zero is simple, and so $\vartheta(0, w) = 0$. Hence, in what follows, it suffices to show the assertion (2) under the condition $\alpha \neq 0$. We now define

$$A_0 := \{z \mid w(z) = w'(z) = 0\}.$$

Clearly, every $\sigma \in A_0$ is a double zero, and we may suppose that A_0 is an infinite set. For a similar argument as to above, define

$$G_0(z) := w'(z)^2 - w(z)^4 - zw(z)^2 - 2\alpha w(z), \tag{10.18}$$

which coincides with (10.13) with $a = 0$. Then,

$$\begin{aligned}
G_0'(z) &= -w(z)^2, \\
G_0''(z) &= -2w(z)w'(z), \\
G_0^{(3)}(z) &= -2w'(z)^2 - 2w(z)w''(z), \\
G_0^{(4)}(z) &= -6w'(z)w''(z) - 2w(z)w^{(3)}(z), \\
G_0^{(5)}(z) &= -8w'(z)w^{(3)}(z) - 6w''(z)^2 - 2w(z)w^{(4)}(z).
\end{aligned}$$

Hence, for every $\sigma \in A_0$, we have $G_0(\sigma) = G_0'(\sigma) = G_0''(\sigma) = G_0^{(3)}(\sigma) = G_0^{(4)}(\sigma) = 0$, $G_0^{(5)}(\sigma) = -6w''(\sigma)^2 = -6\alpha^2 \neq 0$. Therefore, σ is a zero of $G_0(z)$ of multiplicity five. This fact yields

$$N_1(r, 1/w) \le \frac{1}{5}N(r, 1/G_0) + O(\log r) \le \frac{1}{5}T(r, G_0) + O(\log r). \tag{10.19}$$

On the other hand, since all poles of $G_0(z)$ are simple, as seen by differentiating (10.18), $N(r, G_0) = N(r, w)$. Moreover, by (10.18) $m(r, G_0) = O(\log r)$. Substituting these formulas into (10.19), we get

$$N_1(r, 1/w) \le \frac{1}{5}T(r, w) + O(\log r),$$

which implies the desired estimate. □

Remark 2. In the proofs of Theorems 10.3 and 10.4, the argument with the use of the auxiliary functions (10.8) and (10.14) is due to Steinmetz [2]. In Theorem 10.4, by the same method as in the proof of Theorem 10.2, Schubart and Wittich [1] obtained $\le 1/3$ instead of $\le 1/4$.

10.3. Solutions of (P_4). Let $w(z)$ denote an arbitrary transcendental meromorphic solution of

$$w'' = \frac{1}{2}\frac{(w')^2}{w} + \frac{3}{2}w^3 + 4zw^2 + 2(z^2 - \alpha)w + \frac{\beta}{w}. \qquad (P_4)$$

For special combinations of values of (α, β), (P_4) may admit rational solutions as well, see Murata [1], for a complete treatment. See also Chapter 6. The deficiencies and ramification indices for transcendental solutions have been originally estimated by Steinmetz [2].

Theorem 10.5. (1) *We have*

$$m(r, w) = O(\log r) \quad \text{and} \quad \delta(\infty, w) = 0.$$

(2) *If* $(\beta, a) \neq (0, 0)$, *then*

$$m(r, 1/(w - a)) = O(\log r) \quad \text{and} \quad \delta(a, w) = 0.$$

(3) *If* $\beta = 0$ *and if* $w(z)$ *does not satisfy the Riccati differential equation* $w' = \mp(w^2 + 2zw)$, *then*

$$m(r, 1/w) \leq \frac{1}{2}T(r, w) + O(\log r) \quad \text{and} \quad \delta(0, w) \leq \frac{1}{2}.$$

Remark 3. As will be shown in the course of the proof below, (P_4) with $\beta = 0$ admits a family of solutions satisfying $w' = \mp(w^2 + 2zw)$ simultaneously, if and only if $\alpha = \pm 1$. For such solutions, $N(r, 1/w) = 0$, and hence $\delta(0, w) = 1$.

Proof. First recall that all poles of $w(z)$ must be simple, with the residue equal to ± 1. Similarly as to the previous subsections, $m(r, w') = O(\log r)$. By Lemma B.11 and Lemma B.12 again, and the finiteness of the order of $w(z)$, it is sufficient to show the assertion (3). Since $\beta = 0$, we may define

$$\Psi(z) := w'(z)^2/w(z) - w(z)^3 - 4zw(z)^2 - 4(z^2 - \alpha)w(z) \qquad (10.20)$$

to obtain

$$\Psi'(z) = -4w(z)^2 - 8zw(z). \qquad (10.21)$$

Moreover, we define an auxiliary function by

$$\Xi(z) := \Psi(z)/w(z). \qquad (10.22)$$

Differentiating now $\Psi(z) = w(z)\Xi(z)$, (10.21) immediately implies that

$$-4w(z) - 8z = \frac{w'(z)}{w(z)}\Xi(z) + \Xi'(z). \qquad (10.23)$$

Since the poles of $\Psi(z)$ are simple by (10.21) and they appear at the poles z_0 of $w(z)$ only by (10.21), (10.23) immediately results in $\Xi(z_0) = \pm 4$. By the same argument as in the proof of Theorem 10.3, we have

$$m(r, 1/w) \le \frac{1}{2} T(r, w) + O(\log r),$$

and hence $\delta(0, w) \le 1/2$, provided that $\Xi(z) \not\equiv \pm 4$. Therefore, it remains to examine the case where

$$\Xi(z) \equiv \pm 4. \tag{10.24}$$

By (10.20) and (10.22), this is equivalent to

$$\Psi(z) = w'(z)^2/w(z) - w(z)^3 - 4zw(z)^2 - 4(z^2 - \alpha)w(z) \equiv \pm 4w(z). \tag{10.25}$$

Differentiating (10.25), and recalling (10.21), we have

$$\mp w'(z) = w(z)^2 + 2zw(z). \tag{10.26}$$

Substitution of this into (10.25) yields $4\alpha w(z) \equiv \pm 4w(z)$. Hence, (P_4) with $\beta = 0$ admits a family of solutions satisfying (10.26) if and only if $\alpha = \pm 1$. For these solutions only, (10.26) holds. This completes the proof. □

Theorem 10.6. (1) *For every* $a \in \mathbb{C}\setminus\{0\}$,

$$N_1(r, 1/(w-a)) \le \frac{1}{4} T(r, w) + O(\log r) \quad \textit{and} \quad \vartheta(a, w) \le \frac{1}{4}.$$

(2) *If* $\beta \ne 0$, *then*

$$N_1(r, 1/w) = 0 \quad \textit{and} \quad \vartheta(0, w) = 0.$$

On the other hand, if $\beta = 0$, *then*

$$N_1(r, 1/w) = \frac{1}{2} T(r, w) + O(\log r) \quad \textit{and} \quad \vartheta(0, w) = \frac{1}{2}.$$

(3) $N_1(r, w) = 0$ *and* $\vartheta(\infty, w) = 0$.

Proof. (2) and (3) For each transcendental solution $w(z)$ of (P_4), every zero of $w(z)$ is simple if $\beta \ne 0$, and double if $\beta = 0$. Recalling that all poles of $w(z)$ are simple, (2) and (3) immediately follow.

(1). Put $A := \{z \mid w(z) = a,\ w'(z) = 0\}$. Then every $\sigma \in A \setminus \{z_*(a), z_{**}(a)\}$ is a double a-point, where $z_*(a)$, $z_{**}(a)$ are the roots of the quadratic equation $(3/2)a^4 + 4a^3 z + 2(z^2 - \alpha)a^2 + \beta = 0$. This time, we define

$$\Psi(z) := \frac{w'(z)^2}{w(z)} - w(z)^3 - 4zw(z)^2 - 4(z^2 - \alpha)w(z) + \frac{2\beta}{w(z)}.$$

By a direct differentiation, and making use of (P_4), we obtain

$$\Psi'(z) = -4w(z)^2 - 8zw(z).$$

Moreover, we define

$$\begin{aligned} G(z) &:= \frac{w'(z)^2}{w(z)} - (w(z)^3 - a^3) - 4z(w(z)^2 - a^2) \\ &\quad - 4(z^2 - \alpha)(w(z) - a) + 2\beta\left(\frac{1}{w(z)} - \frac{1}{a}\right), \\ &= \Psi(z) + a^3 + 4a^2 z + 4a(z^2 - \alpha) - \frac{2\beta}{a} \end{aligned} \tag{10.27}$$

to consider the function

$$\Theta(z) = G(z)/(w(z) - a). \tag{10.28}$$

Then,

$$\begin{aligned} G'(z) &= -4(w(z)^2 - a^2) - 8z(w(z) - a), \\ G''(z) &= -8(w(z) + z)w'(z) - 8(w(z) - a). \end{aligned}$$

Hence, for every $\sigma \in A \setminus \{z_*(a), z_{**}(a)\}$, $G(\sigma) = G'(\sigma) = G''(\sigma) = 0$. By the same argument as in the proof of Theorem 10.4, (1), we obtain

$$N_1(r, 1/(w - a)) \le \frac{1}{4} T(r, w) + O(\log r),$$

which completes the proof. □

10.4. Solutions of $(\widetilde{P}_3)$. As already mentioned in §3, while considering the modified third Painlevé equation

$$w'' = \frac{(w')^2}{w} + \alpha w^2 + \gamma w^3 + \beta e^z + \frac{\delta e^{2z}}{w}, \tag{$\widetilde{P}_3$}$$

the special cases $\beta = \delta = 0$ and $\alpha = \gamma = 0$ behave symmetrically, reducing back to an equation of the form

$$(v')^2 = \gamma v^2 (v - \tau_1)(v - \tau_2), \quad \gamma, \tau_1, \tau_2 \in \mathbb{C}, \tag{10.29}$$

or, ultimately, to a Riccati differential equation by elementary transformations, see §3 and Hinkkanen and Laine [2], p. 326–327. Therefore, we may omit these special cases here, as the value distribution of (10.29) and of the Riccati equation is well-known, see e.g. Laine [1].

Hence, from now on, we may suppose that

$$(\alpha, \gamma) \ne (0, 0), \quad \text{and} \quad (\beta, \delta) \ne (0, 0). \tag{10.30}$$

Under this supposition, the characteristic functions of the coefficients of $(\widetilde{P}_3)$ are dominated by $T(r, e^z) \asymp O(r)$. A solution $w(z)$ of $(\widetilde{P}_3)$ is said to be *admissible*, if

$$T(r, e^z)/T(r, w) \asymp r/T(r, w) \to 0 \tag{10.31}$$

as $r \to \infty$ outside of a possible exceptional set of finite linear measure. Examples of non-admissible solutions of $(\widetilde{P}_3)$ may easily be constructed by combining the transformation (3.1) with an algebraic solution of (P_3), see §9 and Murata [2] and Gromak [1], [3].

For a concrete example of an admissible solution, see §9.

In what follows, $w(z)$ denotes an arbitrary admissible solution of $(\widetilde{P}_3)$. By Hinkkanen and Laine [2], Theorem 1, $w(z)$ is meromorphic. Concerning their value distribution, we prove the following two theorems, see Shimomura [2]:

Theorem 10.7. *For every $a \in \mathbb{C}$, we have*

$$m(r, 1/(w-a)) = O(r) \quad \text{and} \quad \delta(a, w) = 0.$$

Moreover,

$$m(r, w) = O(r) \quad \text{and} \quad \delta(\infty, w) = 0.$$

Proof. First observe that for every solution $w(z)$ of $(\widetilde{P}_3)$, $T(r, w) = O(\exp(\Lambda r))$, for some $\Lambda \in \mathbb{C}$, without an exceptional set, see §9 and Laine [1], Lemma 1.1.1. Therefore, the Nevanlinna error term $S(r, w)$ here and in the proof of Theorem 10.10 takes the form $O(r)$ without an exceptional set, see Lemma B.13. Using now $(\alpha, \gamma) \neq (0, 0)$, see (10.30), and applying the standard Clunie reasoning to $(\widetilde{P}_3)$, we obtain

$$m(r, w) = O(r) + S(r, w) = O(r).$$

By $(\beta, \delta) \neq (0, 0)$, see (10.30) again, we observe that for every $a \in \mathbb{C} \setminus \{0\}$, $w \equiv a$ is not a solution of $(\widetilde{P}_3)$. By the Mohon'ko–Mohon'ko lemma, see Lemma B.12, we have

$$m(r, 1/(w-a)) = O(r) + S(r, w) = O(r).$$

By the transformation, $W(z) = e^z/w(z)$, $W(z)$ satisfies

$$WW'' = (W')^2 - \beta W^3 - \delta W^4 - \alpha e^z W - \gamma e^{2z}. \tag{10.32}$$

Clearly, $T(r, w) \asymp T(r, W)$. Applying now the Clunie lemma, see Lemma B.11, to (10.32), we obtain

$$m(r, 1/w) = m(r, e^{-z}W) \le m(r, W) + O(r) = S(r, W) + O(r) = O(r).$$

Since $w(z)$ is admissible, the remaining assertions immediately follow. □

To proceed, we first observe that $m(r, w') = O(r)$ by $m(r, w') = m(r, w'/w) + m(r, w) = O(r) + S(r, w) = O(r)$. To prove Theorem 10.10 below, we also need the following elementary lemmas.

Lemma 10.8. *Let $z = z_0$ be an arbitrary pole of $w(z)$. Around $z = z_0$, the Laurent expansion of $w(z)$ starts as*

$$w(z) = \begin{cases} \pm\gamma^{-1/2}(z - z_0)^{-1} + O(1), & \text{if } \gamma \neq 0, \\ (2/\alpha)(z - z_0)^{-2} + O(1), & \text{if } \gamma = 0,\ \alpha \neq 0. \end{cases}$$

Lemma 10.9. *Let $z = c$ be an arbitrary zero of $w(z)$. Around $z = c$, the Taylor expansion of $w(z)$ starts as*

$$w(z) = \begin{cases} \pm(-\delta)^{1/2}e^c(z - c) + O(z - c)^2, & \text{if } \delta \neq 0, \\ -(\beta/2)e^c(z - c)^2 + O(z - c)^4, & \text{if } \delta = 0,\ \beta \neq 0. \end{cases}$$

The expressions in Lemma 10.8 are easily obtained by substituting the Laurent series around z_0 into $(\widetilde{P}_3)$. Using (10.32), we derive Lemma 10.9 from Lemma 10.8.

Theorem 10.10. (1) *Assume that $a \in \mathbb{C}\backslash\{0\}$. If $\delta \neq 0$, then*

$$N_1(r, 1/(w - a)) \leq \frac{1}{4}T(r, w) + O(r) \quad \text{and} \quad \vartheta(a, w) \leq \frac{1}{4}.$$

If $\delta = 0$ and $\beta \neq 0$, then

$$N_1(r, 1/(w - a)) \leq \frac{1}{6}T(r, w) + O(r) \quad \text{and} \quad \vartheta(a, w) \leq \frac{1}{6}.$$

(2) *If $\delta \neq 0$, then*

$$N_1(r, 1/w) = 0 \quad \text{and} \quad \vartheta(0, w) = 0.$$

If $\delta = 0$ and $\beta \neq 0$, then

$$N_1(r, 1/w) = \frac{1}{2}T(r, w) + O(r) \quad \text{and} \quad \vartheta(0, w) = \frac{1}{2}.$$

(3) *If $\gamma \neq 0$, then*

$$N_1(r, w) = 0 \quad \text{and} \quad \vartheta(\infty, w) = 0.$$

If $\gamma = 0$, $\alpha \neq 0$, then

$$N_1(r, w) = \frac{1}{2}T(r, w) + O(r) \quad \text{and} \quad \vartheta(\infty, w) = \frac{1}{2}.$$

Proof. The assertion (2) immediately follows from Lemma 10.8 and Lemma 10.9. For $a \in \mathbb{C} \setminus \{0\}$, consider the sets

$$A := \{z \mid w(z) = a,\ w'(z) = 0\}, \quad \widetilde{A} := \{z \in A \mid w''(z) = 0\}.$$

If $z_* \in \widetilde{A}$, then, substituting $w(z_*) = a$, $w'(z_*) = 0$ into $(\widetilde{P}_3)$, we observe that $\alpha a^3 + \gamma a^4 + \beta e^{z_*} a + \delta e^{2z_*} = 0$. This implies that

$$\widetilde{A} = \{z \in A \mid e^z = \kappa_1, \text{ or } e^z = \kappa_2\}$$

for some $\kappa_1, \kappa_2 \in \mathbb{C}$. Observing that, for every $j \in \mathbb{N}$, $w^{(j)}$ is a polynomial in the variables w, $1/w$, $w^{(s)}$, $s \le j-1$, and e^z, we have that the cardinal number of $\widetilde{A} \cap \{z \mid |z| < r\}$ is at most $O(r)$. Moreover, there exists an integer μ_0 such that, for every a-point belonging to $\widetilde{A}$, the multiplicity of it does not exceed μ_0. If the set $A \setminus \widetilde{A}$ is finite, then we immediately conclude that $N_1(r, 1/(w-a)) = O(r)$, and we are done. Therefore, we may assume that $A \setminus \widetilde{A}$ is an infinite set. By (3.3) and (3.4),

$$\frac{1}{2}V' = \frac{d}{dz}\left(\frac{(w')^2}{w^2} - 2\alpha w - \gamma w^2 + \frac{2\beta e^z}{w} + \frac{\delta e^{2z}}{w^2}\right) = \frac{2\beta e^z}{w} + \frac{2\delta e^{2z}}{w^2}.$$

Therefore, defining now

$$\begin{aligned} H(z) := {} & \frac{w'(z)^2}{w(z)^2} - 2\alpha(w(z)-a) - \gamma(w(z)^2 - a^2) \\ & + 2\beta e^z\left(\frac{1}{w(z)} - \frac{1}{a}\right) + \delta e^{2z}\left(\frac{1}{w(z)^2} - \frac{1}{a^2}\right), \end{aligned} \tag{10.33}$$

we observe by differentiation that

$$H'(z) = 2\beta e^z\left(\frac{1}{w(z)} - \frac{1}{a}\right) + 2\delta e^{2z}\left(\frac{1}{w(z)^2} - \frac{1}{a^2}\right), \tag{10.34}$$

$$H''(z) = 2\beta e^z\left(\frac{1}{w(z)} - \frac{1}{a}\right) - \frac{2\beta e^z w'(z)}{w(z)^2} + 4\delta e^{2z}\left(\frac{1}{w(z)^2} - \frac{1}{a^2}\right) - \frac{4\delta e^{2z} w'(z)}{w(z)^3}.$$

Therefore, it is immediate to observe that for each $\sigma \in A$,

$$H(\sigma) = H'(\sigma) = H''(\sigma) = 0. \tag{10.35}$$

Now we define

$$\Theta(z) := \frac{e^{-z} w(z) H(z)}{w(z) - a}. \tag{10.36}$$

Case I. Suppose that $\delta \ne 0$. By Lemma 10.9, each zero c of $w(z)$ is simple. By (10.34), H has a simple pole at c, and so $\Theta(z)$ is finite at $z = c$. By (10.34) and (10.33), we easily conclude that $\Theta(c) = \pm 2(-\delta)^{1/2} a^{-1}$. By (10.35), for $z_1 \in A \setminus \widetilde{A}$, we have $\Theta(z_1) = 0$, which implies that $\Theta(z)^2 \not\equiv -4\delta a^{-2}$. Hence

$$N(r, 1/w) \le N(r, 1/(\Theta^2 + 4\delta_0 a^{-2})) \le 2T(r, \Theta) + O(1). \tag{10.37}$$

By Theorem 10.7 and its proof, we have

$$m(r, \Theta) \le m(r, w) + m(r, 1/(w-a)) + m(r, H) + O(r) = O(r). \tag{10.38}$$

By (10.35), every double a-point of $w(z)$ is at least a triple zero of $H(z)$. Moreover, for every pole z_0 of $w(z)$, $\Theta(z)$ is analytic at $z = z_0$. Therefore, by (10.38) and the above observations, we derive

$$\begin{aligned} T(r, \Theta) &\leq N(r, \Theta) + O(r) \\ &\leq N(r, 1/(w-a)) - 2N_1(r, 1/(w-a)) + O(r) \\ &\leq T(r, w) - 2N_1(r, 1/(w-a)) + O(r). \end{aligned}$$

Combining this with (10.37) and $N(r, 1/w) = T(r, w) + O(r)$, we obtain

$$N_1(r, 1/(w-a)) \leq \frac{1}{4} T(r, w) + O(r),$$

from which the estimate $\vartheta(a, w) \leq 1/4$ follows.

Case II. Suppose now that $\delta = 0$, $\beta \neq 0$. If now $\Theta(z) \equiv 0$, then $H(z) \equiv 0$, hence $H'(z) \equiv 0$, and a simple computation by (10.34) shows that $w(z)$ is not admissible, a contradiction. Therefore, we may assume that $\Theta(z) \not\equiv 0$. By Lemma 10.9, each zero c of $w(z)$ is now double, and by (10.34), a simple pole of $H(z)$, implying to be a zero of $\Theta(z)$. Moreover, by (10.35) and the definition (10.36), every double a-point of $w(z)$ is a zero of $\Theta(z)$ as well. Hence,

$$(1/2)N(r, 1/w) + N_1(r, 1/(w-a)) \leq N(r, 1/\Theta) + O(r) \leq T(r, \Theta) + O(r).$$

Using this inequality instead of (10.37), by the same argument as in Case I, we obtain

$$N_1(r, 1/(w-a)) \leq \frac{1}{6} T(r, w) + O(r),$$

from which we have $\vartheta(a, w) \leq 1/6$. This completes the proof. □

10.5. Solutions of ($\widetilde{P}_5$). For solutions of ($\widetilde{P}_5$), which are all meromorphic by §5 and Hinkkanen and Laine [3], their deficiencies and ramification indices have been examined in detail only recently, see Shimomura [3]. By Hinkkanen and Laine [3], p. 134–135, the special case $(\gamma, \delta) = (0, 0)$ again reduces, ultimately, back to a Riccati equation. Therefore, it is sufficient to treat ($\widetilde{P}_5$) under the condition

$$(\gamma, \delta) \neq (0, 0), \tag{10.39}$$

only. The necessary considerations are quite similar to those applied for ($\widetilde{P}_3$) in the preceding subsection. Due to more technical details needed for ($\widetilde{P}_5$), we restrict ourselves to give a couple of theorems only, leaving the proofs aside. The interested reader may consult the original reference Shimomura [3]. We first give two one-parameter families of solutions of ($\widetilde{P}_5$); the verification of this proposition is a straightforward computation.

Proposition 10.11. (1) *If* $\alpha = 0$ *and* $-4\beta\delta + (\gamma \pm (-2\delta)^{1/2})^2 = 0$, *then there exists a family of solutions* $\mathcal{V} = \{\chi_{\pm(\gamma,\delta;C)}(z) \mid C \in \mathbb{C}\}$, *where*

$$\chi_{\pm(\gamma,\delta;C)}(z) = \exp(\kappa_{\pm} z \mp (-2\delta)^{1/2} e^z) \cdot \left[C - \kappa_{\pm} \int_0^z \exp(-\kappa_{\pm} s \pm (-2\delta)^{1/2} e^s) ds\right]$$

and

$$\kappa_{\pm} = 1 \pm \gamma(-2\delta)^{-1/2}.$$

(2) *If* $\beta = 0$ *and* $4\alpha\delta + (-\gamma \pm (-2\delta)^{1/2})^2 = 0$, *then there exists a family of solutions* $\mathcal{W} = \{\psi_{\pm(\gamma,\delta;C)}(z) \mid C \in \mathbb{C}\}$, *where*

$$\psi_{\pm(\gamma,\delta;C)}(z) = 1/\chi_{\pm(-\gamma,\delta;C)}(z).$$

We now define an admissible solution $w(z)$ of $(\widetilde{P}_5)$ in the same way as in the case of $(\widetilde{P}_3)$, i.e. $w(z)$ is admissible, provided (10.31) is satisfied as $r \to \infty$ outside of a possible exceptional set of finite linear measure. Recalling, from §9, that $T(r, w) = O(\exp(\Lambda r))$ holds for all solutions $w(z)$ of $(\widetilde{P}_5)$ without any exceptional set, we again have that the Nevanlinna error term $S(r, w)$ will be of the form $O(r)$.

Let now $w(z)$ be an admissible solution of $(\widetilde{P}_5)$. Omitting the proofs, we recall the following two theorems from Shimomura [3]:

Theorem 10.12. *For every* $a \in \mathbb{C} \cup \{\infty\}$, *we have*

$$m(r, 1/(w-a)) = O(r) \quad \text{and} \quad \delta(a, w) = 0,$$

except for the following cases:

(1) *if* $\alpha = 0$, $w \notin \mathcal{V}$, *then*

$$m(r, w) \le \frac{1}{2} T(r, w) + O(r) \quad \text{and} \quad \delta(\infty, w) \le \frac{1}{2};$$

(2) *if* $\beta = 0$, $w \notin \mathcal{W}$, *then*

$$m(r, 1/w) \le \frac{1}{2} T(r, w) + O(r) \quad \text{and} \quad \delta(0, w) \le \frac{1}{2};$$

(3) *if* $\alpha + \beta = 0$, $\gamma = 0$ *and* $\delta \ne 0$, *then*

$$m(r, 1/(w+1)) \le \frac{1}{2} T(r, w) + O(r) \quad \text{and} \quad \delta(-1, w) \le \frac{1}{2};$$

(4) *if* $w \in \mathcal{V}$, *then*

$$m(r, w) = T(r, w) \quad \text{and} \quad \delta(\infty, w) = 1;$$

(5) *if* $w \in \mathcal{W}$, *then*

$$m(r, 1/w) = T(r, w) + O(1) \quad \text{and} \quad \delta(0, w) = 1.$$

Theorem 10.13. *For every $a \in \mathbb{C} \setminus \{0, 1\}$, if $\delta \neq 0$, then*

$$N_1(r, 1/(w-a)) \leq \frac{1}{4}T(r, w) + O(r) \quad \textit{and} \quad \vartheta(a, w) \leq \frac{1}{4}.$$

If $\delta = 0$, $\gamma \neq 0$, then

$$N_1(r, 1/(w-a)) \leq \frac{1}{6}T(r, w) + O(r) \quad \textit{and} \quad \vartheta(a, w) \leq \frac{1}{6}.$$

In particular, if $\alpha + \beta = 0$, $\gamma = 0$, and $\delta \neq 0$, then $N_1(r, 1/(w+1)) = 0$ and $\vartheta(-1, w) = 0$.

§11 The second main theorem for Painlevé transcendents

This section is devoted to proving that the second main theorem reduces to an asymptotic equality for transcendental solutions of (P_1), (P_2), (P_4), $(\widetilde{P}_3)$ and $(\widetilde{P}_5)$, reflecting the surprising regularity of these functions. As a preparation, we begin with

Lemma 11.1. *Let $f(z)$ be an arbitrary transcendental meromorphic function satisfying $m(r, f') = O(r)$, resp. $m(r, f') = O(\log r)$ and $m(r, f/f') = O(r)$, resp. $m(r, f/f') = O(\log r)$. Then*

$$m(r, 1/f) + N(r, 1/f') = N(r, f') + O(r),$$

resp.

$$m(r, 1/f) + N(r, 1/f') = N(r, f') + O(\log r).$$

Proof. By assumption, we have $m(r, 1/f) - m(r, 1/f') \geq -m(r, f/f') = O(r)$. On the other hand, $m(r, 1/f) - m(r, 1/f') \leq m(r, f'/f) = O(r)$. Hence,

$$m(r, 1/f) = m(r, 1/f') + O(r).$$

Using this identity and the assumption $m(r, f') = O(r)$, we get

$$\begin{aligned} m(r, 1/f) + N(r, 1/f') &= T(r, 1/f') + O(r) \\ &= T(r, f') + O(r) \\ &= N(r, f') + O(r). \end{aligned}$$

The respective case will be proved similarly. □

Theorem 11.2. *Let $w(z)$ be an arbitrary solution of (P_1). Then*

$$N(r, 1/w') + N_1(r, w) = 2T(r, w) + O(\log r).$$

Proof. Recall that all solutions of (P_1) are transcendental. By Theorem 10.1, the estimates

$$m(r, 1/w) = O(\log r), \quad m(r, w) = O(\log r), \quad m(r, w') = O(\log r) \tag{11.1}$$

are valid. Since every pole of $w(z)$ is double, see §1,

$$N_1(r, w) = (1/2)N(r, w), \quad N(r, w') = (3/2)N(r, w). \tag{11.2}$$

Differentiating both sides of (P_1), we have

$$w^{(3)} = 12ww' + 1,$$

and so

$$w/w' = ww^{(3)}/w' - 12w^2,$$

which implies $m(r, w/w') \ll m(r, w) + m(r, w^{(3)}/w') \ll \log r$, see Corollary B.6. Then, applying Lemma 11.1 and using (11.1) and (11.2), we get

$$\begin{aligned} N(r, 1/w') + N_1(r, w) &= N(r, w') + O(\log r) + N_1(r, w) \\ &= 2N(r, w) + O(\log r) \\ &= 2T(r, w) + O(\log r). \end{aligned}$$

□

For (P_2) and (P_4), the required asymptotic equality appears to be the same form. Since for every solution of (P_2) with $\alpha \neq 0$, resp. of (P_4) with $\beta \neq 0$, $m(r, 1/w) = O(\log r)$, the asymptotic equality becomes in these cases exactly as in Theorem 11.2.

Theorem 11.3. *Let $w(z)$ be an arbitrary transcendental solution of (P_2), resp. of (P_4). Then*

$$m(r, 1/w) + N(r, 1/w') + N_1(r, w) = 2T(r, w) + O(\log r).$$

Proof. For (P_2), recall that all solutions have simple poles only. Therefore,

$$N_1(r, w) = 0, \quad N(r, w') = 2N(r, w). \tag{11.3}$$

Differentiating (P_2), we obtain

$$w^{(3)} = 6w^2w' + zw' + w,$$

and so

$$w/w' = w^{(3)}/w' - 6w^2 - z.$$

Hence, $m(r, w/w') \ll m(r, w^{(3)}/w') + m(r, w) + \log r \ll \log r$, see Theorem 10.3. Applying Lemma 11.1 and using (11.3), we have

$$\begin{aligned} &m(r, 1/w) + N(r, 1/w') + N_1(r, w) \\ &\qquad = N(r, w') + O(\log r) + N_1(r, w) = 2T(r, w) + O(\log r), \end{aligned}$$

which is the desired identity.

Next write (P_4) in the form

$$\Pi := ww'' - (w')^2/2 - 3w^4/2 - 4zw^3 - 2(z^2-\alpha)w^2 - \beta = 0.$$

By elementary computation of $\frac{1}{w'}\big(\frac{1}{w}\Pi'\big)'$, we obtain

$$4\frac{w}{w'} = \frac{w^{(4)}}{w'} - \frac{w''}{w'}\big(6w^2 + 12zw + 4(z^2-\alpha)\big) - 12(w+z)u' - 20w - 12z.$$

As $w(z)$ is of finite order of growth, we may apply elementary Nevanlinna theory computations to obtain

$$\begin{aligned} m(r, w/w') &\ll m(r, w^{(4)}/w') + m(r, w''/w') \\ &\quad + m(r, w') + m(r, w) + \log r \ll \log r. \end{aligned}$$

Again all poles of $w(z)$ are simple, and so (11.3) remains valid for this case as well. Hence, we may apply Lemma 11.1 exactly as in the case of (P_2) to obtain the asserted conclusion. □

We now proceed to consider solutions $w(z)$ of $(\widetilde{P}_3)$ and $(\widetilde{P}_5)$ under the conditions (10.30) and (10.39), respectively, see Shimomura [2] and [3].

Theorem 11.4. *Let $w(z)$ be an arbitrary transcendental solution of $(\widetilde{P}_3)$. Then*

$$N(r, 1/w') + N_1(r, w) = 2T(r, w) + O(r).$$

Proof. We show that the assumptions of Lemma 11.1 are fulfilled. To this end, we write $(\widetilde{P}_3)$ in the form

$$\Pi := ww'' - (w')^2 - \alpha w^3 - \beta e^z w - \gamma w^4 - \delta e^{2z} = 0.$$

By Theorem 9.1, Theorem 10.7 and Lemma B.13, we have $m(r, w) = O(r)$, $m(r, 1/w) = O(r)$ and $m(r, w^{(k)}) \le m(r, w^{(k)}/w) + m(r, w) = O(r)$ for all $k \in \mathbb{N}$. Moreover, $m(r, w^{(k+1)}/w') = O(r)$ for all $k \in \mathbb{N}$, since $\log T(r, w') = O(r)$. If $\delta \ne 0$, from the relation $(w/w')\big((d/dz)^2 - (d/dz)\big)\Pi(z, w, w', w'') = 0$, we derive

$$\begin{aligned} 2\delta e^{2z}\frac{w}{w'} &= w^2\frac{w^{(4)}}{w'} - w^2\frac{w^{(3)}}{w'} - w\big(w'' + 3\alpha w^2 + 4\gamma w^3 + \beta e^z\big)\frac{w''}{w'} \\ &\quad + ww'' + 3\alpha w^3 - 6\alpha w^2 w' + 4\gamma w^4 - 12\gamma w^3 w' - \beta e^z w. \end{aligned}$$

Using Corollary B.7, this results in $m(r, w/w') = O(r)$. Moreover, whenever $\delta = 0$ and $\beta \ne 0$, then from the relation $(1/w')(d/dz)\Pi(z, w, w', w'') = 0$, it follows that

$$\beta e^z\frac{w}{w'} = w\frac{w^{(3)}}{w'} - w'' - 3\alpha w^2 - 4\gamma w^3 - \beta e^z.$$

Again $m(r, w/w') = O(r)$ in this case. Hence, Lemma 11.1 is applicable, and by $m(r, 1/w) = O(r)$,

$$N\left(r, \frac{1}{w'}\right) = N(r, w') + O(r). \tag{11.4}$$

By Lemma 10.10,

$$N(r, w') = \begin{cases} 2T(r, w) + O(r), & \text{if } \gamma \neq 0, \\ (3/2)T(r, w) + O(r), & \text{if } \gamma = 0, \alpha \neq 0, \end{cases} \tag{11.5}$$

and

$$N_1(r, w) = \begin{cases} 0, & \text{if } \gamma \neq 0, \\ (1/2)T(r, w) + O(r), & \text{if } \gamma = 0, \alpha \neq 0. \end{cases} \tag{11.6}$$

Using (11.4), (11.5) and (11.6), we may apply the same argument as in the proof of Theorem 11.2, to prove the assertion. □

For solutions of $(\widetilde{P}_5)$, the deficiencies of $0, -1$ and ∞ may be strictly positive. Therefore, we now obtain

Theorem 11.5. *Let $w(z)$ be an arbitrary transcendental solution of $(\widetilde{P}_5)$. Then*

$$m(r, w) + m(r, 1/w) + m(r, 1/(w+1)) + N(r, 1/w') + N_1(r, w) = 2T(r, w) + O(r).$$

§12 Value distribution with respect to small target functions

Let $f(z)$ and $g(z)$ be meromorphic functions. Saying that $g(z)$ is small with respect to $f(z)$, we understand that $T(r, g) = S(r, f)$. Then we may define the deficiency and the ramification index for $f(z)$ relative to the small target $g(z)$ by

$$\delta(g, f) := \liminf_{r\to\infty} \frac{m(r, 1/(f-g))}{T(r, f)}, \quad \vartheta(g, f) := \liminf_{r\to\infty} \frac{N_1(r, 1/(f-g))}{T(r, f)}.$$

Clearly, rational functions are small targets with respect to any arbitrary transcendental meromorphic function $f(z)$.

Considering mostly first and second Painlevé transcendents in this section, we prove the following two theorems, essential due to Shimomura [4]. Their proofs below apply the reasoning from Shimomura [4] combined with the finiteness of the order of growth, see Chapter 2.

Theorem 12.1. *Let $w(z)$ be an arbitrary solution of (P_1) and suppose that $T(r, g) = S(r, w)$. Then we have*

$$m(r, 1/(w-g)) \leq \frac{1}{2}T(r, w) + O(\log r + T(r, g)) \quad \text{and} \quad \delta(g, w) \leq \frac{1}{2}.$$

Moreover, if $g(z)$ *satisfies* $\liminf_{r\to\infty}(r^{5/2}/\log r)^{-1}T(r,g) = 0$ *as well, then*

$$m(r, 1/(w-g)) \ll \log r + T(r,g) \quad \textit{and} \quad \delta(g,w) = 0.$$

Proof. Let $w(z)$ be an arbitrary solution of (P_1). The function $u(z) := w(z) - g(z)$ satisfies

$$u''(z) = 6u(z)^2 + 12g(z)u(z) + G(z), \tag{12.1}$$

where

$$G(z) := 6g(z)^2 + z - g''(z).$$

By assumption, $T(r,G) \ll \log r + T(r,g)$, and $\rho(u) \le \rho(w) < +\infty$. Applying Lemma B.11 to (12.1), we have

$$m(r,u) \ll \log r + T(r,g) + T(r,G) \ll \log r + T(r,g). \tag{12.2}$$

By Lemma B.12, if $G(z) \not\equiv 0$, then

$$m(r, 1/(w-g)) = m(r,u) \ll \log r + T(r,g) \quad \text{and} \quad \delta(g,w) = 0. \tag{12.3}$$

Recall from (7.31) that, for every solution $W(z)$ of (P_1),

$$\liminf_{r\to\infty}(r^{5/2}/\log r)^{-1}T(r,W) > 0.$$

If $g(z)$ now satisfies the additional supposition

$$\liminf_{r\to\infty}(r^{5/2}/\log r)^{-1}T(r,g) = 0,$$

then $G(z) \not\equiv 0$, hence we have (12.3), and we obtain the second part of the assertion.

In what follows, we may assume that $G(z) \equiv 0$, meaning that $g(z)$ is a solution of (P_1). Defining now

$$U(z) := u'(z)^2 + 2g'(z)u'(z) - 4u(z)^3 - 12g(z)u(z)^2 - 2(6g(z)^2 + z)u(z), \tag{12.4}$$

it is immediate to observe that

$$U'(z) = -2u(z). \tag{12.5}$$

Now consider the function

$$H(z) := U(z)/u(z).$$

Since $m(r, u'/u) \ll \log r$, we may use (12.4) and (12.2) to see that

$$m(r,H) \ll m(r,u'/u) + m(r,u) + T(r,g) \ll \log r + T(r,g). \tag{12.6}$$

Recall the series expansion of a solution of (P_1) around a pole $z = z_0$: $(z-z_0)^{-2} + O(z-z_0)^2$. Since both $g(z)$ and $w(z)$ are solutions of (P_1), every pole of $u(z)$ is double. By (12.5), these are simple poles of $U(z)$, hence zeros of $H(z)$. Therefore, by (12.6),

$$\frac{1}{2}N(r,u) \le N(r,1/H) \le T(r,H)+O(1) \le N(r,H)+O(\log r+T(r,g)). \quad (12.7)$$

By (12.5) and the definition of $H(z)$, every pole of $H(z)$ is a zero of $u(z)$. Therefore,

$$N(r,H) \le N(r,1/u) = T(r,u) - m(r,1/u) + O(1).$$

Substituting this into (12.7), and using (12.2), we derive

$$\begin{aligned} m(r,1/u) &\le T(r,u) - \frac{1}{2}N(r,u) + O(\log r + T(r,g)) \\ &\le \frac{1}{2}T(r,w) + O(\log r + T(r,g)); \end{aligned}$$

this immediately implies that $\delta(g,w) \leqq 1/2$. □

Remark. It remains open whether (P_1) may admit two solutions $w(z)$, $g(z)$ such that $T(r,g) = S(r,w)$. We believe that this is not the case. If so, then $\delta(g,w) = 0$ holds good for all small target functions.

Theorem 12.2. *Let $w(z)$ be an arbitrary transcendental solution of (P_2). If $T(r,g) = S(r,w)$, then*

$$m(r,1/(w-g)) \le \frac{1}{2}T(r,w) + O(\log r + T(r,g)) \quad \textit{and} \quad \delta(g,w) \le \frac{1}{2}.$$

Proof. In this proof, we use the following special notation: For a set $A \subset \mathbb{C}$, define

$$N(r,f)|_A := \int_0^r \big(n(\rho,f)|_A - n(0,f)|_A\big)\frac{d\rho}{\rho} + n(0,f)|_A \log r,$$

where

$$n(\rho,f)|_A := \sum_{|\tau|\le\rho|A} \mu(\tau,f),$$

where the summation is over all poles τ in $A \cap \{|z| \le \rho\}$, and $\mu(\tau,f)$ stands for the multiplicity of the pole $z=\tau$ of $f(z)$.

Let now $w(z)$ be an arbitrary solution of (P_2). Putting $v(z) := w(z) - g(z)$, we have

$$v''(z) = 2v(z)^3 + 6g(z)v(z)^2 + (6g(z)^2 + z)v(z) + \widetilde{G}(z), \quad (12.8)$$

where

$$\widetilde{G}(z) = 2g(z)^3 + zg(z) + \alpha - g''(z).$$

Observing that $\rho(v) < +\infty$, and making use of the Clunie lemma again, we conclude from (12.8) that

$$m(r,v) \ll \log r + T(r,g). \quad (12.9)$$

In the case $\widetilde{G}(z) \not\equiv 0$, Lemma B.12 implies that $m(r, 1/(w-g)) \ll \log r + T(r, g)$, hence $\delta(g, w) = 0$. Therefore, we may now assume $\widetilde{G}(z) \equiv 0$, meaning that $g(z)$ also satisfies (P_2). Defining now

$$\begin{aligned} V(z) &:= v'(z)^2 + 2g'(z)v'(z) - v(z)^4 - 4g(z)v(z)^3 \\ &\quad - (6g(z)^2 + z)v(z)^2 - (4g(z)^3 + 2zg(z) + 2\alpha)v(z). \end{aligned} \tag{12.10}$$

we readily observe that

$$V'(z) = -v(z)^2 - 2g(z)v(z). \tag{12.11}$$

Consider now the function

$$K(z) := V(z)/v(z). \tag{12.12}$$

By (12.10) and (12.9), we immediately get

$$m(r, K) \ll m(r, v'/v) + m(r, v) + T(r, g) + \log r \ll \log r + T(r, g). \tag{12.13}$$

Recall again that, for each solution of (P_2), every pole is simple. Since both $w(z)$ and $g(z)$ are solutions of (P_2), every pole of $v(z)$ is simple and belongs either to

$$P = \{\tau \mid w(\tau) = \infty,\ g(\tau) \neq \infty\}$$

or to

$$P' = \{\tau \mid g(\tau) = \infty\}.$$

In particular, around $z_0 \in P$, we have $v(z) = \pm(z-z_0)^{-1} + O(1)$ by (2.1). Therefore, by (12.11) and (12.12), it follows that

$$K(z_0) = \pm 1 \tag{12.14}$$

for every $z_0 \in P$.

We proceed to derive the conclusion under the condition that

$$K(z) \not\equiv \pm 1, \tag{12.15}$$

whose validity will be verified afterwards. Note that

$$N(r, v) = N(r, v)|_P + N(r, v)|_{P'} \leq N(r, v)|_P + T(r, g).$$

By (12.14) and (12.15),

$$N(r, v)|_P \leq N(r, 1/(K+1)) + N(r, 1/(K-1)) \leq 2T(r, K) + O(1).$$

Hence, by (12.13),

$$N(r, v) \leq 2N(r, K) + O(\log r + T(r, g)). \tag{12.16}$$

By (12.11), (12.12) and (12.14), we see that each pole z_0 of $K(z)$ belongs to $Z \cup P'$ with $Z := \{\tau' \mid v(\tau') = 0\}$. If now $z_0 \in P' \setminus Z$, and $v(z_0)$ is finite, a contradiction with (12.11) immediately follows. Hence, z_0 is a pole of $v(z)$, and by (12.11) and (12.12), $K(z_0)$ has to be finite, a contradiction. Therefore, every pole of $K(z)$ belongs to Z and, by (12.11), is not a pole of $V(z)$. This fact implies

$$N(r, K) \leq N(r, 1/v) \leq T(r, v) - m(r, 1/v) + O(1). \tag{12.17}$$

From (12.16), (12.17) and (12.9), we conclude that

$$m(r, 1/v) \leq T(r, v) - \frac{1}{2}N(r, v) + O(1) \leq \frac{1}{2}T(r, w) + O(\log r + T(r, g)),$$

which is the desired inequality.

In the remaining part of the proof, we need to show that (12.15) holds. To this end, assume that

$$K(z) \equiv \pm 1. \tag{12.18}$$

Then we have

Lemma 12.3. *If* (12.18) *holds, then both* $g(z)$ *and* $w(z)$ *are solutions of the Riccati differential equation*

$$w' = \mp(w^2 + z/2). \tag{12.19}$$

Proof. Since $V(z) \equiv \pm v(z)$, differentiating and recalling (12.11) results in

$$v'(z) = \mp\big(v(z)^2 + 2g(z)v(z)\big). \tag{12.20}$$

On the other hand, by (12.10),

$$\begin{aligned} &v'(z)^2 + 2g'(z)v'(z) - v(z)^4 - 4g(z)v(z)^3 - (6g(z)^2 + z)v(z)^2 \\ &\quad - (4g(z)^3 + 2zg(z) + 2\alpha)v(z) = \pm v(z). \end{aligned} \tag{12.21}$$

Substituting (12.20) into (12.21), we have

$$F_{\mp}(z)v(z) \pm F'_{\mp}(z) = 0, \quad F_{\mp}(z) = -2g(z)^2 \mp 2g'(z) - z.$$

If $F_{\mp}(z) \not\equiv 0$, then $T(r, v) = T(r, F'_{\mp}/F_{\mp}) \ll T(r, g) + \log r$, which contradicts the supposition $T(r, g) = S(r, w)$. Hence $F_{\mp}(z) \equiv 0$, namely $g(z)$ is a solution of (12.19). Substituting $v(z) = w(z) - g(z)$ into (12.20), we verify that $w(z)$ also satisfies (12.19). □

To derive a contradiction from (12.18), we have to know the growth of solutions of (12.19). To this end, it is well-known that all solutions of Riccati differential equation (12.19) may be expressed, with the respective sign, as $w(z) = \pm\frac{u'}{u}$, where u is a solution of

$$u'' + \frac{z}{2}u = 0. \tag{12.22}$$

By Gundersen [1], Theorem 7, we at once conclude that

$$r^{3/2} \ll N(r, \frac{1}{u}) \le T(r, w) \ll r^{3/2}$$

as the order of $w(z)$ equals to $3/2$. Therefore, as $w(z)$ and $g(z)$ both satisfy (12.19), we have $T(r, g) \ll T(r, w) \ll T(r, g)$, which is a contradiction. This completes the proof of Theorem 12.2. □

For solutions of (P_4), deficiencies relative to small targets can be treated by essentially the same method, see Shimomura [7]. Under certain suppositions, there exist pairs $(\chi(z), g(z))$ of transcendental and rational solutions of (P_4) satisfying $r^2 \ll T(r, \chi) \ll r^2$ and $N(r, 1/(\chi - g)) = 0$. For simplicity, we state here a result for transcendental small targets, in which case the error terms follow from the finiteness of the growth order. For more results and proofs, see Shimomura [7].

Theorem 12.4. *Let $w(z)$ be an arbitrary transcendental solution of (P_4). Suppose that $g(z)$ is transcendental and $T(r, g) = S(r, w)$. Then, if $\beta \neq 0$,*

$$m(r, 1/(w - g)) \le \frac{1}{2} T(r, w) + O(\log r + T(r, g)) \quad \textit{and} \quad \delta(g, w) \le \frac{1}{2},$$

and if $\beta = 0$,

$$m(r, 1/(w - g)) \le \frac{1}{4} T(r, w) + O(\log r + T(r, g)) \quad \textit{and} \quad \delta(g, w) \le \frac{1}{4}.$$

For solutions of (12.1) or (12.8), it is easy to see that the number of zeros with multiplicity ≥ 3 does not exceed $O(T(r, g))$, and hence $\vartheta(g, w) \le 1/2$. As an example of a nontrivial estimate for $\vartheta(g, w)$, see Shimomura [4] for the proof of

Theorem 12.5. *For an arbitrary solution $w(z)$ of (P_1), if $T(r, g) = S(r, w)$, then*

$$N_1(r, 1/(w - g)) \le \frac{5}{12} T(r, w) + O(\log r + T(r, g)) \quad \textit{and} \quad \vartheta(g, w) \le \frac{5}{12}.$$

Concerning ramification indices of Painlevé transcendents relative to small target functions, this topics offers a large number of open research problems. For known results, see Shimomura [4], [7] and Korhonen [1].

Chapter 4
The first Painlevé equation (P_1)

The first Painlevé equation (P_1) is unique among the six classical Painlevé equations in the sense that there are no parameters in the equation itself as is the case for (P_2) through (P_6). Therefore, the results in the three previous chapters on the meromorphic nature, growth and value distribution of the first Painlevé transcendents cover the essential knowledge on them. Some complements have been collected in this chapter. We first offer a rigorous proof for the fact that no first Painlevé transcendent satisfies a first order algebraic differential equation with rational coefficients. Secondly, we present some representations and expansions for solutions of (P_1). Finally, we start our presentation of higher analogues of Painlevé equations, to be continued in the subsequent chapters.

§13 Nonexistence of the first integrals

In this section, we show that the first Painlevé equation does not admit the first algebraic integral. To express this result in a precise form, recall first the standard notation $\mathcal{A}(\mathbb{C}(z))$ for the family of meromorphic functions which satisfy an algebraic differential equation with rational coefficients. The corresponding family related with first order algebraic differential equations will be denoted by $\mathcal{A}_1(\mathbb{C}(z))$. We are now ready to formulate

Theorem 13.1. *Given any solution $w(z)$ of (P_1), then $w(z) \notin \mathcal{A}_1(\mathbb{C}(z))$.*

Proof. Contrary to the assertion, suppose that $w(z) \in \mathcal{A}_1(\mathbb{C}(z))$, hence satisfying a first order algebraic differential equation

$$(w')^m + \sum_{j=1}^{m} P_j(z, w)(w')^{m-j} = 0, \tag{13.1}$$

where $P_j(z, w) \in \mathbb{C}(z)[w]$, $j = 1, \dots, m$, by the second Malmquist theorem, see Erëmenko [1], Theorem 6, and Hotzel [1], Satz 4.5. Consider the corresponding algebraic equation

$$X^m + \sum_{j=1}^{m} P_j(z, w)X^{m-j} = 0. \tag{13.2}$$

Denoting by $\mathbb{A}_z$ the field of algebraic functions of z and decomposing (13.2) in $\mathbb{A}_z[w, X]$ as a polynomial of (w, X), if necessary, we may assume that (13.2) is irreducible in $\mathbb{A}_z[w, X]$. Then (13.2) defines an algebraic function $X = G(z, w)$, which represents a connected algebraic surface S_X in $\widehat{\mathbb{C}} \times \widehat{\mathbb{C}}$, see Ahlfors [1], p. 291–297. Recalling Theorem 10.1, we may take a pole z_0 of $w(z)$ such that all algebraic coefficients of $P_j(z, w)$, $j = 1, \dots m$, assume finite values at z_0. Recalling the series expansion of $w(z)$ around a pole, we observe that $w(z)$ satisfies, locally around z_0, a differential equation

$$w' = g_0(z, w), \tag{13.3}$$

where $g_0(z, w)$ may be expressed around (z_0, ∞) by a series

$$g_0(z, w) = 2\mu_0 w^{3/2} + \frac{\mu_0}{2} z w^{-1/2} - \frac{1}{2} w^{-1} - 7\mu_0 h_0 w^{-3/2} + \cdots \tag{13.4}$$

with $\mu_0^2 = 1$ and with some $h_0 \in \mathbb{C}$, see (1.4). It is not too difficult to show that $g_0(z, w)$ must be one of the branches of $G(z, w)$ around (z_0, ∞). Let now $g_1(z, w), \dots, g_{m-1}(z, w)$ be the other branches of $G(z, w)$ around (z_0, ∞). For each $j \neq 0$, consider now $g_j(z, w)$. In $\widehat{\mathbb{C}} \times \widehat{\mathbb{C}}$, there exists a loop $\Gamma_j \subset \{z_0\} \times \widehat{\mathbb{C}}$ which, by the connectedness of the algebraic surface S_X, possesses the following properties:

(i) Γ_j starts from and ends at (z_0, ∞), and is expressed as $\Gamma_j = \{(z_0, \gamma(t)) \mid 0 \leq t \leq 1\}$, where $\gamma : [0, 1] \to \widehat{\mathbb{C}}$ is continuous satisfying $\gamma(0) = \gamma(1) = \infty$ and $\gamma(t) \neq \infty$ for $0 < t < 1$;

(ii) the analytic continuation of $g_0(z, w)$ along Γ_j results in $g_j(z, w)$, i.e. there exists a function $g(t, z, w)$ such that (a) for each τ, $0 < \tau < 1$, $g(\tau, z, w)$ is analytic at $(z_0, \gamma(\tau))$ and (b) $g(0, z, w) = g_0(z, w)$, $g(1, z, w) = g_j(z, w)$.

By (1.5) and (1.1), (P_1) admits a family of solutions

$$w(z; \sigma) = \frac{1}{(z-\sigma)^2} - \frac{\sigma}{10}(z-\sigma)^2 - \frac{1}{6}(z-\sigma)^3 + h_0(z-\sigma)^4 + \cdots,$$

where h_0 is the same coefficient as in (13.4), such that $w(z; \sigma)$ also satisfies (13.3) in some neighborhood of z_0. We may regard $w(z_0, \sigma)$ as a function of σ, which maps a small open disk, centred at z_0, onto an open set around $w = \infty$. Hence, for every $(z_0, w_0) \in \Gamma_j$ close enough to (z_0, ∞), there exists a solution $w(z; z_0, w_0)$ of (P_1) such that $w(z_0; z_0, w_0) = w_0$, satisfying (13.3) as well. Furthermore, we now conclude that for any point $(z_0, \gamma(\tau))$ with $0 \leq \tau \leq 1$, there exists a solution $w(z; z_0, \gamma(\tau))$ of (P_1) such that $w(z_0; z_0, \gamma(\tau)) = \gamma(\tau)$ and which satisfies

$$w' = g(\tau, z, w) \tag{13.5}$$

around $(z_0, \gamma(\tau))$. To prove this conclusion, define τ_* by

$$\tau_* := \sup\left\{\tau \in [0, 1] \mid \text{the conclusion holds at } (z_0, \gamma(\tau))\right\}.$$

Clearly, $\tau_* > 0$. Suppose that $\tau_* < 1$, and take $(z_0, \gamma(t_0))$ with $t_0 < \tau_*$ sufficiently close to $(z_0, \gamma(\tau_*))$. Then the solution $w(z; z_0, \gamma(t_0))$ of $(P_{\rm I})$ such that $w(z_0; z_0, \gamma(t_0)) = \gamma(t_0)$ satisfies (13.5) with $\tau = t_0$. Since $w(z; z_0, \gamma(t_0))$ is meromorphic in an open neighborhood of $(z_0, \gamma(t_0))$ containing $(z_0, \gamma(\tau_*))$, it satisfies (13.5) with $\tau = \tau_*$ around $(z_0, \gamma(\tau_*))$, a contradiction with the definition of τ_*.

By the above fact, the differential equation

$$w' = g_j(z, w)$$

admits around z_0 a solution $w_j(z)$ such that it also solves $(P_{\rm I})$ and has a pole at z_0. Hence, for every $j = 1, \dots, m-1$,

$$g_j(z, w) = 2\mu_j w^{3/2} + \frac{\mu_j}{2} z w^{-1/2} - \frac{1}{2} w^{-1} - 7\mu_j h_j w^{-3/2} + \cdots$$

around (z_0, ∞) with $\mu_j^2 = 1$ and $h_j \in \mathbb{C}$. Recalling that $g_j(z, w)$, $j = 0, \dots, m-1$, were the branches of the algebraic function $X = G(z, w)$ determined by (13.2), we immediately conclude that

$$P_1(z, w) = -\sum_{j=0}^{m-1} g_j(z, w) = \Big(\sum_{j=0}^{m-1} \mu_j\Big) w^{1/2}\big(2w + \frac{1}{2} z w^{-1}\big) - \frac{m}{2} w^{-1} - \cdots .$$

This is a contradiction, as $P_1(z, w)$ has to be a polynomial in w with rational coefficients. □

Remark. The above result means, intuitively, that the general solution of $(P_{\rm I})$ cannot be an algebraic (or rational) function of the two constants of integration. See Ince [1] for more details.

§14 Representation of solutions as quotient of entire functions

It has been shown in Chapter 1 that any solution $w(z)$ of the equation $(P_{\rm I})$ is a single-valued meromorphic function. Therefore, $w(z)$ may be represented as the quotient of two entire functions, say

$$w(z) = v(z)/u(z). \tag{14.1}$$

To obtain a canonical representation of this type, we shall follow the method proposed in Painlevé [2] and Erugin [1]. For this purpose, we introduce the following function:

$$\eta(z) := (w')^2/2 - 2w^3 - zw. \tag{14.2}$$

Differentiating (14.2) and taking into account $(P_{\rm I})$, we find that $d\eta/dz = -w$. Consequently, $\eta(z)$ satisfies the differential equation

$$\eta''' + 6(\eta')^2 + z = 0. \tag{14.3}$$

Clearly, $\eta(z)$ is a meromorphic function with simple poles only with residue $+1$ exactly at the poles of w. Therefore, there exists an entire function $\xi(z)$ such that $\xi'/\xi = \eta$, see Saks and Zygmund [1], with simple zeros exactly at the poles of $\eta(z)$. Differentiating η, we get the representation

$$w(z) = \frac{(\xi')^2 - \xi\xi''}{\xi^2}, \tag{14.4}$$

of w as a quotient of two entire functions. Substituting $\eta = \xi'/\xi$ into (14.3) we get the following differential equation:

$$\xi\xi^{(4)} - 4\xi'\xi''' + 3(\xi'')^2 + z\xi^2 = 0. \tag{14.5}$$

Therefore, the function $\xi(z)$ has a similar role in the theory of the equation (P_1) as the entire σ-function $\sigma(z)$ in the theory of the Weierstrass $\wp$-function. Note that $\wp(z) = ((\sigma')^2 - \sigma\sigma'')/\sigma^2 = -\varsigma'(z)$, $\varsigma(z) = \sigma'(z)/\sigma(z)$.

We next seek for the solution of the equation (14.5) corresponding to the following initial conditions for solutions of the equation (P_1):

$$w(0) = w_0, \qquad w'(0) = w_1, \tag{14.6}$$

$$w(0) = 0, \qquad w'(0) = 1, \tag{14.7}$$

$$w(0) = 1, \qquad w'(0) = 0, \tag{14.8}$$

$$w(z) \to \infty, \qquad w'(z) \to \infty, \text{ as } z \to z_0. \tag{14.9}$$

Of course, the case (14.9) corresponds to a pole z_0 of a solution $w(z)$.

Consider first the initial conditions for ξ corresponding to the conditions (14.6). This can be done on the basis of the formulas (14.2), (14.4) and of $\xi'/\xi = \eta$. Normalizing first $\xi(0) = 1$, we get $\xi(0) = 1$, $\xi'(0) = w_1^2/2 - 2w_0^3$, $\xi''(0) = (w_1^2/2 - 2w_0^3)^2 - w_0$, $\xi'''(0) = (w_1^2/2 - 2w_0^3)^3 - 3(w_1^2/2 - 2w_0^3)w_0 - w_1$. The solution of the equation (14.5) satisfying these initial conditions will be unique and entire.

In the case of the initial conditions (14.7) we obtain, similarly, $\xi(0) = 1$, $\xi'(0) = 1/2$, $\xi''(0) = 1/4$, $\xi'''(0) = -7/8$. Under the initial conditions (14.7), we get for ξ the series expansion

$$\xi = 1 + \frac{1}{2}z + \frac{1}{8}z^2 - \frac{7}{48}z^3 + \sum_{k=4}^{\infty} \xi_k^{(1)} z^k. \tag{14.10}$$

Respectively, the conditions (14.8) result in $\xi(0) = 1$, $\xi'(0) = -2$, $\xi''(0) = 3$, $\xi'''(0) = -2$ and in the series expansion $\xi = 1 - 2z + \frac{3}{2}z^2 - \frac{1}{3}z^3 - \frac{11}{24}z^4 + \sum_{k=5}^{\infty} \xi_k^{(2)} z^k$.

We now seek for the solution of the equation (14.5) corresponding to the initial conditions (14.9). For this purpose, we shall write the equation (14.5) in the form

$$\xi\xi^{(4)} - 4\xi'''\xi' + 3(\xi'')^2 + (z - z_0)\xi^2 + z_0\xi^2 = 0, \tag{14.11}$$

from which we obtain the solution

$$\xi = \sum_{j=1}^{\infty} \alpha_j (z - z_0)^j, \tag{14.12}$$

as ξ must have a simple zero at z_0. Obviously, we may normalize by $\alpha_1 = 1$. Substituting (14.12) into (14.11), we get the following equations for determining a_j's:

$$\begin{cases} \alpha_2^2 = 2\alpha_3, \\ \alpha_2\alpha_3 = 3\alpha_4, \\ z_0 + (3!)^2\alpha_3^2 - 4!\alpha_2\alpha_4 - 5!\alpha_5 = 0, \\ 3 \cdot 4!\alpha_3\alpha_4 - 5!\alpha_2\alpha_5 - 5!\alpha_6 + 1 + 2z_0\alpha_2 = 0, \\ 72\alpha_4^2 - 2 \cdot 5!\alpha_2\alpha_6 + (2\alpha_3 + \alpha_2^2)z_0 + 2\alpha_2 = 0. \end{cases} \tag{14.13}$$

From the equations in (14.13) we find $\alpha_3, \dots, \alpha_6$ in terms of α_2. Then it is easy to verify that the last equation of (14.13) is satisfied identically. Comparing the coefficients of higher degree we obtain equations from which $\alpha_8, \alpha_9, \dots$ can be defined uniquely provided α_7 is given. Therefore, there exists a two-parameter family of power series (14.12) which satisfy the equation (14.11) formally. Here α_2 and α_7 remain as arbitrary parameters. But we find a relation between the arbitrary constant h in (1.1) and the parameters α_2, α_7. Substituting the expansions (1.1) and (14.12) into (14.4), we obtain

$$h = \frac{1}{4}\alpha_2 + \frac{1}{8}z_0\alpha_2^2 + \frac{1}{24}\alpha_2^6 - 30\alpha_7.$$

In the presence of the regular expansion of $w(z)$ corresponding, for example, to the initial conditions (14.7) and using the series (14.10), it is possible to find, approximately, the root z_0 of the equation $\xi(z_0) = 0$ with the smallest absolute value. This way, given the initial conditions (14.7) for $w(z)$, the pole of $w(z)$ nearest to the origin can be approximately found.

It must be noted that by the definition of the entire function $\xi(z)$ in accordance with (14.2) as to above, we obtain the third order equation (14.3) from the pair

$$\xi'\xi^{-1} = \eta, \quad (\eta'')^2 + 4(\eta')^3 + 2z\eta' - 2\eta = 0. \tag{14.14}$$

Here $\eta \neq Cz + 2C^3$, which corresponds to $w \neq -C$. In fact, we only need to differentiate the second equation of (14.14).

Another method to construct entire functions whose quotient equals to w is given in Gromak and Lukashevich [1]. This method is based on the following observation:

Since

$$F := zw + 2w^3 - \frac{1}{2}w'^2$$

is a primitive function of a solution $w(z)$ of (P_1), then there exists, by Saks and Zygmung [1] again, an entire function $u(z)$ with zeros exactly at the poles of $w(z)$

such that $\frac{u'}{u} = -2F$. As all these zeros are double, $v(z) = u(z)w(z)$ is entire too. As $F' = w$, the system for the entire functions $u(z)$, $v(z)$ takes the form

$$\begin{cases} uu'' - (u')^2 = -2uv \\ v''u^2 + v(u')^2 - 2uu'v' - 4uv^2 - zu^3 = 0 \end{cases}$$

In Davis, Scott, Springer and Resch [1], the question of determining the nearest pole to a holomorphic point of a solution $w(z)$ has also been considered.

§15 Special expansions of solutions

Let $w(z)$ be a solution of the equation (P_1) with the finite initial values w_0', w_0, z_0. Then, by to the Cauchy majorant theorem, this solution is holomorphic in a neighborhood of $z = z_0$. As shown above, all solutions of (P_1) are non-rational meromorphic functions with infinitely many poles accumulating at $z = \infty$. Therefore, the point $z = \infty$ cannot be a pole or a holomorphic point of $w(z)$. Recall that the Laurent expansion (1.1) of $w(z)$ around a pole $z = z_0$ takes the form

$$\begin{aligned} w(z) = (z-z_0)^{-2} - \frac{z_0}{10}(z-z_0)^2 - \frac{1}{6}(z-z_0)^3 \\ + h(z-z_0)^4 + \frac{z_0^2}{300}(z-z_0)^6 + \sum_{k=7}^{\infty} a_k(z-z_0)^k, \end{aligned}$$

where z_0 and h are arbitrary constant parameters and the coefficients a_k, $k \geq 7$, are determined uniquely in terms of z_0 and h. Clearly, $w(z)$ converges in the punctured disc $B(z_0, d) \setminus \{z_0\}$, where d is the distance from z_0 to the nearest pole $z_0' \neq z_0$ of $w(z)$.

This radius of convergence of the series (1.1) can be estimated directly, see Hille [2], p. 442. Actually, let us choose the number $M > 1$ such that $|z_0| < 10M$, $|h| < M^3$. Looking at the expansion (1.1), it is evident that

$$|a_k| < M^k, \quad 1 \leq k \leq 6. \tag{15.1}$$

The coefficients a_k, $k > 6$, are given by the recurrence relation $[k(k-1) - 12]a_k = 6\sum_{j=2}^{k-4} a_j a_{k-2-j}$. Let us assume that the inequality (15.1) holds for $k < n$. Then for $|a_n|$ the estimate $|a_n| \leq (6M^{n-2}/(n+3) < M^n$, follows. Therefore, (15.1) holds for all $k \in \mathbb{N}$, and so (1.1) converges in the domain $0 < |z - z_0| < M^{-1}$.

Writing (P_1) in the modified form

$$y'' = -z + 6y^2, \tag{15.2}$$

we observe that (15.2) has a formal solution

$$y(z) = \frac{z^{1/2}}{\sqrt{6}} \sum_{k=0}^{\infty} \frac{a_k}{(z^{1/2})^{5k}}, \tag{15.3}$$

where $z^{1/2}$ denotes a fixed branch of the square root, $a_0 = 1$ and the coefficients a_{k+1}, $k \geq 0$, are given by the recurrence relation

$$a_{k+1} = \frac{25k^2 - 1}{8\sqrt{6}} a_k - \frac{1}{2} \sum_{m=1}^{k} a_m a_{k+1-m}, \quad k \geq 0. \tag{15.4}$$

This can be verified by the direct substitution of (15.3) into (15.2). The series (15.3) has an asymptotic character, as already observed by Boutroux [2]. For a modern proof of this property, see Joshi and Kitaev [1], we have to rely on the following

Theorem 15.1 (Wasow [1], Theorem 12.1). *Let S be an open sector in the complex plane with vertex at the original and of a positive opening not exceeding $\pi/(q+1)$ for some $q \in \mathbb{N}$. Let $f(z, \vec{w})$ be an r-dimensional vector function of z and of an r-dimensional vector $\vec{w}$ with the following properties:*

(a) *The vector $f(z, \vec{w})$ is a polynomial in the components $w_1, \ldots, w_r$ with coefficients holomorphic in the region*

$$0 < z_0 \leq |z| < \infty, \quad z \in S.$$

(b) *The coefficients of the polynomial $f(x, \vec{w})$ have asymptotic series in powers of z^{-1}, as $z \to \infty$ in S.*

(c) *If $f_j(z, \vec{w})$ denotes the components of $f(z, \vec{w})$, then all eigenvalues $\lambda_1, \ldots \lambda_r$ of the Jacobian matrix*

$$\left\{ \lim_{S \ni z \to \infty} \left(\frac{\partial f_j}{\partial w_k} |_{\vec{w}=0} \right) \right\}$$

are different from zero.

(d) *The differential equation*

$$z^{-q} \vec{w}' = f(z, \vec{w}) \tag{15.5}$$

is formally satisfied by a power series of the form

$$\sum_{k=1}^{\infty} b_k z^{-k}.$$

Then there exists, for sufficiently large z in S, a solution $\vec{w} = \phi(z)$ of (15.5) such that, in every proper subsector of S,

$$\phi(z) \sim \sum_{k=1}^{\infty} b_k z^{-k}, \quad as\ z \to \infty.$$

Here the notation $\sim$ is used in the usual asymptotic sense.

Theorem 15.2. *For any sector of Ω of opening less than $4\pi/5$ and with vertex at the origin, there exists a solution of* (15.2) *whose asymptotic behavior as $|z| \to \infty$, $z \in \Omega$, is given by the asymptotic series* (15.3).

Proof. Let $y(z)$ be a solution of (15.2). Define a transformation

$$w(\zeta) = \frac{y(z)}{z^{1/2}} - \frac{1}{\sqrt{6}}, \quad \zeta = \frac{4z^{5/4}}{5}, \tag{15.6}$$

where $z^{1/4} = (z^{1/2})^{1/2}$. Then the function $w(\zeta)$ solves the differential equation

$$w'' = 2\sqrt{6}w + 6w^2 - \frac{w'}{\zeta} + \frac{4(w + 1/\sqrt{6})}{25\zeta^2}, \tag{15.7}$$

where w' now stands for the differentiation with respect to ζ. The series (15.3) becomes

$$w(\zeta) \sim \frac{1}{\sqrt{6}} \sum_{n=1}^{\infty} a_n \left(\frac{16}{25}\right)^n \zeta^{-2n}. \tag{15.8}$$

By denoting $w_1 = w$, $w_2 = w'$, we may consider (15.7) as a pair of differential equations,

$$\begin{cases} w_1' = w_2, \\ w_2' = 2\sqrt{6}w_1 + 6w_1^2 - \dfrac{w_2}{\zeta} + \dfrac{4(w_1 + 1/\sqrt{6})}{25\zeta^2}. \end{cases} \tag{15.9}$$

Theorem 14.1 now applies for (15.9) with $q = 0$, $b_{2j} = a_j(16/25)^j/\sqrt{6}$, $b_{2j+1} = 0$, $r = 2$, and $\lambda_1 = -\lambda_2 = \sqrt{2\sqrt{6}}$. □

Boutroux also showed that the five rays $\arg(x) = 2\pi n/5$, $n = 0, \pm 1, \pm 2$ play an important role in the asymptotic description of solutions. According to his analysis, the general solution of (15.2) has sequences of poles which are asymptotic to these rays. The solutions determined by Theorem 15.2 have the property that the sequences of poles asymptotic to one of the these rays are truncated for large z. In fact, by the asymptotic behavior of these solutions $y(z) \to \left(\frac{z}{6}\right)^{1/2}$, as $|z| \to \infty$, $x \in \Omega$.

Asymptotic properties of solutions of the Painlevé equations and related problems have been investigated by the method of isomonodromic deformations of linear systems, see. e.g. Kitaev [2], Fokas and Its [1] and Its and Novokshenov [1].

Substitute now in the equation (P_1),

$$w = z^3 v, \quad z = \exp(\tau/5). \tag{15.10}$$

Denoting now $dv/d\tau$ by v', we obtain

$$25v'' + 25v' + 6v = 1 + 6v^2 \exp\tau. \tag{15.11}$$

The solution of the equation (15.11) is now sought for in the form of a special series

$$v(\tau) = \sum_{n=0}^{\infty} v_n \exp(n\tau). \tag{15.12}$$

Substituting (15.12) into (15.11) and comparing the coefficients, we obtain $v_0 = 1/6$,

$$(25n^2 + 25n + 6)v_n = 6 \sum_{j=0}^{n-1} v_j v_{n-1-j}, \quad n \geq 1. \tag{15.13}$$

It is immediate that (15.13) determines uniquely all coefficients v_n. Consequently, the formal expansion of the solution which satisfies the initial conditions $w(0) = w'(0) = 0$ exists and has the form (15.10), (15.12). Denote now $\sigma_n := 25n^2 + 25n + 6$. From (15.13) we obtain $v_1 = \frac{1}{6\sigma_1}$, $v_2 = \frac{1}{3\sigma_1\sigma_2}$, $v_3 = \frac{4\sigma_1+\sigma_2}{6\sigma_1^2\sigma_2\sigma_3}$, $v_4 = \frac{\sigma_2+4\sigma_1+2\sigma_3}{3\sigma_1^2\sigma_2\sigma_3\sigma_4}$. Thus, we can determine, uniquely, all the coefficients of the formal solution $w(z)$ satisfying the initial conditions $w(0) = w'(0) = 0$:

$$w(z) = z^3 \sum_{n=0}^{\infty} v_n z^{5n}. \tag{15.14}$$

If we seek for the solution of the equation (15.11) in the form $v(\tau) = \exp(-\tau) + \sum_{n=0}^{\infty} v_n \exp(n\tau)$, then we have $v_0 = -1/6$ and

$$(25n^2 + 25n - 6)v_n = 6 \sum_{j=0}^{n-1} v_j v_{n-1-j}, \quad n \geq 1. \tag{15.15}$$

Determining v_n from (15.15), we verify that the equation (P_1) has a formal solution of the form

$$w(z) = z^{-2} + z^3 \sum_{n=0}^{\infty} v_n z^{5n}. \tag{15.16}$$

It is not difficult to prove the convergence of the series involved in (15.14) and (15.16). For example, the convergence of (15.14) follows from the above proof of the convergence of series (1.1). Moreover, in this case, the domain of the convergence of the series (15.14) is at least $|z| < 1$, and $0 < |z| < 1$ for the series (15.16).

The asymptotic formulas of solutions (15.14) and (15.16) have been obtained by Kapaev [1] and Kitaev [3], respectively. In particular, see Kapaev [1], as $|z| \to \infty$ on the rays $\arg(z) = \pi + 2\pi k/5$, $k = 0, 1, 2, 3, 4$, the solution (15.14) behaves asymptotically as

$$
\begin{aligned}
w(z) &= \exp(-i\pi k)(z\exp(-i\pi))^{1/2}v(t), \quad t = \frac{4}{5}\exp\Big(-\frac{i\pi k}{2}\Big)(z\exp(-i\pi))^{5/4} \\
v(t) &\sim -\frac{1}{\sqrt{6}} + \frac{2\alpha}{\sqrt{5t}}\cos(h+\varphi), \\
h &= 2\Big(\frac{3}{2}\Big)^{1/4} t + \rho\ln(5t) + \frac{11}{4}\rho\ln 2 + \frac{5}{4}\rho\ln 3 + \frac{3\pi}{4}, \\
\varphi &= -\pi - \arg\Gamma(i\rho) - \arg(ip), \\
\alpha &= \Big(\frac{2}{3}\Big)^{1/8}(-\rho)^{1/2} > 0, \quad \rho = \frac{\ln p}{2\pi} < 0, \ p^2 + p = 1.
\end{aligned}
$$

§16 Higher order analogues of (P_1)

There are several approaches to construct higher order differential equations with properties similar to those of Painlevé equations. Historically, the first approach is the method of using Korteweg–de Vries equations, considered below in the case of (P_1). Examples of other approaches are the method of the isomonodromy deformation of linear systems via Garnier systems, see Kimura and Okamoto [1], Iwasaki, Kimura, Shimomura and Yosida [1] and the direct method of higher order Hamiltonian systems, Takasaki [1].

It is well-known, see Lax [1], that the Korteweg–de Vries equation for $u(x,t)$, written in the form

$$\frac{\partial u}{\partial t} = 6u\frac{\partial u}{\partial x} - \frac{\partial^3 u}{\partial x^3} \tag{KdV}$$

can be interpreted as the compatibility condition for two linear operators L and A. Let L be the Schrödinger operator $-\partial^2/\partial x^2 + u$ and $A = 4\frac{\partial^3}{\partial x^3} - 3u\frac{\partial}{\partial x} - 3\frac{\partial}{\partial x}u$. Consider now the equation

$$\frac{\partial}{\partial t}L = [L, A] = LA - AL. \tag{16.1}$$

Here $\frac{\partial}{\partial t}L$ denotes differentiation of the coefficients of the operator L in x, i.e. $\frac{\partial}{\partial t}L$ is the operator of multiplying by u_t. By simple computation, the commutator $[L, A]$ is the operator of multiplying by $6uu_x - u_{xxx}$. Therefore, for L and A as chosen here, (16.1) implies that $u(x,t)$ satisfies the (KdV) equation. The commutator representation (16.1) is called the Lax representation, and the operators L, A are an example of a Lax pair. We now seek for an operator A_m of order $2m+1$ with coefficients depending of $u(x)$ and its derivatives such that this operator equals to $\frac{1}{2}P(\frac{1}{i}\frac{\partial}{\partial x})$, if $u \equiv 0$, and satisfies the condition

$$\Big[\frac{\partial^2}{\partial x^2} - u(x), A_m\Big] = F\Big(u, \frac{\partial u}{\partial x}, \ldots, \frac{\partial^{2m+1}u}{\partial x^{2m+1}}\Big),$$

where F is the operator of multiplying by a function. Such an operator A_m exists, and it is possible to obtain recursion relations for its coefficients, see Lax [1]. In this case the equations

$$\frac{\partial u}{\partial t} = F\left(u, \frac{\partial u}{\partial x}, \dots, \frac{\partial^{2m+1} u}{\partial x^{2m+1}}\right) \tag{16.2}$$

are called higher analogues of the Korteweg–de Vries equation (KdV).

Consider now the KdV hierarchy in the form

$$\frac{\partial \omega}{\partial t} + \frac{\partial}{\partial x} B^{n+1}(\omega) = 0, \tag{16.3}$$

where $B^n(\omega)$ is determined by the Lenard recursion relations, see Lax [2], as follows:

$$B^1(\omega) = \omega, \quad \frac{\partial}{\partial x} B^{n+1}(\omega) = \left(D^3 + 2\omega D + \frac{\partial \omega}{\partial x}\right) B^n(\omega), \quad n \in \mathbb{N},$$

where $D := \frac{\partial}{\partial x}$. If $n = 1$ in (16.3), then we obtain (KdV) in the form $\omega_t + 3\omega\omega_x + \omega_{xxx} = 0$.

We also consider the singular manifold equation for the KdV hierarchy, see Weiss [2],

$$\frac{\partial z}{\partial t} + \frac{\partial z}{\partial x} B^n\big(\{z; x\}\big) = 0, \tag{16.4}$$

where

$$\{z; x\} := \frac{\dfrac{\partial^3 z}{\partial x^3}}{\dfrac{\partial z}{\partial x}} - \frac{3}{2} \frac{\left(\dfrac{\partial^2 z}{\partial x^2}\right)^2}{\left(\dfrac{\partial z}{\partial x}\right)^2}$$

is the Schwarzian derivative. Now, if $z(x, t)$ is a solution of (16.4), then $\omega(x, t) = \{z, x\}$ is a solution of (16.3), see Weiss [2], Theorem 1. We now seek the solutions in the form

$$z = z(\xi), \quad \xi = x(\lambda t)^m,$$

where λ is a parameter to be determined, see Kudryashov [1]. Then

$$\{z; x\} = \{z; \xi/(\lambda t)^m\} = (\lambda t)^{2m}\{z; \xi\} = (\lambda t)^{2m} F(\xi),$$

where $F(\xi) = \{z(\xi); \xi\}$.

We introduce an operator d by

$$d^1(F) = F, \quad \frac{\partial}{\partial \xi} d^{n+1}(F) = (D^3 + 2FD + F_\xi) d^n(F), \quad n \in \mathbb{N},$$

where $D = \frac{\partial}{\partial \xi}$. Then $B^n\big(\{z; x\}\big) = (\lambda t)^{2mn} d^n(F)$. In fact, for $n = 1$ we have

$$B^1(\{z; x\}) = \{z; x\} = \{z; \xi/(\lambda t)^m\} = (\lambda t)^{2m} F(\xi) = (\lambda t)^{2m} d^1 F(\xi).$$

To complete the induction, we have

$$\begin{aligned}\frac{\partial}{\partial x}B^{n+1}(\{z;x\}) &= (D_x^3+2\{z;x\}D_x+(\partial/\partial x)\{z;x\})B^n(\{z;x\})\\ &= (D_x^3+2\{z;x\}D_x+(\partial/\partial x)\{z;x\})(\lambda t)^{2mn}d^n(F)\\ &= (\lambda t)^{3m}(D_\xi^3+2\{z;\xi\}D_\xi+(\partial/\partial\xi)\{z;\xi\})(\lambda t)^{2mn}d^n(F)\\ &= (\lambda t)^{2m(n+1)}(\lambda t)^m(D_\xi^3+2FD_\xi+F_\xi)d^n(F)\\ &= (\lambda t)^{2m(n+1)}(\lambda t)^m\frac{\partial}{\partial\xi}d^{n+1}(F)=\frac{\partial}{\partial x}((\lambda t)^{2m(n+1)}d^{n+1}(F)),\end{aligned}$$

which implies

$$B^{n+1}(\{z;x\}) = (\lambda t)^{2m(n+1)}d^{n+1}(F).$$

Assuming now $m=-1/(2n+1)$, the equation (16.4) takes the form

$$d^n(F)-\frac{\lambda}{2n+1}\xi=0. \tag{16.5}$$

To prove (16.5), let $Z(\xi)$ be a function (of ξ) such that $z=Z((\lambda t)^m x)$ is a solution of (16.4). To show that $F(\xi)=\{Z(\xi);\xi\}$ satisfies (16.5), substitute $z=Z((\lambda t)^m x)$ in (16.4). Observing that $\{z;x\}=(\lambda t)^{2m}F(\xi)|_{\xi=(\lambda t)^m x}$, we obtain

$$\begin{aligned}\frac{\partial z}{\partial t}+\frac{\partial z}{\partial x}B^n(\{z;x\}) &= m\lambda^m t^{m-1}xZ'(\xi)|_{\xi=(\lambda t)^m x}\\ &\quad+(\lambda t)^m Z'(\xi)(\lambda t)^{2mn}d^n(F(\xi))|_{\xi=(\lambda t)^m x}\\ &= mt^{-1}x(\lambda t)^m Z'(\xi)|_{\xi=\cdots}+(\lambda t)^{(2n+1)m}Z'(\xi)d^n(F(\xi))|_{\xi=\cdots}\\ &= mt^{-1}\xi Z'(\xi)|_{\xi=(\lambda t)^m x}+(\lambda t)^{-1}Z'(\xi)d^n(F(\xi))|_{\xi=(\lambda t)^m x}=0,\end{aligned}$$

since $m=-1/(2n+1)$. Hence we have $\lambda^{-1}d^n(F(\xi))+m\xi=0$, proving (16.5).

If $n=1$, we obtain the degenerated equation $F-\frac{1}{3}\lambda\xi=0$. If $n=2$, then we have the first Painlevé equation in a modified form $F''+\frac{3}{2}F^2-\frac{1}{5}\lambda\xi=0$. In order to obtain the first Painlevé equation in the standard form we take $\lambda=-4(2n+1)$ and apply the scaling transformation

$$F\mapsto -4w,\quad \xi\mapsto z.$$

Then we get

$$d^{n+1}(w)+4z=0, \tag{$_{2n}P_1$}$$

where

$$d^1(w)=-4w,\quad \frac{d}{dz}d^{n+1}(w)=(D^3-8wD-4w_z)d^n(w),\quad n\in\mathbb{N}, \tag{16.6}$$

and where now $D = \frac{d}{dz}$. Then for $n = 1, n = 2$ and $n = 3$ we have, correspondingly,

$$w'' = 6w^2 + z, \qquad ({}_2P_1)$$

$$w^{(4)} = 20ww'' + 10(w')^2 - 40w^3 + z, \qquad ({}_4P_1)$$

$$w^{(6)} = 42(w'')^2 + 56w'w''' + 28ww^{(4)} - 280w((w')^2 + ww'' - w^3) + z. \quad ({}_6P_1)$$

The equation $({}_2P_1)$ is the first Painlevé equation in the standard form. We shall call $({}_{2n}P_1)$ with $n \geq 2$ the higher analogues of the first Painlevé equation. Clearly, the order of the equation $({}_{2n}P_1)$ is $2n$. As we know by §1, all solutions of $({}_2P_1)$ are meromorphic functions with double poles only. This remains open for $({}_{2n}P_1), n \geq 2$, in general. For $({}_4P_2)$, a proof by using isomonodromic deformations may be found in Shimomura [5], and we conjecture that all solutions of $({}_{2n}P_1)$ are meromorphic functions. We now consider solutions of $({}_4P_1)$.

Theorem 16.1. *Given $z_0 \in \mathbb{C}$, there exist local meromorphic solutions of $({}_4P_1)$ which have a double pole at z_0.*

Proof. We first observe, by substituting the formal Laurent expansion $w(z) = c(z - z_0)^k + \cdots$ into $({}_4P_1)$ that whenever a local meromorphic solution has a pole at z_0, then $k = -2$ and either $c = 1$ or $c = 3$.

Consider first the case $c = 1$. In order to find the powers which can be connected with arbitrary parameters of integration, called resonances, let us put $t := z - z_0$ and substitute $w \sim t^{-2} + \beta t^r$ into the dominant monomials of the equation $({}_4P_1)$. Then we obtain

$$\beta\big[r(r-6)(r+3)(r-3)\big]t^{r-4} = 0,$$

which follows by looking at terms of order one in β. Therefore, the nonnegative resonances are $r = 0, r = 3, r = 6$. Hence, a formal solution in a neighborhood of z_0 must be of the form

$$w = \frac{1}{t^2} + h_1 + a_2t^2 + h_2t^3 + a_4t^4 + a_5t^5 + h_3t^6 + a_7t^7 + \sum_{k=8}^{\infty} a_kt^k, \qquad (16.7)$$

where $t = z - z_0$, and h_1, h_2, h_3 are arbitrary constants. Moreover, it is immediate to get

$$a_2 = -3h_1^2, \quad a_4 = -10h_1^3 - \frac{1}{56}z_0, \quad a_5 = \frac{3}{2}h_1h_2 - \frac{1}{80}, \quad a_7 = -\frac{1}{140}h_1.$$

To prove the convergence of the formal series (16.7) in a disc $|z - z_0| < \rho, \rho > 0$, we transform the initial equation into a Briot–Bouquet system. To this end, rewrite $({}_4P_1)$ in the form of the equivalent system

$$\begin{cases} w_1' = w_2, \\ w_2' = w_3, \\ w_3' = w_4, \\ w_4' = 20w_1w_3 + 10w_2^2 - 40w_1^3 + z, \end{cases} \qquad (16.8)$$

where $w_1 = w$. In (16.8) we make a change of variables by

$$y_1 := t^2 w_1 - 1, \quad y_2 := t^3 w_2 + 2, \quad y_3 := t^4 w_3 - 6, \quad y_4 := t^5 w_4 + 24, \quad (16.9)$$

where $t := z - z_0$, resulting in a Briot–Bouquet system

$$\begin{cases} ty_1' = 2y_1 + y_2, \\ ty_2' = 3y_2 + y_3, \\ ty_3' = 4y_3 + y_4, \\ ty_4' = -40y_2 + 20y_3 + 5y_4 + 20y_1y_3 + 10y_2^2 - 40y_1^3 - 120y_1^2 + (t + z_0)t^6 \end{cases} \quad (16.10)$$

with eigenvalues $\lambda_1 = -1$, $\lambda_2 = 2$, $\lambda_3 = 5$, $\lambda_4 = 8$ of the Wronskian matrix. From this situation the existence of a three-parametric family of locally holomorphic solutions of (16.10) with the property $y_j \to 0$, $j \in \{1, 2, 3, 4\}$, as $t \to 0$ follows, see Theorem A.12. Thus, due to (16.8) and (16.9), the expansion (16.7) gives the local representation for solutions of $({}_4P_1)$ meromorphic in a neighborhood of an arbitrary pole.

Consider next the case $c = 3$. The equation for resonances now takes the form, with the same notations as in the previous case,

$$\beta(r + 5)(r - 6)(r + 3)(r - 8) = 0.$$

Therefore, we now have two positive resonances $r = 6$, $r = 8$. The formal pole expansion in a neighborhood of z_0 takes the form

$$w = \frac{3}{t^2} + a_4t^4 + a_5t^5 + h_1t^6 + h_2t^8 + a_{10}t^{10} + a_{11}t^{11} + a_{12}t^{12} + \sum_{k=13}^{\infty} a_k t^k, \quad (16.11)$$

where h_1, h_2 are arbitrary parameters and

$$a_4 = z_0/504, \quad a_5 = 1/240, \quad a_{10} = a_4^2/39, \quad a_{11} = 2a_4a_5/21,$$
$$a_{12} = 5h_1a_4/51 + 29a_5^2/612.$$

We now make the transformation

$$y_1 = t^2 w_1 - 3, \quad y_2 = t^3 w_2 + 6, \quad y_3 = t^4 w_3 - 18, \quad y_4 = t^5 w_4 + 72,$$

in the system (16.8). This time we obtain a Briot–Bouquet system

$$\begin{cases} ty_1' = 2y_1 + y_2, \\ ty_2' = 3y_2 + y_3, \\ ty_3' = 4y_3 + y_4, \\ ty_4' = -720y_1 - 120y_2 + 60y_3 + 5y_4 + 20y_1y_3 + 10y_2^2 \\ \qquad - 40y_1^3 - 360y_1^2 + (t + z_0)t^6 \end{cases} \quad (16.12)$$

with eigenvalues $\lambda_1 = -1$, $\lambda_2 = -3$, $\lambda_3 = 8$, $\lambda_4 = 10$ of the Wronskian matrix. Therefore, the system (16.12) has a two-parametric family of locally holomorphic solutions with the property $y_j \to 0$, $j \in \{1, 2, 3, 4\}$, as $t \to 0$. These solutions generate polar solutions of $({}_4P_1)$ of the form (16.11). Thus the assertion holds. □

As shown in §14 above, any solution of (P_1) has a representation $w(z) = v(z)/u(z)$, where the entire functions $v(z)$, $u(z)$ satisfy (14.4). We now proceed to obtain a similar representation for meromorphic solutions of $({}_{2n}P_1)$.

Lemma 16.2. *For any meromorphic solution $w(z)$ of $({}_{2n}P_1)$ with a pole at $z_0 \in \mathbb{C}$, the pole is double and the residues of $(z - z_0)w(z)$ at z_0 are integers.*

Proof. Suppose that $({}_{2n}P_1)$ admits a solution $w = b\zeta^{-p} + \cdots$, $\zeta := z - z_0$, $b \neq 0$, $p \geq 3$. Then $d^k(w) = b_k\zeta^{-pk} + \cdots$, $b_k \neq 0$ for every $k \geq 1$, implying that $d^{n+1}(w) + 4z = b_{n+1}\zeta^{-p(n+1)} + \cdots \not\equiv 0$, a contradiction. Hence, every pole is at most double and so $p \leq 2$. We use induction to show that all poles are in fact double, recalling that this is true for $n = 1$. Suppose now that $b_k \neq 0$ for some k. Then

$$\begin{aligned}\frac{d}{dz}d^{k+1}(w) &= (D^3 - 8wD - 4w_z)d^k(w)\\ &= (-8(-pk) - 4(-p))bb_k\zeta^{-pk-p-1} + \cdots\\ &= 4(2k+1)pbb_k\zeta^{-p(k+1)-1} + \cdots,\end{aligned}$$

and so $d^{k+1}(w) = -4(2k+1)(k+1)^{-1}bb_k\zeta^{-p(k+1)} + \cdots$, which implies $b_{k+1} \neq 0$.

To determine the required residues, we write $d^k(w) = A_k\zeta^{-2k} + \cdots$, $k \geq 1$, denoting $A_1 =: c$. Then

$$\begin{aligned}\frac{d}{dz}d^{k+1}(w) &= (D^3 - 8wD - 4w_z)d^k(w)\\ &= (D^3 - 8(c\zeta^{-2} + \cdots)D + 8c\zeta^{-3} + \cdots)(A_k\zeta^{-2k} + \cdots)\\ &= [(-2k)(-2k-1)(-2k-2) - 8c(-2k) + 8c](a_k\zeta^{-2k-3} + \cdots)\\ &= 8(c - k(k+1)/2)(2k+1)A_k\zeta^{-2k-3} + \cdots,\end{aligned}$$

and hence

$$d^{k+1}(w) = -4(k+1)^{-1}(2k+1)(c - k(k+1)/2)A_k\zeta^{-2(k+1)} + \cdots.$$

This implies $A_{k+1} = -4(k+1)^{-1}(2k+1)(c - k(k+1)/2)A_k$, and we obtain

$$A_{n+1} = C_{n+1}\prod_{j=0}^{n}(c - j(j+1)/2), \quad C_{n+1} \neq 0, \quad n \geq 1. \tag{16.13}$$

In particular, for $n = 1$, we have $c = 1$, while for $n = 2$, the possible residue values are $c = 1$ and $c = 3$, and for $n = 3$, $c = 1$, $c = 3$ and $c = 6$. □

Theorem 16.3. *Any meromorphic solution $w(z)$ of the equation $({}_{2n}P_1)$ may be represented as the quotient of two entire functions $u(z)$, $v(z)$ such that*

$$uu'' = (u')^2 - 2uv, \quad d^{n+1}(v/u) + 4z = 0. \tag{16.14}$$

Proof. For any meromorphic solution $w(z)$ of $({}_{2n}P_1)$, we have

$$\begin{aligned}\frac{d}{dz}d^{n+2}(w) &= (D^3 - 8wD - 4w_z)d^{n+1}(w)\\ &= (D^3 - 8wD - 4w_z)(-4z) = \frac{d}{dz}(16zw) + 16w,\end{aligned}$$

implying that $w(z)$ admits a meromorphic primitive function $\eta(z) = d^{n+2}(w)/16 - zw$. Moreover, by Lemma 16.2, every pole of $\eta(z)$ is simple. By Saks and Zygmund [1], there exists an entire function $u(z)$ such that $u'/u = -2\eta$. Clearly, zeros of $u(z)$ appear exactly at the poles of $w(z)$. Moreover, a pole of $w(z)$ of type $w(z) = c_j(z-z_0)^{-2} + \cdots$ corresponds to a $2c_j$-fold zero of $u(z)$. Therefore, $v(z) = u(z)w(z)$ is entire as well, and we have a representation $w(z) = v(z)/u(z)$ of the meromorphic solution $w(z)$ of $({}_{2n}P_1)$. From $(u'/u)' = -2w$ we immediately obtain (16.14). □

In (16.14), the derivatives $u^{(n)}$, $n \geq 2$, in the second equation of (16.14) are replaced by the corresponding expressions from the first equation. For $n = 1$, (16.14) reduces back to the case of §14. For $n = 2$, (16.14) takes the form

$$\begin{cases} uu'' = (u')^2 - 2uv, \\ v^{(4)}u^4 - 4v^{(3)}u^3u' - 4u(u')^3v' + 6u^2v''(u')^2 + 16u'u^2vv' \\ \quad - 2u^3(v')^2 - 6u^3vv'' + v(u')^4 - 8u(u')^2v^2 + 16v^3u^2 - zu^5 = 0. \end{cases}$$

We add a final remark concerning the system (16.14). A solution of (16.14) will be determined by the initial values

$$u(z_0) = u_0, \quad u'(z_0) = u_0', \quad v(z_0) = v_0, \quad v'(z_0) = v_0', \ldots, \quad v^{(2n-1)}(z_0) = v_0^{(2n-1)}, \tag{16.15}$$

where z_0, u_0, u_0', v_0, v_0', ..., $v^{(2n-1)} \in \mathbb{C}$. If $u(z_0) = 0$, then the initial values (16.15) become singular. Hence, we shall assume that $u(z_0)v(z_0) \neq 0$. By a direct substitution of (16.16) below into (16.14), we obtain

Lemma 16.4. *If $(v(z), u(z))$ is a solution of the system* (16.14), *then*

$$(\widetilde{v}, \widetilde{u}) = (\lambda(z)v, \lambda(z)u)), \quad \lambda(z) \not\equiv 0, \tag{16.16}$$

is a solution of (16.14) *if and only if* $\lambda(z) = \exp(az + b)$, $a, b \in \mathbb{C}$.

Remark. By Lemma 16.4, the representation of a meromorphic solution $w(z)$ of $({}_{2n}P_1)$ in Theorem 16.3 is uniquely determined up to a factor of the form $\lambda(z) = \exp(az + b)$, $a, b \in \mathbb{C}$. Conversely, an arbitrary entire solution (v, u) of (16.14) with $u \not\equiv 0$ determines a meromorphic solution $w = v/u$ of $({}_{2n}P_1)$.

Chapter 5

The second Painlevé equation (P_2)

The second Painlevé equation (P_2) is the first of the six classical Painlevé equations with a complex parameter in the equation. This means, of course, that several phenomena may appear which depend on the value of the parameter. Since (P_2) has one parameter only, it may be considered as a pilot case to these aspects. In fact, the reasoning remains more straightforward as is the case for (P_3) through (P_6). Typical parameter dependent phenomena in the case of (P_2) are the existence of rational solutions and of subnormal solutions which satisfy an algebraic first-order differential equation. A powerful tool, in this and the subsequent chapters, to consider such phenomena is the use of Bäcklund transformations. These transformations, usually related to a parameter change, will be intensively applied in all remaining chapters.

§17 Canonical representation of solutions

The second Painlevé equation

$$w'' = 2w^3 + zw + \alpha, \tag{P_2}$$

has one free parameter $\alpha \in \mathbb{C}$. As originally attempted to prove by Painlevé [1], all solutions of (P_2) are meromorphic functions. This classical result has been made rigorous in Hinkkanen and Laine [1] recently. See also the proof in Chapter 1. Hence, poles are the only singularities of solutions of (P_2). Let now $w(z)$ be a solution, with a pole at z_0. By (P_2), the Laurent expansion of $w(z)$ around z_0 takes the form

$$w(z) = \frac{\varepsilon}{\tau} - \frac{1}{6}\varepsilon z_0 \tau - \frac{\alpha + \varepsilon}{4}\tau^2 + h\tau^3 + \frac{3\alpha + \varepsilon}{72} z_0 \tau^4 + O(\tau^5), \tag{17.1}$$

where $\tau := z - z_0$, $\varepsilon^2 = 1$ and h is an arbitrary complex constant. Any solution $w(z)$ of (P_2) may be expressed as the quotient of two entire functions,

$$w(z) = \frac{v(z)}{u(z)}. \tag{17.2}$$

Of course, this representation is not unique. However, we seek for a certain canonical representation of type (17.2). To this end, we observe from (17.1) that

$$w(z)^2 = \frac{1}{(z - z_0)^2} - \frac{1}{3} z_0 + O(z - z_0).$$

There exists a meromorphic function $W(z)$ such that $W' = w^2$. In fact, $W(z) := zw(z)^2 - w'(z)^2 + w(z)^4 + 2\alpha w(z)$ has the required property by (P_2). Clearly, all poles of $W(z)$ are simple with residue -1, located exactly at poles of $w(z)$. By Saks and Zygmund [1], there exists an entire function $u(z)$ such that $u' = -Wu$. Of course, u may be represented as

$$u(z) = \exp\Big(- \int^z ds \int^s w^2(t)\, dt\Big),$$

where the path of integration avoids the poles of w. Given now a pole z_0 of $w(z)$, $u(z_0) = 0$. Moreover, multiplying by a constant, we may also assume that $u'(z_0) = 1$. Define now $v(z) := w(z)u(z)$. Since all zeros of $u(z)$ are simple and exactly at the poles of $w(z)$, $v(z)$ is entire as well. By (P_2), it is not difficult to see that $u(z)$, $v(z)$ satisfy the following pair of differential equations:

$$\begin{cases} uu'' = (u')^2 - v^2 \\ v''u^2 + v(u')^2 - 2u'v'u = v^3 + zvu^2 + \alpha u^3. \end{cases} \tag{17.3}$$

In fact,

$$\frac{u''}{u} = \left(\frac{u'}{u}\right)' + \left(\frac{u'}{u}\right)^2 = -W' + \left(\frac{u'}{u}\right)^2 = -w^2 + \left(\frac{u'}{u}\right)^2 = -\left(\frac{v}{u}\right)^2 + \left(\frac{u'}{u}\right)^2.$$

Moreover, by (P_2),

$$w'' = \left(\frac{v}{u}\right)'' = \frac{u^2v'' - uu''v - 2uu'v' + 2v(u')^2}{u^3} = 2\left(\frac{v}{u}\right)^3 + z\frac{v}{u} + \alpha.$$

Substituting uu'' from the first equation of (17.3) now results in the second equation in (17.3). We now proceed to investigate (17.3) in more detail. The next two lemmas follow by direct computation.

Lemma 17.1. *If $\alpha = 0$ in (P_2), then* (17.3) *has a solution*

$$\begin{cases} v(z) = 0 \\ u(z) = \exp(az + b) \end{cases} \tag{17.4}$$

with arbitrary parameters $a, b \in \mathbb{C}$.

Lemma 17.2. *If (v, u) is a solution of* (17.3), *and $\lambda(z) \not\equiv 0$, then*

$$\big(\widetilde{v}(z), \widetilde{u}(z)\big) = \big(\lambda(z)v(z), \lambda(z)u(z)\big)$$

is a solution of (17.3) *if and only if $\lambda(z) = \exp(az + b)$, where $a, b \in \mathbb{C}$.*

The solution (v, u) of (17.3) will be determined by the initial conditions

$$u(z_0) = u_0, \quad u'(z_0) = u_0', \quad v(z_0) = v_0, \quad v'(z_0) = v_0', \tag{17.5}$$

where $z_0, u_0, u_0', v_0, v_0' \in \mathbb{C}$, see Theorem A.3. If $u(z_0) = 0$, then (17.5) may be called singular for (P_2).

Theorem 17.3. *All solutions of* (17.3) *are pairs of entire functions.*

Proof. If $u = 0$, then $v = 0$ as well. Hence, we may assume that $u \not\equiv 0$, and we may fix the initial values (17.5) to be non-singular for (P_2), i.e. $u(z_0) \neq 0$. Fix a solution $w(z)$ of (P_2) by the initial conditions

$$w(z_0) = \frac{v_0}{u_0}, \quad w'(z_0) = \frac{u_0 v_0' - v_0 u_0'}{u_0^2}. \tag{17.6}$$

By a straightforward computation with (17.3), $\frac{v}{u}$ satisfies (P_2) with the initial values (17.5). Hence, $w = \frac{v}{u}$ by the standard local uniqueness argument. Let now $\widetilde{u}, \widetilde{v}$ be the canonical entire solution of (17.3), constructed for $w(z)$ above, meaning that $\frac{v}{u} = \frac{\widetilde{v}}{\widetilde{u}}$. Rewriting the first equation of (17.3), we get

$$-\left(\frac{u'}{u}\right)' = \left(\frac{v}{u}\right)^2 = \left(\frac{\widetilde{v}}{\widetilde{u}}\right)^2 = -\left(\frac{\widetilde{u}'}{\widetilde{u}}\right)'.$$

Therefore, $u(z) = e^{az+b}\widetilde{u}(z)$ for some $a, b \in \mathbb{C}$, and so u has to be entire, and the same applies for $v(z)$ as well. □

Theorem 17.4. *Let (v, u) be any nonzero solution of* (17.3) *for a given value α of the parameter, (v, u) being different from $\big(0, \exp(az + b)\big)$. Then the quotient* (17.2) *is a solution of (P_2) with the same parameter value α. Conversely, any solution w of (P_2) can be expressed as $w = v/u$, where (v, u) is a solution of* (17.3) *determined uniquely apart of a factor* $\exp(az + b)$, $a, b \in \mathbb{C}$.

Proof. Let (v, u) be a solution (17.3) as assumed, and define $w = \frac{v}{u}$. By (17.3) w satisfies (P_2) with the same parameter.

Conversely, let w be a solution of (P_2). If $\alpha = 0$ and $w = 0$, we may take (v, u) as in (17.4). So we may assume that $w \not\equiv 0$. Fix now $z_0 \in \mathbb{C}$ such that $w(z_0) = w_0 \neq 0, \infty$. Define a solution (v, u) of (17.3) with the initial values $u(z_0) = 1$, $v(z_0) = w_0$ and $u_0' = u'(z_0)$, $v_0' = v'(z_0)$ such that $v_0' - w_0 u_0' = w_0'$. Since $\frac{v}{u}$ solves (P_2), it clearly coincides with w. By Lemma 17.2, (v, u) is uniquely defined apart of a factor $\exp(az + b)$, $a, b \in \mathbb{C}$. □

Remark 1. Note that the reasoning implies a one-to-one correspondence between the solutions w of (P_2) and the quotient of solutions $\frac{u}{v}$ of (17.3) with initial values nonsingular for (P_2). Note that this does not hold for the pair u, v described in Lukashevich [3]:

$$\begin{cases} uu'' = (u')^2 - v^2 \\ (v')^2 - vv'' = uu' + \alpha uv. \end{cases}$$

Here, for $\alpha = -1$, the solution $u(z) = az + b$, $v(z) = a$ with $ab \neq 0$ does not determine a solution of (P_2) by (17.2).

Finally, we remark that the question of representing the solutions of (P_2) as a quotient of two entire functions has been investigated by Yablonskiĭ [2].

Remark 2. We close this section by a remark on solutions of (17.3) with initial data singular for (P_2), rewriting the first equation of (17.3) as

$$w^2 = -\left(\frac{u'}{u}\right)' = \left(\frac{v}{u}\right)^2 \tag{17.7}$$

again. Let now z_0 be a zero of u of multiplicity k. By (17.7),

$$w(z)^2 = \frac{k}{(z - z_0)^2} + \Phi(z)$$

around z_0, with a holomorphic $\Phi(z)$. Hence, $w(z)$ has a simple pole at z_0, and so $k = 1$. Therefore, $u(z) = (z - z_0)\phi(z)$ with $\phi(z)$ holomorphic in a neighborhood of z_0. This implies that $v(z)$ has to be holomorphic around z_0 and $v(z_0) = \varepsilon\phi(z_0)$, where ε is the residue of $w(z)$ at z_0. Moreover, $u'(z_0) = \phi(z_0)$ and $v'(z_0) = \varepsilon\phi'(z_0)$. Therefore, the initial value problem for (17.3) with $u(z_0) = 0$, $u'(z_0) = u_0'$, $v(z_0) = v_0$, $v'(z_0) = v_0'$ has a solution in the case $v_0 = \varepsilon u_0' \neq 0$ only. This solution is not unique, since (P_2) permits a one-parameter family of solutions with a pole at z_0.

§18 Poles of second Painlevé transcendents

Applying the standard Clunie reasoning as in the case of (P_1), we observe that all non-rational solutions $w(z)$ of (P_2) have infinitely many poles, all simple by the preceding section. By (17.1), the residue of $w(z)$ at any pole z_0 has to be either $+1$ or -1. We may now prove

Theorem 18.1. *Any non-rational solution of (P_2) has infinitely many poles with residue $+1$, and infinitely many with residue -1, provided $\alpha \neq \pm\frac{1}{2}$.*

To prepare the proof of Theorem 18.1, fix ε such that $\varepsilon^2 = 1$, and consider the pair of differential equations

$$\begin{cases} w' = \varepsilon w^2 + \dfrac{1}{2}\varepsilon z + v \\ v' = \alpha - \dfrac{\varepsilon}{2} - 2\varepsilon wv. \end{cases} \tag{18.1}$$

Suppose (w, v) is a solution of (18.1). Eliminating v from (18.1) shows at once that w is a solution of (P_2). Conversely, given a solution $w(z)$ of (P_2) and ε such that $\varepsilon^2 = 1$, define v by the first equation of (18.1). Differentiating, and making use of (P_2) immediately proves that (w, v) is a solution of (18.1). Moreover, a simple computation now results in the following formulas:

$$v = w' - \varepsilon w^2 - \frac{1}{2}\varepsilon z, \tag{18.2}$$

$$2vw = \varepsilon(\alpha - \frac{1}{2}\varepsilon - v'), \tag{18.3}$$

$$(v')^2 - 2vv'' = 2zv^2 + 4\varepsilon v^3 + \left(\alpha - \frac{\varepsilon}{2}\right)^2. \tag{18.4}$$

Proof (of Theorem 18.1). Let $w(z)$ be a non-rational solution of (P_2) with finite initial values $w(z_0) = w_0$, $w'(z_0) = w_0'$. Moreover, determine, for both values $\varepsilon = \pm 1$, $v(z, \varepsilon)$ by (18.1), hence

$$\begin{cases} v(z_0, \varepsilon) = w_0' - \varepsilon w_0^2 - \dfrac{1}{2}\varepsilon z_0 \\ v'(z_0, \varepsilon) = \alpha - \dfrac{\varepsilon}{2} - 2\varepsilon w_0 v(z_0, \varepsilon). \end{cases} \tag{18.5}$$

Of course, the meromorphic functions $v(z, \varepsilon)$ are solutions of (18.4), with initial values (18.5) at z_0. Moreover, by (18.3), both of $v(z, 1)$, $v(z, -1)$ are non-vanishing functions as $\alpha \neq \pm\frac{1}{2}$. Finally, by (18.3) again, $v(z, \varepsilon)$ has to be non-rational, as $w(z)$ is non-rational.

Looking at the Laurent expansions of $v(z, \varepsilon)$ at the poles z_0 of $w(z)$, we observe by (17.1) that the principal term of $v(z, 1)$ has to be $-2(z - z_0)^{-2}$ at z_0, while $v(z, -1)$ has to be holomorphic around z_0, provided z_0 is a pole of $w(z)$ with residue $+1$. Similarly, the principal term of $v(z, -1)$ has to be $2(z - z_0)^2$ at z_0, while $v(z, 1)$ is holomorphic around z_0, whenever z_0 is a pole of $w(z)$ with residue -1. Therefore, as $v(z, \varepsilon)$ has to be holomorphic, whenever $w(z)$ is holomorphic, the poles of $w(z)$ with residue $+1$, resp. -1, are exactly the poles of $v(z, 1)$, resp. $v(z, -1)$. Writing now (18.4) in the form

$$4\varepsilon v^3 = -2zv^2 + (v')^2 - 2vv'' - \left(\alpha - \frac{\varepsilon}{2}\right)^2$$

and making use of the Clunie reasoning again, we conclude that $v(z, \varepsilon)$, $\varepsilon = \pm 1$, has infinitely many poles, and we are done. □

Theorem 18.2. *Let $w(z)$ be a non-rational solution of (P_2) with $\alpha = \frac{\varepsilon}{2}$, where $\varepsilon^2 = 1$. Then all poles of w are with residue $-\varepsilon$ if and only if w satisfies the Riccati differential equation*

$$w' = \varepsilon w^2 + \frac{1}{2}\varepsilon z. \tag{18.6}$$

Proof. By the Clunie reasoning, w has infinitely many poles. Suppose first w also satisfies (18.6). By an elementary local consideration at a pole z_0 of w, the residue has to be $-\varepsilon$. On the other hand, if w does not satisfy (18.6), then $v = v(z, \varepsilon) \not\equiv 0$. If now $v(z, -\varepsilon) = 0$, then $w' = -\varepsilon w^2 - \frac{1}{2}\varepsilon z$ and we obtain $w'' = 2w^3 + zw - \frac{\varepsilon}{2}$ by differentiation, a contradiction. Since $v(z, \varepsilon)$ and $v(z, -\varepsilon)$ both are non-vanishing functions, the preceding proof may be repeated to show that w has infinitely many poles with residue $-\varepsilon$ and infinitely many with residue $+\varepsilon$. □

Remark 1. If w satisfies (18.6), then w is a solution of (P_2) with $\alpha = \varepsilon/2$. All poles of $-\varepsilon w$ are with residue $(-\varepsilon)^2 = 1$. Hence there exists an entire function u such that $u'/u = -\varepsilon w$. By a simple computation,

$$u'' + \frac{z}{2}u = 0. \tag{18.7}$$

This is the well-known Airy equation. For the zero distribution of Airy functions, see Gundersen and Steinbart [1].

Remark 2. The first equation of (18.1) is the Riccati equation with respect to w and it may be considered as the Miura transformation of (P_2), see Miura [1]. In this case the equation (18.4) is the result of the Miura transformation. The pair (18.1) is a Hamiltonian system with the Hamiltonian

$$H(z, w, v) := \frac{v^2}{2} + \varepsilon(w^2 + z/2)v - \left(\alpha - \frac{\varepsilon}{2}\right) w. \tag{18.8}$$

Denoting now $h(z) := H\big(z, w(z), v(z)\big)$ and $\tau(z) := \exp \int h(z)\,dz$, we observe that $\tau(z)$ is an entire function and $h(z)$ satisfies the equation

$$(h'')^2 + 4(h')^3 + 2h'(zh' - h) - \frac{1}{4}\left(\alpha + \frac{1}{2}\right)^2 = 0.$$

This may be verified by direct computation. Moreover,

$$4h'w = 2h'' + \alpha + \frac{1}{2}, \quad v = -2h', \quad \varepsilon = -1.$$

§19 The Bäcklund transformations of (P_2)

The Bäcklund transformations, originally introduced by A. Bäcklund in 1870's, see e.g. Lamb [1], in relation to investigating surfaces of constant negative curvature, form a powerful device to studying solutions of (P_2). To this end, denote by $P_2(\alpha)$ the family of solutions of (P_2) with the parameter value $\alpha \in \mathbb{C}$, and by w_α a solution in $P_2(\alpha)$. Moreover, define $\mathcal{F} := \{(P_2(\alpha)) \mid \alpha \in \mathbb{C}\}$. Then the Bäcklund transformations are certain mappings from $\mathcal{F}$ into $\mathcal{F}$.

We first observe that $S : \mathcal{F} \to \mathcal{F}$ defined by

$$S(w_\alpha) = -w_\alpha \in P_2(-\alpha) \tag{19.1}$$

is a transformation of required type. Moreover, $S \circ S = I$, the identity on $\mathcal{F}$.

To define another transformation, take $w \in P_2(\alpha)$, and define

$$y = T^{-1}(w) := \begin{cases} -w, & \text{if } \alpha = \frac{1}{2} \text{ and } w' = w^2 + \frac{z}{2}, \\ -w + \frac{\alpha - 1/2}{w' - w^2 - z/2}, & \text{otherwise}, \end{cases} \tag{19.2}$$

see Lukashevich [3]. We proceed to show that $T^{-1}(w) \in P_2(\alpha - 1)$, hence defining again a required transformation. Clearly we may assume that $\alpha \neq \frac{1}{2}$. Fix now $\varepsilon = 1$, and use (18.1) to define

$$v = w' - w^2 - \frac{z}{2}.$$

By (19.2) and (P_2), it is a straightforward computation to show that

$$-\left(y' + y^2 + \frac{z}{2}\right) = v. \tag{19.3}$$

Differentiating (19.3), and making use of (18.1) for v', we obtain

$$\begin{aligned} y'' &= -2yy' - \frac{1}{2} - v' \\ &= -2y\left(-y^2 - \frac{z}{2} - v\right) - \frac{1}{2} - \alpha + \frac{1}{2} + 2v\left(-y + \frac{\alpha - 1/2}{w' - w^2 - z/2}\right) \\ &= 2y^3 + zy + 2yv - \alpha - 2vy + 2\left(\alpha - \frac{1}{2}\right) \\ &= 2y^3 + zy + \alpha - 1, \end{aligned}$$

as required.

By a similar computation, for $y \in P_2(\alpha - 1)$,

$$w = T(y) := \begin{cases} -y, & \text{if } \alpha = \frac{1}{2} \text{ and } y' = -y^2 - \frac{z}{2}, \\ -y - \frac{\alpha - 1/2}{y' + y^2 + z/2}, & \text{otherwise}. \end{cases} \tag{19.4}$$

defines the inverse transformation such that $T(y) \in P_2(\alpha)$. Moreover, it is immediate to see that S and T generate a transformation group G_2 in $\mathcal{F}$. By the notations

$\tau(\alpha) := -\alpha$, $\tau_+(\alpha) := \alpha + 1$, for transformations $\tau : \mathbb{C} \to \mathbb{C}$, $\tau_+ : \mathbb{C} \to \mathbb{C}$ a transformation group $\widetilde{G}_2$ in $\mathbb{C}$ will be defined. Hence, we are now ready to state the following theorem, see Okamoto [4]:

Theorem 19.1. *The Bäcklund transformations S, T defined in* (19.1), (19.2) *form in a transformation group G_2 on $\mathcal{F}$. Moreover, the transformation groups G_2, $\widetilde{G}_2$ in $\mathbb{C}$ and $W(A_1)$, where $W(A_1)$ is the affine Weyl group associated with the Lie algebra of type A_1 are isomorphic.*

We now proceed to construct another transformation formula related to the mapping $T : \mathcal{F} \to \mathcal{F}$. To this end, assume that $\alpha \neq \pm\frac{1}{2}$, choose $w_{\alpha-1} \in P_2(\alpha - 1)$, and denote now $w_\alpha = T(w_{\alpha-1})$ and $w_{\alpha+1} = T(w_\alpha)$. By (19.4),

$$w_{\alpha+1} = -w_\alpha - \frac{\alpha + 1/2}{w'_\alpha + w_\alpha^2 + z/2} \tag{19.5}$$

and by (19.2),

$$w_{\alpha-1} = -w_\alpha + \frac{\alpha - 1/2}{w'_\alpha - w_\alpha^2 - z/2}. \tag{19.6}$$

Solving w'_α from (19.6) and substituting into (19.5) we obtain an algebraic relation between $w_{\alpha-1}, w_\alpha, w_{\alpha+1}$ as follows:

$$w_{\alpha+1} = -w_\alpha - \frac{\alpha + 1/2}{2w_\alpha^2 + z + \dfrac{\alpha - 1/2}{w_\alpha + w_{\alpha-1}}}. \tag{19.7}$$

By a similar computation, a relation between $w_{\alpha-2}, w_{\alpha-1}, w_\alpha, w_{\alpha+1}$ follows:

$$w_{\alpha-2} = -w_{\alpha-1} + \frac{(2\alpha - 3)(w_\alpha + w_{\alpha+1})}{2\alpha + 1 + 4(w_\alpha^2 - w_{\alpha-1}^2)(w_\alpha + w_{\alpha+1})}. \tag{19.8}$$

Another, less obvious transformation connects $P_2\left(\frac{\varepsilon}{2}\right)$, $\varepsilon^2 = 1$ and $P_2(0)$. In order to construct this transformation, consider $w \in P_2\left(\frac{\varepsilon}{2}\right)$ such that the corresponding $v = v(z, \varepsilon)$ in (18.1) does not vanish identically. Therefore, by (18.1) and (18.4), $v = w' - \varepsilon w^2 - \frac{1}{2}\varepsilon z \neq 0$ and

$$v'' = \frac{(v')^2}{2v} - zv - 2\varepsilon v^2. \tag{19.9}$$

We now define an algebroid function $u(z)$ by

$$v = \lambda u^2 \tag{19.10}$$

with $\lambda = -\varepsilon \cdot 2^{1/3}$. Differentiating (19.10) twice and making use of (19.9) we get

$$u'' = \frac{1}{2\lambda u} v'' - \frac{(u')^2}{u} = \frac{1}{2\lambda u}\left(\frac{(v')^2}{2v} - zv - 2\varepsilon v^2\right) - \frac{(u')^2}{u}.$$

By (19.10) and $v' = 2\lambda uu'$ it is straightforward to verify that

$$u'' = -\varepsilon\lambda u^3 - \frac{1}{2}zu. \tag{19.11}$$

We now define a change of variable by $z = -2^{1/3}\tau$. Defining $y(\tau) := u(-2^{1/3}\tau) = u(z)$, it is immediate to conclude that

$$y''(\tau) = 2y(\tau)^3 + \tau y(\tau),$$

hence $y(\tau) \in P_2(0)$, and so y has to be meromorphic by Theorem 2.1. As an obvious consequence, u is meromorphic as well. Combining the definition of $y(\tau)$ above, (19.10) and the pair (18.1) of differential equations we immediate obtain the asserted connection:

Theorem 19.2. *For $y(\tau) \in P_2(0)$ and $w(z) \in P_2(\frac{\varepsilon}{2})$, $\varepsilon^2 = 1$, $w' \neq \varepsilon w^2 + \frac{1}{2}\varepsilon z$, it is true that*

$$\begin{cases} -2^{1/3}\varepsilon y(\tau)^2 = w'(z) - \varepsilon w(z)^2 - \dfrac{1}{2}\varepsilon z, \\ w(z) = 2^{-1/3}\varepsilon \dfrac{y'(\tau)}{y(\tau)}. \end{cases} \tag{19.12}$$

Corollary 19.3. *To determine solutions of (P_2) for arbitrary values of the parameter α, it suffices to determine solutions of (P_2) for all α satisfying*

$$0 \leq \operatorname{Re}\alpha \leq \frac{1}{2}. \tag{19.13}$$

Remark 1. The strip (19.13) is a fundamental domain of the parameter space, called the Weyl chamber.

Remark 2. As a consequence of the above reasoning, see (19.10), we observe that all zeros and poles of v are of even multiplicity whenever $w \in P_2(\frac{\varepsilon}{2})$, $\varepsilon^2 = 1$, and $v = v(z, \varepsilon) \neq 0$ is determined by the first equation of (18.1).

Remark 3. From (19.2) with $\alpha \neq \frac{1}{2}$, it is a direct computation to verify that

$$w' - w^2 + y' + y^2 = 0. \tag{19.14}$$

Now, (19.14) is a Riccati equation for w with a particular solution $w = -y$. This enables us to solve the equation (19.14). With the initial data $w(z_0) = w_0$, $w'(z_0) = w_0'$, we obtain

$$w_{\alpha-1} = -w_\alpha + \frac{(\alpha - 1/2)J(w_\alpha)}{w_0' - w_0^2 - z_0/2 + (\alpha - 1/2)\int_{z_0}^{z} J(w_\alpha)\,dz}, \tag{19.15}$$

where $J(w_\alpha)$ is an entire function such that $J'(w_\alpha)/J(w_\alpha) = 2w_\alpha$ and $J(w_\alpha)(z_0) = 0$. The formula (19.15) is called the integral form of the Bäcklund transformation.

Remark 4. The formulas (19.7) and (19.8) determine a principle of nonlinear superposition for the equation (P_2). The transformations (19.2) and (19.4) permit us to construct the solutions for the different values of the parameters. Note that if $\alpha \in \mathbb{N}$, then (19.2) and (19.4) may be considered as the discrete analogues of (P_2) in the parameter α. The discrete analogues of the Painlevé equations have recently been derived using either a method based on the singularity confinement or the orthogonal polynomials, see Grammaticos, Nijhoff and Ramani [1]. Relations (19.2) and (19.4) present the additional method of deriving the nonautonomous integrable discrete equations, based on the use of the Bäcklund transformations for the continuous Painlevé equations.

The transformations have been applied to construct solutions of (P_2) with a change of the parameter α. However, we may apply these transformations to obtain modified solutions with the same parameter α as in the initial solution. Such transformations are called auto-Bäcklund transformations.

A simple example of an auto-Bäcklund transformation is the transformation S with $\alpha = 0$. Another example is the transformation $S_j : \mathcal{F} \to \mathcal{F}$ defined by

$$S_j\big(w_\alpha(z)\big) := \mu w_\alpha(\mu z)$$

with $\mu = e^{2\pi i j/3}$, $j \in \mathbb{Z}$. In general, $w_\alpha(z)$ and $S_j\big(w_\alpha(z)\big)$ are distinct solutions. Indeed, let $w_\alpha(z) \in P_2(\alpha)$ with a pole at $z = 0$ and the Laurent expansion at $z = 0$

$$w_\alpha(z) = \frac{\varepsilon}{z} - \frac{\alpha + \varepsilon}{4} z^2 + hz^3 + O(z^4),$$

see (17.1). Then it is easy to verify the corresponding expansion for $S_j \circ w_\alpha$:

$$S_j\big(w_\alpha(z)\big) = \frac{\varepsilon}{z} - \frac{\alpha + \varepsilon}{4} z^2 + \mu hz^3 + O(z^4).$$

Hence, $w_\alpha(z)$, $S_j\big(w_\alpha(z)\big)$ are distinct provided that $h \neq 0$.

Theorem 19.4. *Let $w_\alpha(z)$, $\alpha \in \mathbb{Z}$, be a solution in (P_2). Then*

$$\widetilde{w}_\alpha(z) = T^\alpha S T^{-\alpha} w_\alpha(z) \tag{19.16}$$

is also a solution of (P_2) and $\widetilde{w}_\alpha(z)$, $w_\alpha(z)$ are distinct solutions, provided $w_\alpha(z)$ is non-rational.

Proof. Indeed, from (19.2), (19.4) it follows that (19.16) is a solution. Noncoincidence of the solutions $w_\alpha(z)$ and $\widetilde{w}_\alpha(z)$ follows from the fact that $w_\alpha(z)$ is nonrational solution and in this case $ST^{-\alpha}w_\alpha(z) = -T^{-\alpha}w_\alpha(z) \not\equiv 0$. □

As an example, we construct the auto-Bäcklund transformation TST^{-1} for $\alpha = 1$ explicitly.To this end, let $w_1(z)$ be a nonrational solution of (P_2) with the parameter

$\alpha = 1$. Then, in accordance with Theorem 19.4, the "new" solution $\widetilde{w}_1(z)$ with parameter $\alpha = 1$ may be constructed by the scheme

$$w_1(z) \overset{T^{-1}}{\to} w_0(z) \overset{S}{\to} \widetilde{w}_0(z) \overset{T}{\to} \widetilde{w}_1(z), \tag{19.17}$$

where $\widetilde{w}_0(z) = Sw_0(z) = -w_0(z)$. Since $\widetilde{w}_1(z) = T\big(\widetilde{w}_0(z)\big) = T\big(-w_0(z)\big)$, we have

$$\widetilde{w}_1(z) = -\widetilde{w}_0(z) - \frac{1/2}{\widetilde{w}_0'(z) + \widetilde{w}_0^2(z) + z/2} = w_0(z) - \frac{1/2}{-w_0'(z) + w_0^2(z) + z/2}. \tag{19.18}$$

However, from (19.2) with $\alpha = 1$ it follows that

$$w_0(z) = -w_1(z) + \frac{1/2}{w_1'(z) - w_1^2(z) - z/2}. \tag{19.19}$$

Substituting (19.19) into (19.18) and keeping in mind that $w_1''(z) \equiv 2w_1^3(z) + zw_1(z) + 1$, we obtain the expression of the "new" solution $\widetilde{w}_1(z)$ by means of the "old" solution $w_1 := w_1(z)$:

$$\widetilde{w}_1(z) = -w_1 + \frac{1}{2\theta} - \frac{\theta^2}{(2w_1' + 2w_1^2 + z)\theta^2 - 4w_1\theta + 1}, \tag{19.20}$$

where $\theta := w_1' - w_1^2 - z/2$. Substituting $w_1'(z)$ from (19.19) into (19.20), we obtain an algebraic relation between the solutions $w_0 := w_0(z), w_1 = w_1(z)$, $\widetilde{w}_1 := \widetilde{w}_1(z)$:

$$\frac{1}{\widetilde{w}_1 - w_0} + \frac{1}{w_1 + w_0} + 2z + 4w_0^2 = 0. \tag{19.21}$$

§20 Rational solutions of (P_2)

Rational solutions of (P_2) have been investigated in detail, e.g. by Murata [1] and by Yablonskiĭ [1] and Vorob'ev [1]. We offer here a short characterization of these solutions.

Lemma 20.1. *(P_2) admits a rational solution if and only if (17.3) admits a polynomial solution* $(v, u) = (P, Q) \neq (P, 0)$.

Proof. Let (P, Q) be a polynomial solution of (17.3). By Theorem 17.4, P/Q solves (P_2).

Suppose next w is a rational solution of (P_2). Let W be a meromorphic function such that $W' = w^2$. Since $w(z) = O(z^{-1})$ around $z = \infty$, W is rational and so with finitely many poles, all of them being simple and with residue equal to -1. Therefore, there exists a polynomial Q such that $Q' = -WQ$. Obviously, $P := wQ$ is a polynomial as well and (P, Q) is a solution of (17.3). □

Theorem 20.2. (P_2) *admits a uniquely determined rational solution if and only if* $\alpha \in \mathbb{Z}$.

Proof. We first observe that (P_2) admits at least one rational solution for every $\alpha \in \mathbb{Z}$. In fact, $w = 0$ trivially satisfies (P_2) with $\alpha = 0$. Applying (19.4) inductively, we may construct a rational solution of (P_2) for every $\alpha \in \mathbb{N}$, extending this to $\alpha \in \mathbb{Z}$ by an application of (19.1).

We next show that $w = 0$ is the unique rational solution for

$$w'' = 2w^3 + zw. \tag{20.1}$$

Suppose $w \neq 0$ is a rational solution of (20.1), which may be written as

$$\frac{w''}{w} = 2w^2 + z. \tag{20.2}$$

Obviously, $2w^2 + z \to \infty$ as $|z| \to \infty$. On the other hand, writing $w = P/Q$ with P, Q polynomials, we obtain

$$\frac{w''}{w} = \frac{P''}{P} - \frac{Q''}{Q} - 2\frac{Q'}{Q}\frac{P'}{P} + 2\left(\frac{Q'}{Q}\right)^2.$$

Hence, as $z \to \infty$, then $w''/w \to 0$, a contradiction.

Suppose now $\alpha \neq 0$, and let $w = \frac{P}{Q} \neq 0$ be a rational function in $P_2(\alpha)$. By Lemma 20.1 and (17.3),

$$P^2 = (Q')^2 - QQ''. \tag{20.3}$$

Let q, resp. p, denote the degree of the polynomial Q, resp. P. By (20.3), $2p = 2q-2$, hence $p = q - 1$. Since $\alpha \neq 0$, w must be non-constant. Write now

$$\begin{cases} Q(z) = q_n z^n + q_{n-1} z^{n-1} + \cdots + q_0, \quad q_n \neq 0, \\ P(z) = p_{n-1} z^{n-1} + p_{n-2} z^{n-2} + \cdots + p_0. \end{cases} \tag{20.4}$$

Substitute (20.4) into (20.3) and into

$$P''Q^2 + P(Q')^2 - 2QQ'P' = P^3 + zPQ^2 + \alpha Q^3, \tag{20.5}$$

see (17.3). Comparing the coefficients of the leading terms in (20.3) and (20.5) results in

$$nq_n^2 = p_{n-1}^2, \quad p_{n-1} = -\alpha q_n$$

and so

$$n = \alpha^2. \tag{20.6}$$

Therefore $|\alpha| \geq 1$. Applying now the Bäcklund transformation T for $\alpha \leqq -1$, resp. T^{-1} for $\alpha \geq 1$, finitely many times, we observe that $\alpha \in \mathbb{Z}$. Since $T^k(w) \equiv 0$, resp.

$(T^{-1})^k(w) \equiv 0$, we conclude that for a given $\alpha \in \mathbb{Z}$, $P_2(\alpha)$ contains exactly one rational element. □

Starting from $w = 0$ for $\alpha = 0$, and applying the Bäcklund transformations S, T, it is immediate to apply mathematical software to compile the following list of first rational solutions in $P_2(\pm\alpha)$, $\alpha = 1, \ldots, 7$, see the table below.

Table 20.1. List of the first rational solutions of (P_2).

α	$w_\alpha(z)$
0	0
± 1	$\mp\frac{1}{z}$
± 2	$\pm\frac{4-2z^3}{4z+z^4}$
± 3	$\mp\frac{3z^2(160+8z^3+z^6)}{-320+24z^6+z^9}$
± 4	$\mp\frac{4(-224000-112000z^3-22400z^6+1000z^9+50z^{12}+z^{15})}{z(-80+20z^3+z^6)(11200+60z^6+z^9)}$
± 5	$\mp(5(14049280000-14049280000z^3+1404928000z^6+213248000z^9$ $+6899200z^{12}+35840z^{15}+7840z^{18}+152z^{21}+z^{24}))/$ $((11200z+60z^7+z^{10})$ $(-6272000-3136000z^3+78400z^6+2800z^9+140z^{12}+z^{15}))$
± 6	$\mp(6z^2(-121160990720000000-60580495360000000z^3$ $-2704486400000000z^6+29503488000000z^9-9904742400000z^{12}$ $-313474560000z^{15}-131712000z^{18}+103488000z^{21}+2352000z^{24}+44800$ $z^{27}+350z^{30}+z^{33}))/((-6272000-3136000z^3+78400z^6+2800$ $z^9+140z^{12}+z^{15})(-38635520000+19317760000z^3+1448832000z^6$ $-17248000z^9+627200z^{12}+18480z^{15}+280z^{18}+z^{21}))$
± 7	$\mp(7(1707652696086937600000000+1707652696086937600000000z^3$ $+512295808826081280000000z^6-56413526567157760000000z^9$ $-2418712625545216000000z^{12}-462433422409728000000z^{15}$ $+46999213735936000000z^{18}+1426812844441600000z^{21}$ $+50732301312000000z^{24}+471353344000000z^{27}+1733079040000z^{30}$ $+87193344000z^{33}+2099955200z^{36}+25356800z^{39}+186240z^{42}$ $+688z^{45}+z^{48}))/((-38635520000+19317760000z^3+1448832000z^6$ $-17248000z^9+627200z^{12}+18480z^{15}+280z^{18}+z^{21})$ $(-30939324416000000z-49723914240000z^7-828731904000z^{10}$ $+13039488000z^{13}+62092800z^{16}+5174400z^{19}+75600z^{22}+504z^{25}+z^{28}))$

Recalling that all poles are simple for rational solutions of (P_2) as well, with residue ± 1, we add the following theorem, which parallels to Theorem 15.1 and Theorem 15.2 in the case of rational solutions.

Theorem 20.3. *Let w_α be the rational solution of (P_2) with the parameter $\alpha \neq 0$, and let l_+, resp. l_-, denote the number of poles with the residue $+1$, resp. -1. Then*

$$l_+ = \frac{\alpha(\alpha-1)}{2}, \quad l_- = \frac{\alpha(\alpha+1)}{2}.$$

Proof. Recalling (20.4) and (20.6) from the preceding proof, we first observe that

$$l_+ + l_- = n = \alpha^2. \tag{20.7}$$

By elementary properties of rational functions and their residues, see e.g. Hahn and Epstein [1], p. 151–155,

$$-\operatorname{Res}(\infty, w_\alpha) = \lim_{z\to\infty} z w_\alpha(z) = \lim_{z\to\infty} \sum_{j=1}^{n} \pm \frac{z}{z - z_j} = l_+ - l_-,$$

as this limit exists (and so w_α has a simple zero at $z = \infty$). Denoting $f(t) := w_\alpha\left(\frac{1}{t}\right)$, we get $w'_\alpha\left(\frac{1}{t}\right) = -t^2 f'(t)$ and $w''_\alpha\left(\frac{1}{t}\right) = t^4 f''(t) + 2t^3 f'(t)$. Substituting into (P_2) results in

$$t^4 f''(t) + 2t^3 f'(t) = 2f(t)^3 + \frac{1}{t} f(t) + \alpha. \tag{20.8}$$

Since $f(t) = \beta_1 t + \beta_2 t^2 + \cdots$ has a simple zero at $t = 0$, (20.8) implies that $\beta_1 = -\alpha$, hence $\operatorname{Res}(\infty, w_\alpha) = \alpha$. Therefore

$$l_+ - l_- = -\alpha. \tag{20.9}$$

Now, (20.7) and (20.9) yield the assertion. □

We close this section by an immediate consequence of the Bäcklund transformations:

Theorem 20.4. *For all rational solutions $w(z)$ of (P_2), there exists a rational function $R(z)$ such that*

$$z w(z) = R(z^3). \tag{20.10}$$

Proof. Clearly, it suffices to show that the Bäcklund transformations S and T preserve the form (20.10). For S, this is trivial. Assume now that $zw(z) = R(z^3)$, and consider $y = T^{-1}(w)$. Since $zw'(z) + w(z) = 3z^2 R'(\tau)$, where $\tau := z^3$, we get

$$\begin{aligned} zy(z) &= -zw(z) + \frac{(\alpha - 1/2)z^2}{zw'(z) - zw^2(z) - z^2/2} \\ &= -R(\tau) + \frac{(\alpha - 1/2)\tau}{3\tau R'(\tau) - zw(z) - z^2 w^2(z) - \tau/2} \\ &= -R(\tau) + \frac{(\alpha - 1/2)\tau}{3\tau R'(\tau) - R(\tau) - R^2(\tau) - \tau/2} \end{aligned}$$

which is of the required form. □

§21 The Airy solutions of (P_2)

By the second Malmquist theorem, see again Erëmenko [1], Theorem 6, and Hotzel [1], Satz 4.5, if a transcendental solution $w(z)$ of (P_2) satisfies a first order algebraic differential equation $P(z, w, w') = 0$, this equation reduces to

$$(w')^m + \sum_{j=1}^{m} \Big(\sum_{i=0}^{n(j)} q_{ij}(z) w^i \Big) (w')^{m-j} = 0, \tag{21.1}$$

where $n(j) \leq 2j$ and the coefficients $q_{ij}(z)$ are rational functions. Such transcendental solutions $w(z) \in \mathcal{A}_1(\mathbb{C}(z))$ of (P_2), hence satisfying (21.1), are called *Airy solutions* in what follows, see Remark 1 following Theorem 18.2. It is immediate to observe that all solutions $w(z)$ of the Riccati differential equation

$$w' = \varepsilon w^2 + \frac{1}{2}\varepsilon z, \quad \varepsilon^2 = 1, \tag{21.2}$$

also satisfy (P_2) with $\alpha = \frac{\varepsilon}{2}$. Recall again that all poles of $w(z)$ are simple and with residue $-\varepsilon$. Therefore, there exists an entire function g such that $w = -\varepsilon \frac{g'}{g}$ and that $g'' + \frac{z}{2} g = 0$. By Laine [1], Proposition 5.1, $\rho(g) = \frac{3}{2}$. Since $m(r, w) = S(r, w)$ by the Clunie reasoning, $\rho(w) = \frac{3}{2}$ for all such solutions of (P_2). Invoking now the Bäcklund transformations S, T, T^{-1} defined in §19, it becomes immediate to construct, for each $\alpha = \frac{2m+1}{2}$, $m \in \mathbb{Z}$, at least one solution of (P_2) satisfying a first order differential equation $P(z, w, w') = 0$, necessarily being of the form (21.1). Denoting these solutions by $w_\alpha(z)$, whenever $w_\alpha \in P_2(\alpha)$, we recall by §19 that

$$w_{\alpha+1} := T(w_\alpha) = -w_\alpha - \frac{\alpha + 1/2}{w'_\alpha + w_\alpha^2 + z/2}, \tag{21.3}$$

$$w_\alpha := T^{-1}(w_{\alpha+1}) = -w_{\alpha+1} + \frac{\alpha + 1/2}{w'_{\alpha+1} - w_{\alpha+1}^2 - z/2}. \tag{21.4}$$

Therefore, all solutions of (P_2) constructed from the Riccati differential equations (21.2) via the Bäcklund transformations, and which satisfy a first order algebraic differential equation, being necessarily of type (21.1), must be of order $\rho(w) = \frac{3}{2}$, as seen from (21.3) and (21.4) by elementary order considerations.

Example. Considering the Riccati differential equation (21.2) with $\varepsilon = 1$, i.e.

$$w' = w^2 + \frac{z}{2}, \tag{21.5}$$

and denoting by $w_{1/2}(z)$ its solutions, then

$$y = w_{3/2} := -w_{1/2} - \frac{1}{w'_{1/2} + w_{1/2}^2 + z/2}. \tag{21.6}$$

Making use of the inverse transformation,

$$w_{1/2} = -y + \frac{1}{y' - y^2 - z/2}, \tag{21.7}$$

and substituting (21.7) into (21.5), we conclude that all functions of type (21.6) satisfy

$$(y')^3 - (y^2 + \frac{z}{2})(y')^2 - (y^4 + zy^2 + 4y + \frac{z^2}{4})y' + y^6 + \frac{3}{2}zy^4 + 4y^3 + \frac{3}{4}z^2y^2 + 2yz + 2 + \frac{z^3}{8} = 0.$$

Recalling from §18 that $w_{1/2} = U := -u'/u$ where u is a solution of the Airy equation, we obtain the following table of the first Airy solutions described above:

Table 21.1. List of the first Airy solutions of (P_2).

α	$w_\alpha(z)$
$\pm\frac{1}{2}$	$\pm U$
$\pm\frac{3}{2}$	$\mp\frac{2U^3 + zU - 1}{2U^2 + z}$
$\pm\frac{5}{2}$	$\pm\frac{1 + 6U^3 + 3Uz - 4U^4z - 4U^2z^2 - z^3}{(2U^2 + z)(1 + 4U^3 + 2Uz)}$
$\pm\frac{7}{2}$	$\pm(-96U^7 - 152U^5z - 80U^3z^2 + 16U^6z^2 + 24U^4(-1 + z^3) + z^2(-5 + 2z^3) + 2U^2z(-11 + 6z^3) - U(1 + 14z^3))/((1 + 4U^3 + 2Uz)(-3 - 16U^3 - 8Uz + 8U^4z + 8U^2z^2 + 2z^3))$
$\pm\frac{9}{2}$	$\pm(27 + 312U^2z^2 + 96U^8z^2 - 9z^3 - 2z^6 - 64U^9(-27 + 2z^3) - 24U^5z^2(-45 + 8z^3) + 16U^6(85 + 8z^3) + 4U^4z(335 + 12z^3) + U^7(2448z - 256z^4) + U^3(338 + 108z^3 - 64z^6) + U(171z - 18z^4 - 8z^7))/((-3 - 16U^3 - 8Uz + 8U^4z + 8U^2z^2 + 2z^3)(72U^5 + 9z + 88U^3z + 26Uz^2 - 16U^4z^2 - 4z^4 + U^2(14 - 16z^3)))$
$\pm\frac{11}{2}$	$\pm(128U^{11}z^2(-243 + 8z^3) + 32U^{10}z(3321 + 40z^3) + 16U^8z^2(13969 + 328z^3) + 16U^5z^2(271 - 4146z^3 + 80z^6) + 16U^7z(-4879 - 6972z^3 + 160z^6) + 16U^9(-4779 - 5828z^3 + 160z^6) + 16U^6(-2660 + 10711z^3 + 456z^6) + 8U^4z(-5001 + 6719z^3 + 584z^6) + 2U^2z^2(-3564 + 1881z^3 + 712z^6) + Uz(-4050 + 6399z^3 - 2292z^6 + 32z^9) + 3(-162 + 378z^3 - 261z^6 + 56z^9) + 4U^3(-1971 + 6231z^3 - 4894z^6 + 80z^9))/((72U^5 + 9z + 88U^3z + 26Uz^2 - 16U^4z^2 - 4z^4 + U^2(14 - 16z^3))(-54 - 488U^3 - 252Uz + 9z^3 + 8z^6 + 8U^6(-135 + 8z^3) + 12U^4z(-91 + 8z^3) + 6U^2z^2(-43 + 8z^3)))$

On the other hand, the process above covers all Airy solutions of (P_2). In fact, we proceed to prove

Theorem 21.1. *The second Painlevé equation (P_2) admits no other Airy solutions than those obtained from the solutions of the Riccati differential equations* (21.2) *by repeated applications of the Bäcklund transformations. Therefore, an Airy solution of (P_2) exists if and only if $\alpha \in \frac{1}{2} + \mathbb{Z}$. All Airy solutions $w(z)$ of (P_2) are of order $\rho(w) = \frac{3}{2}$.*

Proof. We first prove that no solutions of (21.1) appear, if $|\operatorname{Re}\alpha| \le \frac{1}{2}$ and $\alpha \ne \frac{1}{2}$. Recalling the transformation S in §19, we may assume that $0 \le \operatorname{Re}\alpha \le \frac{1}{2}$ and $\alpha \ne \frac{1}{2}$. Suppose now, contrary to the assertion, that a transcendental solution $\varphi(z)$ of (P_2) also satisfies a first order equation (21.1). Consider now a pole z_0 of $\varphi(z)$. As the pole is simple, and so $\varphi(z)$ is locally univalent around z_0, it is not difficult to obtain the expansion

$$\varphi' = \varepsilon\varphi^2 + \frac{1}{2}\varepsilon z + \left(\alpha\varepsilon - \frac{1}{2}\right)\varphi^{-1} + \sum_{j=2}^{\infty} A_{-j}\varphi^{-j}, \tag{21.8}$$

where A_{-2} may be arbitrary, valid in some neighborhood of z_0. For a construction of this expansion, see the proof of Theorem 2.1 and the formula (2.6) in particular.

Write now the equation (21.1) as

$$P(z, w, w') = (w')^m + P_1(z, w)(w')^{m-1} + \cdots + P_m(z, w) = 0, \tag{21.9}$$

where each $P_j(z, w)$ is a polynomial in w with rational coefficients and where $\deg_w P_j(z, w) \le 2j$. The case $m = 1$ is trivial. In fact, (21.9) then reduces to a Riccati differential equation

$$w' = a_0(z) + a_1(z)w + a_2(z)w^2$$

with rational coefficients by the Malmquist theorem, see Laine [1], Theorem 10.2. Differentiating we get

$$w'' = b_0(z) + b_1(z)w + b_2(z)w^2 + b_3(z)w^3$$

with some rational coefficients. Since this has to be equal to (P_2), it is elementary to show that $b_0 = \alpha = \pm\frac{1}{2}$, a contradiction. Hence, we may assume that $m \ge 2$, although this is actually not needed below.

Consider the (generalized) algebraic equation

$$X^m + P_1(z, w)X^{m-1} + \cdots + P_m(z, w) = 0. \tag{21.10}$$

We may assume that (21.10) is irreducible in $\mathbb{A}_z[w, X]$. Clearly, (21.10) has a solution

$$X = S(z, w),$$

where $S(z, w)$ is algebraic with respect to w and defines a connected algebraic curve. Consider now $S(z, w)$ around $w = \infty$. We first observe that (21.9) is equivalent to

$$w' = S(z, w). \tag{21.11}$$

There exists a branch $S_1(z, w)$ of $X = S(z, w)$ such that the solution $\varphi(z)$ satisfies $w' = S_1(z, w)$ around the pole $z = z_0$. Then $S_1(z, w)$ may be expressed in the form

$$S_1(z, w) = \varepsilon w^2 + \frac{1}{2}\varepsilon z + \left(\alpha\varepsilon - \frac{1}{2}\right)\frac{1}{w} + \sum_{j=2}^{\infty} A_{-j} w^{-j} \tag{21.12}$$

around $w = \infty$. Let $S_2(z, w), \ldots, S_m(z, w)$ be the other local branches of $S(z, w)$ around $w = \infty$, expressed by Puiseux series, see, e.g. Ahlfors [1], p. 287–290. Consider next the analytic continuation with respect to w from $S_1(z, w)$ to $S_2(z, w)$, both locally around $w = \infty$. Then, by the same argument as in the proof of Theorem 13.1, the equation

$$w' = S_2(z, w)$$

admits a solution $\varphi_2(z)$ satisfying (P_2) simultaneously, and at $z = z_0$, $\varphi_2(z)$ has a pole. Hence, around $w = \infty$, $S_2(z, w)$ has an expansion of the same form as (21.12), possibly with a different sign of ε, and with a different A_{-2}. Continuing we obtain similar expansions for all branches $S_2(z, w), \ldots, S_m(z, w)$, denoted by $S_j(z, w, \varepsilon, A_{-2,j})$ to indicate the values of ε and $A_{-2,j}$ in (21.12). Denote now by l_+, resp. l_-, the number of branches with $\varepsilon = 1$, resp. $\varepsilon = -1$, in the expansion (21.12). Then

$$P(z, w, w') = \prod_{j=1}^{l_+}(w' - S_j(z, w, 1, A_{-2,j})) \prod_{k=1}^{l_-}(w' - S_k(z, w, -1, A_{-2,k})). \tag{21.13}$$

Expanding now (21.13), we obtain

$$P(z, w, w') = (w')^m - \left(\sigma_2 w^2 + \sigma_1 w + \sigma_0 + \sum_{j=1}^{\infty} \sigma_{-j} w^{-j}\right)(w')^{m-1} + \cdots,$$

where $\sigma_2 = l_+ - l_-$, $\sigma_1 = 0$, $\sigma_0 = \frac{1}{2}(l_+ - l_-)z$ and $\sigma_{-1} = \alpha(l_+ - l_-) - \frac{1}{2}m$. Since (21.13) has to be a polynomial in w, w', we have $\sigma_{-j} = 0$ for $j \geq 1$. Therefore

$$\begin{cases} l_+ + l_- = m \\ \alpha(l_+ - l_-) = \dfrac{1}{2}m. \end{cases} \tag{21.14}$$

From (21.14) we first observe that $\alpha \in \mathbb{R}$, and since $2\alpha < 1$, we get $l_+ + l_- = m$, $l_+ - l_- > m$, which is not possible for $l_+ \geq 0$, $l_- \geq 0$.

It remains to show that for $\alpha = \frac{1}{2}$ all solutions of (P_2) also solving a first-order equation of type (21.1) are solutions of the Riccati differential equation (21.2) with

$\varepsilon = 1$. But (21.14) now implies that $l_+ = m$, $l_- = 0$. In this case, around each pole $\tilde{z}_0$ of $\varphi(z)$, the function $\varphi(z)$ satisfies either of the equations

$$w' = S_j(z, w) = w^2 + \frac{1}{2}z + \cdots, \qquad j = 1, \dots, m,$$

since $l_+ = m$. Therefore, every pole of $\varphi(z)$ has the same residue -1. By Theorem 18.2, $\varphi(z)$ satisfies the Riccati equation (21.2). □

§22 Higher order analogues of (P_2)

Similarly as to §19 for (P_1), we apply the method of using higher order Korteweg–de Vries equations to obtain higher order analogues of (P_2). To this end, we consider the Korteweg–de Vries hierarchy for a function $u(x, t)$ of two variables, sufficiently smooth, as follows:

Let D stand, for a while, for the partial differentiation operator $\frac{\partial}{\partial x}$, and define

$$L_u := 2u + 2DuD^{-1} - D^2, \tag{22.1}$$

where D^{-1} stands for the inverse operator of D. We now consider the equations

$$(2m-1)\frac{\partial u}{\partial t} = X_m u, \tag{$_m$KdV}$$

where

$$X_1 u = Du = D\frac{\partial H_1}{\partial u}, \qquad X_m u = L_u X_{m-1} u = D\frac{\partial H_m}{\partial u}, \qquad m \geq 2. \tag{22.2}$$

If $m = 2$, then we obtain the Korteweg–de Vries equation

$$3\frac{\partial u}{\partial t} = X_2 u = (2u + 2DuD^{-1} - D^2)\frac{\partial u}{\partial x} = 6u\frac{\partial u}{\partial x} - \frac{\partial^3 u}{\partial x^3}. \tag{KdV}$$

Similarly, for $m = 3$, we get

$$5\frac{\partial u}{\partial t} = 30u^2\frac{\partial u}{\partial x} - 20\frac{\partial u}{\partial x}\frac{\partial^2 u}{\partial x^2} - 10u\frac{\partial^3 u}{\partial x^3} + \frac{\partial^5 u}{\partial x^5}.$$

By a change of variables,

$$z := xt^{-1/(2m-1)}, \quad q(z) := t^{2/(2m-1)}u(x, t), \tag{22.3}$$

($_m$KdV) reduces to the ordinary differential equation

$$X_m q + 2q + zq' = 0, \tag{22.4}$$

where L_q and $X_m q$ are defined by (22.1), (22.2) with u replaced by q, x replaced by z and $'$ denoting the differentiation with respect to z. For the convenience, we also use now D to denote the same as $'$.

Given a meromorphic w, denote now $q := w' + w^2$ and define an operator S_w by

$$S_w := 4w^2 + 4w'D^{-1}w - D^2.$$

It is easy to verify that

$$(2w + D)S_w = L_q(2w + D). \tag{22.5}$$

Moreover,

$$X_m q = (2w + D)S_w^{m-1}(w'). \tag{22.6}$$

In fact, since $X_1 q = Dq = (2w + D)(w') = (2w + D)S_w^0(w')$, (22.6) holds for $m = 1$. Supposing (22.6) holds for m, we get

$$\begin{aligned} X_{m+1}q &= L_q X_m q = L_q(2w + D)S_w^{m-1}(w') \\ &= (2w + D)S_w(S_w^{m-1}(w')) = (2w + D)S_w^m(w'). \end{aligned}$$

Hence we can integrate the equation (22.4) to obtain

$$D^{-1}S_w^n(w') + zw + \alpha = 0, \tag{$_{2n}P_2$}$$

where $n = m - 1$, $n = 1, 2, \dots$ and α is a constant of integration. In fact, this follows by using (22.6) and differentiating $({}_{2n}P_2)$. Clearly, the order of the equation $({}_{2n}P_2)$ is $2n$.

For $n = 2$, hence $m = 2$, the equations $({}_m\mathrm{KdV})$ and $({}_{2n}P_2)$ become the Korteweg–de Vries equation (KdV), already computed above, and the second Painlevé equation

$$w'' = 2w^3 + zw + \alpha, \tag{P_2}$$

with the connection

$$u(x, t) = t^{-2/3}(w'(z) + w^2(z)), \quad z = xt^{-1/3}$$

between the solutions. This is a well-known result due to Ablowitz and Segur [1]. For $n = 2$, $n = 3$, $n = 4$ in $({}_{2n}P_2)$ we obtain, correspondingly,

$$w^{(4)} = 10w^2w'' + 10w(w')^2 - 6w^5 - zw - \alpha. \tag{$_4P_2$}$$

$$\begin{aligned} w^{(6)} &= 14w^2w^{(4)} + 56ww'w^{(3)} + 42w(w'')^2 - 70\big(w^4 - (w')^2\big)w'' \\ &\quad - 140w^3(w')^2 + 20w^7 + zw + \alpha, \end{aligned} \tag{$_6P_2$}$$

$$\begin{aligned} w^{(8)} &= 18w^2w^{(6)} + 108ww'w^{(5)} - 6\big(21w^4 - 35(w')^2 - 38ww''\big)w^{(4)} \\ &\quad - 138w\big(w^{(3)}\big)^2 - 252w'(4w^3 - 3w'')w^{(3)} + 182(w'')^3 \\ &\quad - 756w^3(w'')^2 + 84w^2\big(5w^4 - 37(w')^2\big)w'' - 798w(w')^4 \\ &\quad + 1260w^5(w')^2 - 70w^9 - zw - \alpha. \end{aligned} \tag{$_8P_2$}$$

Presently, there is no rigorous proof to show that all solutions of ($_{2n}P_2$) would be meromorphic functions, although we conjecture that this is the case. Therefore, in what follows, we restrict ourselves to considering meromorphic solutions of ($_{2n}P_2$), which may, of course, be represented as the quotient $w(z) = v(z)/u(z)$ of two entire functions $v(z)$, $u(z)$. We establish a one-to-one correspondence between the meromorphic solutions of the equation ($_{2n}P_2$) and the entire solutions of a system constructed below.

To this end, we first observe, by ($_{2n}P_2$) and the definition of the operator S_w that $w(z)$ has the Laurent expansion

$$w(z) = \alpha_{-1}(z - z_0)^{-1} + \alpha_1(z - z_0) + \varphi_n(z - z_0) \tag{22.7}$$

at a pole $z = z_0$ with α_{-1} taking one of the values $\pm 1, \pm 2, \dots, \pm n$ and $\varphi_n(z - z_0)$ being analytic in a neighborhood of z_0. Similarly as to (P_2) in §17, we may construct a meromorphic function $W(z)$ such that $W' = w^2$, see Gromak [8], [13] and an entire function $u(z)$ such that $u' = -Wu$, with a zero of multiplicity α_{-1}^2 at $z = z_0$. Again, u may be represented as

$$u(z) = \exp\Big(- \int^z ds \int^s w^2(t)\,dt \Big),$$

where the path of integration avoids the poles of w. Defining $v(z) := w(z)u(z)$, we obtain the required representation. In fact, differentiating $u' = -Wu$ and looking at ($_{2n}P_2$), we obtain

$$\begin{cases} uu'' - (u')^2 = -v^2, \\ D^{-1}S^n_{vu^{-1}}(v'u^{-1} - vu'u^{-2}) + zvu^{-1} + \alpha = 0. \end{cases} \tag{22.8}$$

For example, for the equation ($_4P_2$) we have

$$\begin{cases} uu'' - (u')^2 = -v^2, \\ v^{(4)}u^4 - 4v^{(3)}u'u^3 + 6v''(u')^2u^2 - 2v''v^2u^2 - 4v'(u')^3u \\ + 4v'u'v^2u + v(u')^4 - 2v^3(u')^2 + v^5 + zvu^4 + \alpha u^5 = 0. \end{cases}$$

We now proceed to consider the pair (22.8) in more detail.

We first observe that the initial conditions

$$\begin{aligned} &u(z_0) = u_0, \quad u'(z_0) = u_0', \\ &v(z_0) = v_0, \quad v'(z_0) = v_0', \dots, v^{(2n-1)}(z_0) = v_0^{(2n-1)}, \end{aligned} \tag{22.9}$$

where $z_0, u_0, u_0', v_0, v_0', \dots, v_0^{(2n-1)} \in \mathbb{C}$, determine the solution of (22.8), provided $u(z_0) \neq 0$. If $u(z_0) = 0$, the initial conditions (22.9) are singular.

Lemma 22.1. *The pair* (22.8) *has a solution*

$$v = 0, \quad u = \exp(az + b) \tag{22.10}$$

for any $a, b \in \mathbb{C}$, *provided* $\alpha = 0$.

Lemma 22.2. *If* (v, u) *is a solution of the pair* (22.8) *and* $\lambda(z) \not\equiv 0$, *then*

$$(\widetilde{v}, \widetilde{u}) = \big(\lambda(z)v, \lambda(z)u\big), \quad \lambda(z) \not\equiv 0 \tag{22.11}$$

is a solution of the pair (22.8) *if and only if* $\lambda(z) = e^{az+b}$, $a, b \in \mathbb{C}$.

Similarly as to Lemma 22.1 and Lemma 22.2, these lemmas follow by direct substitution of (22.10) and (22.11) into (22.8).

Theorem 22.3. *Let* (v, u) *be an arbitrary non-zero entire solution of the pair* (22.8) *for some fixed value of parameter* α *different from the special solution* (22.10). *Then the quotient* $v(z)/u(z)$ *represents a meromorphic solution of the equation* $({}_{2n}P_2)$ *with the same parameter value* α.

Proof. Let (v, u) be a non-zero entire solution of (22.8) with the initial conditions (22.9). If $u = 0$, then $v = 0$. Therefore $u \not\equiv 0$. Then there exists z_0 such that $u(z_0) = u_0 \neq 0$. Let us take a solution $w(z)$ of equation $({}_{2n}P_2)$ with the initial conditions $w(z_0) = v_0/u_0$, $w'(z_0) = v_0'/u_0 - v_0u_0'/u_0^2, \ldots, w^{(2n-1)}(z_0) = (v/u)^{(2n-1)}(z_0)$ such that $u''(z_0) = ((u_0')^2 - v_0^2)/u_0$. We may now construct a solution of (22.8) as

$$\begin{aligned} u_1(z) &:= u_0 \exp\Big((z - z_0)\frac{u_0'}{u_0} - \int_{z_0}^{z} dz \int_{z_0}^{z} w^2(z)\, dz\Big), \\ v_1(z) &:= u_1(z)w(z), \end{aligned} \tag{22.12}$$

where the path of integration avoids poles of $w(z)$. It is not difficult to see that the solution $(v_1(z), u_1(z))$ satisfies the same initial conditions as $(v(z), u(z))$. Hence, by the local uniqueness, $(v_1(z), u_1(z)) \equiv (v(z), u(z))$. The assertion follows. □

Theorem 22.4. *Any meromorphic solution* $w(z)$ *of the equation* $({}_{2n}P_2)$ *may be represented in the form* $w(z) = v(z)/u(z)$, *where* $(v(z), u(z))$ *is an entire solution of* (22.8), *determined up to factor* $\exp(az + b)$.

Proof. If $w(z) = 0$ with $\alpha = 0$, we shall take the entire solution $((v(z), u(z)) = (0, \exp(az + b))$ of the pair (22.8). Let $w(z) \not\equiv 0$. Then there exists a z_0 that $w(z_0) \neq 0$. Let us take $u(z)$ and $v(z) = w(z)u(z)$ as in the construction in (22.8). In this case, $v(z)$ and $u(z)$ are entire solutions of (22.8) and $w(z) = v(z)/u(z)$. However, by (22.11), $w(z)$ may also be expressed in terms of the solution

$$(\widetilde{v}(z), \widetilde{u}(z)) = (v(z)\exp(az + b), u(z)\exp(az + b)),$$

and the assertion follows. □

Theorems 22.3 and 22.4 establish the required correspondence between the meromorphic solutions of the equation $({}_{2n}P_2)$ and the entire solutions of the pair (22.8).

Let now

$$w(z) = P(z)/Q(z) \tag{22.13}$$

be a rational solution of ($_{2n}P_2$), where $P(z)$ and $Q(z)$ are polynomials with no non-trivial common factors. By direct substitution of (22.13) into ($_{2n}P_2$) it follows that $\deg P(z) = \deg Q(z) - 1$. Therefore, any rational solution of ($_{2n}P_2$) may be represented in the form

$$w(z) = \sum_{k=1}^{n} k\left[\sum_{j_k=1}^{l_k^+} \frac{1}{z - z_{j_k}} - \sum_{i_k=1}^{l_k^-} \frac{1}{z - z_{i_k}}\right] = \sum_{k=1}^{n} k\left(\frac{P_k'(z)}{P_k(z)} - \frac{Q_k'(z)}{Q_k(z)}\right),$$

where $l_k^{\pm}$ is a number of poles with the residue k and $-k$, respectively, and

$$P_k(z) = \prod_{j_k=1}^{l_k^+} (z - z_{j_k}), \quad Q_k(z) = \prod_{i_k=1}^{l_k^-} (z - z_{i_k}).$$

Therefore,

$$Q(z) = \prod_{s=1}^{n} P_s Q_s \quad P(z) = \sum_{k=1}^{n} k\widetilde{P}_k(P_k' Q_k - P_k Q_k'),$$

where

$$\widetilde{P}_k = \prod_{\substack{s=1 \\ s\neq k}}^{n} P_s Q_s.$$

Then the pair (22.8) admits a solution

$$u(z) = N(z)\exp(g(z)), \quad v(z) = M(z)\exp(g(z)), \tag{22.14}$$

where $g(z)$ is a polynomial and

$$N(z) = \prod_{k=1}^{n} (P_k Q_k)^{k^2}, \quad M(z) = P(z)\prod_{k=1}^{n} (P_k Q_k)^{k^2-1}. \tag{22.15}$$

Substituting (22.15) into (22.8), we get a pair of differential equations of the form (22.8) for $M(z)$ and $N(z)$ with u, v replaced by N, M. It is easy to observe that $g''(z) = 0$. As the second equation of (22.8) results by substituting $w(z) = v(z)/u(z)$ into ($_{2n}P_2$), the following theorem follows:

Theorem 22.5. *For the existence of a rational solution of* ($_{2n}P_2$) *it is necessary and sufficient that* (22.8) *with u, v replaced by N, M has a polynomial solution* $\big(M(z), N(z)\big)$, $N(z) \neq 0$, *under the condition that at all poles of $w(z)$ with residue $\pm k$, the polynomials $M(z)$ and $N(z)$ have zeros of multiplicity $k^2 - 1$ and k^2, respectively.*

Denote now

$$M(z) = \sum_{j=0}^{N-1} p_j z^{N-1-j}, \quad N(z) = \sum_{j=0}^{N} q_j z^{N-j}, \quad q_0 = 1. \tag{22.16}$$

Clearly

$$N = \deg N(z) = \sum_{k=1}^{n} (l_k^+ + l_k^-) k^2, \tag{22.17}$$

where l_k^+, l_k^- are the numbers of poles of rational solutions with residues equal to k and $-k$. Substituting (22.16) into the first equation of (22.8) we observe that $p_0^2 = N$. From the second equation of (22.8) it follows that $p_0 = -\alpha$. Therefore $N = \alpha^2$. For a rational solution of $({}_{2n}P_2)$, $z = \infty$ is a holomorphic point. Substituting $w(z) = \sum_{j=0}^{\infty} a_j z^{-j}$ into $({}_{2n}P_2)$ yields $a_0 = 0$, $a_1 = -\alpha$. Now from the total sum of the residues of a single-valued function in the complex plane, see Hahn and Epstein [1], we obtain

$$\sum_{k=1}^{n} (l_k^+ - l_k^-) k = -\alpha, \quad \sum_{k=1}^{n} (l_k^+ + l_k^-) k^2 = \alpha^2. \tag{22.18}$$

Therefore, we easily obtain

Theorem 22.6. *For the existence of a rational solution of $({}_{2n}P_2)$, it is necessary and sufficient that $\alpha \in \mathbb{Z}$. For every integer α, the equation $({}_{2n}P_2)$ has a unique rational solution.*

Proof. The necessity follows from the first equation of (22.18), as the left-hand side of this equation is an integer. The sufficiency follows from the Bäcklund transformations of $({}_{2n}P_2)$ constructed below. They permit us to construct the rational solutions for all integer parameters α from the seed solution $w = 0$ for $\alpha = 0$. □

Theorem 22.7. *The pair of differential equations*

$$w' + w^2 = q, \quad w\Big(2\frac{\partial H_n}{\partial q} + z\Big) + \alpha = X_n q \tag{22.19}$$

is equivalent to the higher order Painlevé equation $({}_{2n}P_2)$.

Proof. The equivalence between $({}_{2n}P_2)$ and (22.19) immediately follows from the identity

$$2w\frac{\partial H_n}{\partial q} - X_n q = D^{-1} S_w^n(w'), \quad q = w' + w^2,$$

which is verified as follows:

$$\begin{aligned}
2w\frac{\partial H_n}{\partial q} &- X_n q - D^{-1}S_w^n(w') \\
&= 2wD^{-1}X_n q - X_n q - D^{-1}S_w^n(w') \\
&= \big(2wD^{-1}(2w+D) - (2w+D)\big)S_w^{n-1}(w') - D^{-1}S_w^n(w') \\
&= (4wD^{-1}w - D)S_w^{n-1}(w') - D^{-1}S_w^n(w') \\
&= \big(4wD^{-1}w - D - D^{-1}(4w^2 + 4w'D^{-1}w - D^2)\big)S_w^{n-1}(w') \\
&= D^{-1}\big(D(4wD^{-1}w - D) - (4w^2 + 4w'D^{-1}w - D^2)\big)S_w^{n-1}(w') \\
&= 0.
\end{aligned}$$

□

Using the equivalent system (22.19), we obtain the Bäcklund transformations for ($_{2n}P_2$), see Airault [1] and Gromak [8].

Theorem 22.8. *Let $w(z)$ be a meromorphic solution of ($_{2n}P_2$) with parameter α such that*

$$z + 2\frac{\partial H_n}{\partial r} \neq 0, \quad r := w^2 - w'.$$

Then the transformations

$$T^{-1} : w \mapsto \widetilde{w} = -w - (2\alpha - 1)\left(z + 2\frac{\partial H_n}{\partial r}\right)^{-1}, \tag{22.20}$$

$$T : \widetilde{w} \mapsto w = -\widetilde{w} - (2\widetilde{\alpha} + 1)\left(z + 2\frac{\partial H_n}{\partial \widetilde{q}}\right)^{-1}, \tag{22.21}$$

$$S : w_\alpha(z) \mapsto -w_{-\alpha}(z), \tag{22.22}$$

where $\widetilde{q} = \widetilde{w}^2 + \widetilde{w}'$, determine Bäcklund transformations of ($_{2n}P_2$) such that $T^{-1}(w)$ is a solution of ($_{2n}P_2$) with parameter $\widetilde{\alpha} = \alpha - 1$.

Proof. The pair (22.19) shows that a solution $w_\alpha(z)$ of ($_{2n}P_2$) with parameter α generates a solution $q(z)$ of (22.4). Conversely, a fixed solution $q(z)$ forms a solution $w_\alpha(z)$ of ($_{2n}P_2$) with parameter α. Denote now

$$Q := 2\frac{\partial H_n}{\partial q} + z, \quad P := X_n q. \tag{22.23}$$

Then from (22.19) we get $w = (P - \alpha)/Q$ and

$$Q' = 2D\frac{\partial H_n}{\partial q} + 1 = 2X_n q + 1 = 2P + 1. \tag{22.24}$$

Suppose now that a solution $q(z)$ can be obtained from another solution $\widetilde{w}_{\widetilde{\alpha}}(z)$ of ($_{2n}P_2$). Then we have

$$\widetilde{w}' + \widetilde{w}^2 = q, \quad \widetilde{w}Q + \widetilde{\alpha} = P. \tag{22.25}$$

Therefore

$$\widetilde{w}' + \widetilde{w}^2 = w' + w^2, \quad \widetilde{w}Q + \widetilde{\alpha} = P \tag{22.26}$$

and

$$\widetilde{w} = \frac{P - \widetilde{\alpha}}{Q} = \frac{wQ + \alpha - \widetilde{\alpha}}{Q} = w + \frac{\alpha - \widetilde{\alpha}}{Q}. \tag{22.27}$$

Substituting $\widetilde{w}$ from (22.27) and $w = (P - \alpha)/Q$ into the first equation of (22.26) we find that

$$(\alpha - \widetilde{\alpha})Q' = 2(\alpha - \widetilde{\alpha})(P - \alpha) + (\alpha - \widetilde{\alpha})^2. \tag{22.28}$$

Substituting Q' from (22.24) gives two possibilities for $\widetilde{\alpha}$: either $\widetilde{\alpha} = \alpha$ with $\widetilde{w} = w$, or $\widetilde{\alpha} = -\alpha - 1$ with $\widetilde{w} = w + (2\alpha + 1)/Q$. Finding w by means of $\widetilde{w}$ yields the inverse transformation. We note that if w is a solution of $({}_{2n}P_2)$ with the parameter α, then $-w$ is a solution with the parameter $-\alpha$. Thus $w \mapsto -w$ implies $\alpha \mapsto -\alpha, q \mapsto r = -w' + w^2$. This completes the proof. □

From (22.19) it follows that $({}_{2n}P_2)$ has a $(2n - 1)$-parameter family of solutions when $\alpha = m + 1/2, m \in \mathbb{Z}$. This family is generated by the generic solution of the equation

$$2\frac{\partial H_n}{\partial r} + z = 0. \tag{22.29}$$

Finally, we remark that the fundamental domain of equation $({}_{2n}P_2)$ is also $0 \leq \mathrm{Re}(\alpha) \leq 1/2$ and the following statement on the auto-Bäcklund transformations holds:

Theorem 22.9. *Let $w_\alpha(z)$ be a meromorphic solution of $({}_{2n}P_2)$. Then, for $\alpha \in \mathbb{Z}$, the function*

$$\widetilde{w}_\alpha(z) = T^\alpha S T^{-\alpha} w_\alpha(z)$$

is a solution of $({}_{2n}P_2)$ as well and $\widetilde{w}_\alpha \neq w_\alpha$, if $w_\alpha(z)$ is non-rational.

Proof. Similar as the proof of Theorem 19.4. □

To close this chapter, we shortly consider the case of $({}_4P_2)$. The pair of differential equations equivalent to $({}_4P_2)$, corresponding to (22.19), takes the form

$$q = w' + w^2, \quad w(6q^2 + z - 2q'') = 6qq' - q''' - \alpha, \tag{22.30}$$

and the Bäcklund transformations of $({}_4P_2)$ are

$$T^{-1} : w \mapsto \widetilde{w} = -w - (2\alpha - 1)R^{-1}(r),$$
$$T : \widetilde{w} \mapsto w = -\widetilde{w} - (2\widetilde{\alpha} + 1)R^{-1}(\widetilde{q}),$$

where $R(r) := z + 6r^2 - 2r'', r := w^2 - w', \widetilde{q} := \widetilde{w}' + \widetilde{w}^2$. The poles of the solutions $w(z)$ are simple and have residues $-1, +1$ or $-2, +2$. A rational solution exists if and only if α is an integer. Applying the Bäcklund transformations to the seed solution $w = 0$ for $\alpha = 0$, we obtain the following list of the first rational solutions of $({}_4P_2)$:

Table 22.1. List of the first rational solutions of ($_4P_2$).

α	$w_\alpha(z)$
0	0
± 1	$\mp\frac{1}{z}$
± 2	$\mp\frac{2}{z}$
± 3	$\pm\left(\frac{2}{z} - \frac{5z^4}{144+z^5}\right)$
± 4	$\pm\frac{-6967296 - 870912z^5 - 288z^{10} - 4z^{15}}{z(144+z^5)(-48384+1008z^5+z^{10})}$

It follows from the equivalent pair (22.30) that ($_4P_2$) has a 3-parameter family of solutions (called 3-solutions), defined by the system

$$w' = -\varepsilon w^2 + y, \quad y'' = 3\varepsilon y^2 + \varepsilon z/2, \quad \varepsilon^2 = 1, \quad \alpha = -\varepsilon/2. \tag{22.31}$$

It is not difficult to deduce that any five solutions of the same Bäcklund hierarchy are algebraically dependent. This results in a nonlinear superposition formula between these solutions, see Gromak [15].

The second equation of (22.31) is, essentially, the first Painlevé equation. Taking a meromorphic function u such that $w = \varepsilon u'/u$, we obtain

$$u'' - \varepsilon y u = 0, \quad y'' = 3\varepsilon y^2 + \varepsilon z/2. \tag{22.32}$$

The first equation in (22.32) is linear. Therefore, the coefficient of this linear equation is determined by a solution of the first Painlevé equation. In this sense, this result is similar to the Lamé equation

$$u'' + (a\wp(z) + b)u = 0,$$

where $\wp(z)$ is the Weierstrass elliptic function.

Remark. Value distribution of meromorphic solutions of ($_{2n}P_2$) has been recently investigated by Li an He [1]. Connections between the higher order analogues of (P_1) and of (P_2) has been investigated by Kudryashov [1].

Chapter 6

The fourth Painlevé equation (P_4)

The fourth Painlevé equation (P_4) is characterized, along with (P_1) and (P_2), by the property that all of its solutions are meromorphic functions, as proved in §4. However, as (P_4) has two complex parameters, several phenomena appear to be more complicated as is the case for (P_2). In particular, this concerns the analysis of the existence of rational solutions and the same applies for Bäcklund transformations. Therefore, the fundamental parameter domain too becomes less transparent than in the simple case of (P_2). On the other hand, the polar behavior is quite similar to the corresponding behavior for (P_2). A new phenomenon is the connection of (P_4) with the complementary error function.

§23 Preliminary remarks

The fourth Painlevé equation

$$w'' = \frac{(w')^2}{2w} + \frac{3}{2}w^3 + 4zw^2 + 2(z^2 - \alpha)w + \frac{\beta}{w} \tag{P_4}$$

contains two free parameters $\alpha, \beta \in \mathbb{C}$, in contrast to the first two Painlevé equations (P_1) and (P_2). As shown in §4, see also Steinmetz [4], all solutions of (P_4) are meromorphic functions. From the proof of this fact in §4, we recall that all poles of any solution $w(z)$ of (P_4) have to be simple with the residue ± 1. In fact, around a pole z_0, the Laurent expansion of $w(z)$ reads as

$$w(z) = \frac{\mu}{z - z_0} - z_0 + \frac{\mu}{3}(z_0^2 + 2\alpha - 4\mu)(z - z_0) + h(z - z_0)^2 + O(z - z_0)^3, \tag{23.1}$$

where $\mu^2 = 1$, $h \in \mathbb{C}$ is arbitrary and all other coefficients are uniquely determined in terms of α, β, z_0 and h.

Very much similarly as in §17 for (P_2), we may determine entire functions $u(z)$, $v(z)$ such that any solution $w(z)$ equals to their quotient $v(z)/u(z)$. Indeed, by (23.1) it is immediate to see that $w^2 + 2zw = (w + z)^2 - z^2$ may be represented as

$$w^2 + 2zw = \frac{1}{(z - z_0)^2} + \phi(z, z_0) \tag{23.2}$$

around z_0, where $\phi(z, z_0)$ is analytic in a neighborhood of z_0. By (23.2), $w^2 + 2zw$ has double poles only. Hence, exactly as in §17, there exists a meromorphic function

$W'(z) = w^2 + 2zw$, and an entire function $u(z)$ such that $u' = -Wu$. Again, u may be expressed as

$$u(z) = \exp\Big(-\int^z ds \int^s \big(w^2(t) + 2tw(t)\big)\,dt\Big),$$

where the path of integration avoids the poles of w. We may assume that $u(z_0) = 0$ and $u'(z_0) = 1$. Defining again $v(z) := w(z)u(z)$, we obtain the desired quotient representation.The entire functions $u(z)$, $v(z)$ now satisfy the following pair of differential equations:

$$\begin{cases} uu'' - (u')^2 + v^2 + 2zuv = 0 \\ 2u^2vv'' + v^2(u')^2 - 2uvu'v' - u^2(v')^2 \\ \quad = v^4 + 4zuv^3 + 4(z^2 - \alpha)u^2v^2 + 2\beta u^4. \end{cases} \tag{23.3}$$

The first equation in (23.2) follows from

$$\begin{aligned} \frac{u''}{u} &= \Big(\frac{u'}{u}\Big)' + \Big(\frac{u'}{u}\Big)^2 = -W' + \Big(\frac{u'}{u}\Big)^2 = -w^2 - 2zw + \Big(\frac{u'}{u}\Big)^2 \\ &= -\Big(\frac{v}{u}\Big)^2 - 2z\frac{v}{u} + \Big(\frac{u'}{u}\Big)^2, \end{aligned}$$

while the second one can be obtained by first substituting $w(z) = v(z)/u(z)$ into (P_4), and then replacing uu'' from the first equation above.

Before proceeding, we make the following remark which describes an interesting symmetry property of (P_4). Namely, let $P_4(\alpha, \beta)$ denote the family of solutions of (P_4) with parameters α, β. Let now λ be such that $\lambda^4 = 1$, and let α, β be given. If $w(z) \in P_4(\alpha, \beta)$, then it is an elementary calculation to show that if $\widetilde{w}(z) := \lambda^{-1}w(\lambda z)$, then $\widetilde{w}(z) \in P_4(\alpha\lambda^2, \beta)$.

We now proceed to show how a certain one-parameter family of solutions of (P_4) may be constructed. To this end, observe that by a straightforward computation, all solutions of the Riccati differential equation

$$w' = \mu w^2 + 2\mu zw - 2(1 + \alpha\mu), \tag{23.4}$$

where $\mu^2 = 1$, also satisfy (P_4) with $\beta = -2(1 + \alpha\mu)^2$. As is well-known, all poles of w are simple with residue $-\mu$. Therefore, there exists an entire function u such that $w = -\mu\frac{u'}{u}$ implying that u satisfies the Weber–Hermite linear differential equation

$$u'' - 2\mu zu' - 2(\alpha + \mu)u = 0. \tag{23.5}$$

We remark that (23.5) can be further reduced to the Whittaker differential equation

$$4\zeta^2\frac{d^2y}{d\zeta^2} = (\zeta^2 - 4k\zeta + 4m^2 - 1)y, \tag{23.6}$$

where $y = y(k, m, \zeta)$ depends on the parameters k, m. It is immediate to check that

$$u(z) = z^{-1/2} \exp\left(\frac{1}{2}\mu z^2\right) y\left(\frac{1}{4} + \frac{1}{2}\alpha\mu, \frac{1}{4}, -\mu z^2\right).$$

If we now substitute $z = \tau\sqrt{\mu}$, assuming that $-1 - \alpha\mu = n \in \mathbb{N}$, then $\beta = -2n^2$ in (P_4) and all solutions of (23.5) may be expressed as $u(z) = U(z/\sqrt{\mu})$ with

$$U(\tau) = \exp(\tau^2)\frac{d^n}{d\tau^n}\Big(\exp(-\tau^2)(C_1 + C_2\int^{\tau} \exp(t^2)\,dt)\Big).$$

In particular, (23.5) admits solutions expressible in terms of the Hermite polynomials

$$U(\tau) = H_n(\tau) := (-1)^n \exp(\tau^2)\frac{d^n \exp(-\tau^2)}{d\tau^n} = \sum_{\nu=0}^{[n/2]}(-1)^{\nu}\binom{n}{2\nu}\frac{(2\nu)!}{\nu!}(2\tau)^{n-2\nu}. \tag{23.7}$$

We remark, according to Bassom, Clarkson and Hicks [1], that the transformation

$$u(z) = \eta(\xi)\exp\left(\frac{1}{4}\mu\xi^2\right), \quad \xi := \sqrt{2}z$$

reduces (23.5) into the parabolic cylinder equation

$$\frac{d^2\eta}{d\xi^2} = \left(\frac{1}{4}\xi^2 + \alpha + \frac{1}{2}\mu\right)\eta. \tag{23.8}$$

All solutions of (23.8) may be expressed in the form

$$\eta(\xi) = c_1 D_\nu(\xi) + c_2 D_\nu(-\xi), \tag{23.9}$$

where $\nu := -\alpha - \frac{1}{2}(1+\mu)$, and where $D_\nu(\xi)$ denotes a parabolic cylinder function satisfying

$$\frac{d^2 D_\nu}{d\xi^2} = \left(\frac{1}{4}\xi^2 - \nu - \frac{1}{2}\right)D_\nu$$

and having the asymptotic behavior

$$D_\nu(\xi) \sim \xi^\nu \exp(-\xi^2/4), \quad \xi \to +\infty,$$

$$D_\nu(\xi) \sim -\frac{\sqrt{2\pi}}{\Gamma(-\nu)}e^{\nu\pi i}\xi^{-\nu-1}\exp(\xi^2/4), \quad \xi \to -\infty,\ \arg\xi = \pi,$$

provided $\nu \notin \mathbb{Z}$. If $\nu \in \mathbb{N}$, then $D_\nu(\xi) = 2^{-\nu/2}H_\nu(\xi/\sqrt{2})\exp\left(-\frac{1}{4}\xi^2\right)$. For further properties of Hermite polynomials, see e.g. Abramowitz and Stegun [1], p. 687–691 and Rainville [1], p. 187–198.

We also point out that (P_4) is equivalent to as non-linear autonomous system, see Adler [1]. In fact, consider the autonomous system

$$\begin{cases} g_1' = g_1(g_2 - g_3) + \alpha_1 \\ g_2' = g_2(g_3 - g_1) + \alpha_2 \\ g_3' = g_3(g_1 - g_2) + \alpha_3. \end{cases} \tag{23.10}$$

Provided $\alpha_1 + \alpha_2 + \alpha_3 \neq 0$, a linear change of variables in (23.10) results in a normalization

$$g_1 + g_2 + g_3 = -2z, \quad \alpha_1 + \alpha_2 + \alpha_3 = -2.$$

Then it is immediate to see that the first component g_1 also satisfies (P_4) with the parameter values $\alpha = \frac{1}{3}(\alpha_3 - \alpha_2)$, $\beta = -\frac{1}{2}\alpha_1^2$, see also Noumi and Yamada [1].

§24 Poles of fourth Painlevé transcendents

We shall consider the following pair of differential equations, where $\mu^2 = 1$:

$$\begin{cases} w' = \mu w^2 + 2\mu z w + \xi \\ 2w\xi' = \xi^2 - 2\mu w^2(2(1+\alpha\mu) + \xi) + 2\beta. \end{cases} \tag{24.1}$$

By an elementary computation, (24.1) is equivalent with (P_4), as observed by Lukashevich [1]. Making use of (24.1), we may proceed similarly as in §18 for (P_2) to prove the following

Theorem 24.1. *All nonrational solutions of* (P_4), *not satisfying the Riccati differential equation* (23.4), *have infinitely many poles with residue* $= 1$ *and, respectively, with residue* $= -1$.

Proof. Differentiating the second equation of (24.1), substituting for w' the first equation of (24.1), and dividing by $2w$, we obtain

$$\xi'' + 2\mu z\xi' + 2\mu w\xi' + 2\mu(\mu w^2 + 2\mu z w + \xi)(2(1+\alpha\mu) + \xi) = 0.$$

Substituting now for $2w\xi'$ the second equation of (24.1), we further get

$$\xi'' + 2\mu z\xi' + \mu\xi^2 + 2\mu\beta + (2(1+\alpha\mu) + \xi)(4zw + 2\mu\xi) = 0. \tag{24.2}$$

As w does not satisfy (23.4), we have $\xi + 2(1+\alpha\mu) \not\equiv 0$, and so we may solve w from (24.2) to obtain

$$w = -\frac{\mu\xi'' + 2z\xi' + 3\xi^2 + 4(1+\alpha\mu)\xi + 2\beta}{4\mu z(\xi + 2(1+\alpha\mu))}. \tag{24.3}$$

Substituting further (24.3) in the second equation of (24.1), we obtain, after some computation

$$\begin{aligned} D(\xi,\mu) := {}& 9\xi^4 + 6\mu\xi^2\xi'' + 8\big(3(1+\alpha\mu) - \mu z^2\big)\xi^3 + (\xi'')^2 + 8(\alpha+\mu)\xi\xi'' \\ & - 4z^2(\xi')^2 + \big(12\beta + 16(\alpha+\mu)^2 - 16(\alpha+\mu)z^2\big)\xi^2 + 4\mu\beta\xi'' \\ & + 16\beta(1+\alpha\mu-\mu z^2)\xi + 4\beta^2 - 32\beta(\alpha+\mu)z^2 = 0, \end{aligned} \tag{24.4}$$

which is an algebraic differential equation to determine $\xi(z)$. The solutions $\xi(z)$ of (24.4) will be expressed in terms of the fourth Painlevé transcendents by the formula

$$\xi(z) = w'(z) - \mu w(z)^2 - 2\mu z w(z), \tag{24.5}$$

while the solutions $w(z)$ of (P_4) will be expressed in terms of the solutions $\xi(z)$ of (24.4) by (24.3), provided $w(z)$ does not satisfy the Riccati differential equation (23.4).

By (24.5), it is immediate to verify that the solutions $\xi(z)$ of $D(\xi, 1)$ have a pole at z_0 if and only if $w(z)$ has a pole at z_0 with residue $= 1$. Similarly, the solutions $\xi(z)$ of $D(\xi, -1)$ have a pole at z_0 if and only if $w(z)$ has a pole at z_0 with residue $= -1$.

Now, independently of μ, the equation (24.4) has exactly one dominant term $9\xi^4$ in the sense of the Clunie lemma, Lemma B.11. By this lemma, ξ must have infinitely many poles, and we are done. □

Remark. In case a fourth Painlevé transcendent $w(z)$ also satisfies the Riccati differential equation (23.4) with $\mu = +1$, resp. $\mu = -1$, all poles of $w(z)$ must be with residue $-\mu$.

§25 Connection formulae between solutions of (P_4)

To construct the Bäcklund transformations for the fourth Painlevé transcendents, we observe that the following pair of two linked Riccati differential equations is equivalent to (P_4):

$$\begin{cases} w' = q + 2\mu z w + \mu w^2 + 2\mu w u \\ u' = p - 2\mu z u - \mu u^2 - 2\mu w u, \end{cases} \tag{25.1}$$

where $q^2 = -2\beta$, $p = -1 - \alpha\mu - \frac{1}{2}q$ and $\mu^2 = 1$. In fact, putting $\xi = 2\mu w u + q$ in (24.1), we immediately see that w satisfies (P_4) with parameters (α, β). On the other hand, by the symmetry of (25.1), u also satisfies (P_4) with parameters $(\widetilde{\alpha}, \widetilde{\beta})$ such that $p^2 = -2\widetilde{\beta}$, $q = -1 + \widetilde{\alpha}\mu - \frac{1}{2}p$. Hence,

$$\widetilde{\alpha} := \frac{1}{4}(2\mu - 2\alpha + 3\mu q), \quad \widetilde{\beta} := -\frac{1}{2}\left(1 + \alpha\mu + \frac{1}{2}q\right)^2. \tag{25.2}$$

Therefore, we obtain the following theorem, due to Lukashevich [1]:

Theorem 25.1. *Let $w = w(z, \alpha, \beta) \not\equiv 0$ be a solution of (P_4) with parameters (α, β). Then the transformations*

$$T : w \mapsto \widetilde{w}, \quad \Lambda : (\alpha, \beta) \mapsto (\widetilde{\alpha}, \widetilde{\beta}), \tag{25.3}$$

where

$$\widetilde{w}(z) := \frac{R(w)}{2\mu w} = \frac{1}{2\mu w}(w' - \mu w^2 - 2\mu z w - q), \quad \mu^2 = 1, \quad q^2 = -2\beta, \tag{25.4}$$

determine a solution $\widetilde{w} = \widetilde{w}(z, \widetilde{\alpha}, \widetilde{\beta})$ of (P_4) with parameters $\widetilde{\alpha}$, $\widetilde{\beta}$, given in (25.2).

Proof. This is an immediate consequence of the equivalence between (P_4) and the pair of differential equations (25.1), with u denoted by $\widetilde{w}$. □

We note that from the system (25.1) the inverse transformation to T can also be obtained. For this purpose, we express w in terms of u from the second equation of (25.1). This results in

$$T^{-1} : \widetilde{w} \mapsto w, \quad \Lambda^{-1} : (\widetilde{\alpha}, \widetilde{\beta}) \mapsto (\alpha, \beta), \tag{25.5}$$

where

$$w(z) = \frac{\widetilde{R}(w)}{2\mu\widetilde{w}} = -\frac{1}{2\mu\widetilde{w}}(\widetilde{w}' - \widetilde{q} + 2\mu z\widetilde{w} + \mu\widetilde{w}^2) \tag{25.6}$$

and

$$\begin{aligned} \widetilde{q}^2 &= -2\widetilde{\beta}, & \alpha &= (-2\mu - 2\widetilde{\alpha} - 3\mu\widetilde{q})/4, \\ q &= \widetilde{\alpha}\mu - 1 - \widetilde{q}/2, & \beta &= -q^2/2. \end{aligned} \tag{25.7}$$

If in (25.5) and (25.6), $\sqrt{-2\widetilde{\beta}} = p$, then it is easy to show that $TT^{-1} = T^{-1}T = I$, $\Lambda\Lambda^{-1} = \Lambda^{-1}\Lambda = I$, where I denotes the identity transformation.

Thus, the above relations yield a one-to-one correspondence between the solutions $w(z, \alpha, \beta)$ for which $R(w) \not\equiv 0$ and the solutions $\widetilde{w}(z, \widetilde{\alpha}, \widetilde{\beta})$ under the given choice of μ and q.

Bearing in mind that our goal is to generate solution hierarchies, it is important to consider the result of applying several consecutive Bäcklund transformations. It allows us to deduce the general form of α and β after an arbitrary number of the Bäcklund transformations. During repeated applications of the recursion relation (25.3), the choice of μ and q at the jth step will be denoted by μ_j and ν_j, respectively, where $\mu_j^2 = \nu_j^2 = 1$. Then, (25.2) is written in the form $\widetilde{\alpha} = \frac{1}{4}(2\mu_j - 2\alpha + 3\mu_j\nu_j q)$, $\widetilde{\beta} = -\frac{1}{2}(1 + \alpha\mu_j + \frac{1}{2}\nu_j q)^2$, respectively, where the value of $q, \widetilde{q}$ is fixed in a suitable manner.

We now turn to the problem of finding one-parameter families of solutions of (P_4), called 1-solutions. To this end, we again recall the Riccati differential equation (23.4), the solutions of which also satisfy (P_4) with parameters $(\alpha, \beta) = (\alpha, -2(1 + \alpha\mu)^2)$.

Taking, as above, q such that $q^2 = -2\beta$, then we may take σ such that $\sigma^2 = 1$ and $q\sigma = -2(1 + \alpha\mu)$, hence (23.4) may be written as

$$R_1(w) := w' - \mu w^2 - 2\mu z w - q\sigma = 0. \tag{25.8}$$

This now determines a one-parameter family of solutions of (P_4) with parameter values

$$q\sigma + 2(1 + \alpha\mu) = 0, \quad \mu^2 = \sigma^2 = 1, \quad q^2 = -2\beta. \tag{25.9}$$

We now apply the (T, Λ)-transformation to a solution $w(z)$ of (25.8) with parameter values (25.9), choosing them so that $(\mu_1, \nu_1) \neq (\mu, \sigma)$. If $\nu_1 = -\sigma$, $\mu_1 = -\mu$, say, then the first transformation step results in

$$(\widetilde{w}')^2 + 4\widetilde{w}' - \widetilde{w}^4 - 4z\widetilde{w}^3 - 4(z^2 - \widetilde{\alpha})\widetilde{w}^2 + 4 = 0, \tag{25.10}$$

which determines a one-parameter family of solutions of (P_4) with $\widetilde{\beta} = -2$ and $\widetilde{\alpha}$ arbitrary. By (25.4) and (25.8), the solutions of (25.10) may be expressed in terms of the solutions of (25.8) by means of the formula

$$\widetilde{w}(z) = \frac{2(\alpha + \mu)}{w(z)} - 2z - w(z), \quad \widetilde{\alpha} = -2\alpha - 2\mu,$$

i.e. in terms of the Weber–Hermite functions.

Remark 1. If $\alpha = 0$, then the equation (25.10) factorizes into two Riccati equations:

$$(w' + w^2 + 2zw + 2)(w' - w^2 - 2zw + 2) = 0.$$

Remark 2. After an application of the Bäcklund transformation (25.3) with $\mu = 1$ to solutions of (25.10) we obtain the equation

$$(w')^3 + P_1(z, w)(w')^2 + P_2(z, w)w' + P_3(z, w) = 0,$$

where

$$\begin{aligned}
P_1(z, w) &:= w^2 + 2zw + 6 - 2\alpha, \\
P_2(z, w) &:= -w^4 - 4zw^3 - 4w^2(z^2 - 1 - \alpha) - 8zw(1 - \alpha) - (6 - 2\alpha)^2, \\
P_3(z, w) &:= -w^6 - 6zw^5 + (-12z^2 - 2 + 6\alpha)w^4 - 4z(2z^2 + 6 - 6\alpha)w^3 \\
&\quad + 4(6\alpha z^2 - 10z^2 - 3\alpha^2 + 10\alpha - 11)w^2 \\
&\quad - 8z(15 - 14\alpha + 3\alpha^2)w - (6 - 2\alpha)^3.
\end{aligned}$$

This equation determines solutions of (P_4) for $\beta = -2(\alpha - 3)^2$. The next step of Bäcklund transformations gives

$$(w')^4 + 8(w')^3 + Q_2(z, w)(w')^2 + Q_3(z, w)w' + Q_4(z, w) = 0,$$

where

$$Q_2(z,w) := -2w^2(w^2+4zw+4z^2-4\alpha)$$
$$Q_3(z,w) := -8(w^4+4zw^3+4w^2(z^2-\alpha)+16)$$
$$Q_4(z,w) := w^8+8zw^7+8(3z^2-\alpha)w^6+32(z^3-\alpha z)w^5 + 16(1+z^4-2\alpha z^2+\alpha^2)w^4+64zw^3-256,$$

determining solutions of (P_4) for $\beta = -8$ and α arbitrary.

The following two theorems on Bäcklund transformations of (P_4) originate from Gromak [10]. For an illustration of Theorem 25.2, see Figure 26.1, p. 141.

Theorem 25.2. *The equation (P_4) with parameter values (α, β) such that either*

$$\beta = -2(\alpha\mu + 2n - 1)^2, \quad n \in \mathbb{N}, \quad \mu^2 = 1, \tag{25.11}$$

or

$$\beta = -2n^2, \quad n \in \mathbb{N}, \tag{25.12}$$

has one-parameter families of solutions expressed in terms of the Weber–Hermite functions.

Proof. Indeed, if $n = 1$ in (25.11) and (25.12), then we have the one-parameter families of solutions of equation (25.8) and (25.10), respectively. Let the statement now be valid for $n = m$ in (25.11). Then applying (T, Λ)-transformations with $\mu_1 = -\mu, \nu_1 = \nu$ to solutions with parameters (25.11) in which $q = 2\nu(\alpha\mu+2m-1)$, we obtain the case (25.12) with $n = m$. If $\mu_1 = \mu$, $\nu_1 = -\nu$, then we have the case (25.12) with $n = m - 1$. If $\mu_1\nu_1 = \mu\nu$, then one obtains the previous case (25.11) with $n = m$. By (25.2) with (μ_1, ν_1), $\widetilde{\beta} = -2(1-\mu_1\widetilde{\alpha}+\nu_1 q)^2$. Assuming the validity of the statement for $n = m$ in (25.12) and $q = 2m\nu$, and applying (25.3), we obtain (25.11), where

(a) $n = m + 1$, if $\mu_1 = -\mu$, $\nu_1 = \nu$;

(b) $n = m$, if $\mu_1 = \mu$, $\nu_1 = -\nu$;

(c) $n = -m$, if $\mu_1 = \mu$, $\nu_1 = \nu$;

(d) $n = -(m - 1)$, if $\mu_1 = -\mu$, $\nu_1 = -\nu$.

Therefore, the proof is completed. □

Let us now assume that a solution $w(z, \alpha, \beta)$ does not belong to the one-parameter families of solutions which have been obtained in Theorem 25.2. To this end, it is sufficient that α, β be inconsistent with the parameters (25.11), (25.12). It is clear from (25.4) that this assumption makes possible to apply successive Bäcklund transformations to a solution $w(z, \alpha, \beta)$.

Theorem 25.3. *Successive applications of the* (T, Λ)*-transformation to a solution* $w(z, \alpha, \beta)$ *of* (P_4) *lead to a solution* $\widetilde{w}(z, \widetilde{\alpha}, \widetilde{\beta})$*, where* $(\widetilde{\alpha}, \widetilde{\beta})$ *is expressible as either*

$$\widetilde{\alpha} = \alpha + n_1, \quad \widetilde{\beta} = -\frac{1}{2}(2n_2 + \sqrt{-2\beta})^2 \tag{25.13}$$

or

$$\begin{cases} \widetilde{\alpha} = n_1 + \dfrac{1}{2}(\mu - \alpha) + \dfrac{3}{4}\mu\sqrt{-2\beta}, \\ \widetilde{\beta} = -\dfrac{1}{2}\Big(2n_2 + 1 + \mu\alpha + \dfrac{1}{2}\sqrt{-2\beta}\Big)^2, \end{cases} \tag{25.14}$$

with $(n_1, n_2) \in \mathbb{Z} \times \mathbb{Z}$ *satisfying* $n_1 + n_2 \in 2\mathbb{Z}$*. Moreover, for every pair* (n_1, n_2) *of such integers, there exists a composition of* (T, Λ)*-transformations which carries* (α, β) *to* $(\widetilde{\alpha}, \widetilde{\beta})$ *as in* (25.13) *or* (25.14).

Proof. According to (25.2), after the first application of the Λ-transformation, we have

$$_1\Lambda_{\mu_1,\nu_1} : \left(\alpha, -\frac{q^2}{2}\right) \mapsto \left\{-\frac{\alpha}{2} + \frac{3}{4}\mu_1\nu_1 q + \frac{\mu_1}{2},\ -\frac{1}{2}\Big(\alpha\mu_1 + \frac{\nu_1}{2}q + 1\Big)^2\right\}.$$

By a suitable choice of operators, we obtain

$$\begin{aligned} {}_2\Lambda_{\mu_2,\mu_2}(\mu_1, \nu_1) : \left(\alpha, -\frac{q^2}{2}\right) \mapsto \Bigg\{ & \frac{\alpha}{4}(1 + 3\mu_1) + \frac{3}{8}\nu_1(1 - \mu_1)q + \frac{3}{4} - \frac{\mu_1}{4} + \frac{\mu_2}{2}, \\ & -\frac{1}{2}\left[\frac{\mu_2}{2}\alpha(\mu_1 - 1) + \frac{\mu_2\nu_1}{4}(1 + 3\mu_1)q + \frac{\mu_1\mu_2}{2} + \frac{\mu_2}{2} + 1\right]^2\Bigg\} \end{aligned}$$

and

$$\begin{aligned} {}_2\Lambda_{\mu_2,-\mu_2}(\mu_1, \nu_1) : \left(\alpha, -\frac{q^2}{2}\right) \mapsto \Bigg\{ & \frac{\alpha}{4}(1 - 3\mu_1) - \frac{3}{8}\nu_1(1 + \mu_1)q - \frac{3}{4} - \frac{\mu_1}{4} + \frac{\mu_2}{2}, \\ & -\frac{1}{2}\left[-\frac{\mu_2}{2}\alpha(\mu_1 + 1) + \frac{\mu_2\nu_1}{4}(-1 + 3\mu_1)q + \frac{\mu_1\mu_2}{2} - \frac{\mu_2}{2} + 1\right]^2\Bigg\}. \end{aligned}$$

Let us now consider the transformations:

$$\begin{aligned} S_1 &= {}_2\Lambda_{1,1}(1, 1) : (\alpha, -q^2/2) \mapsto (\alpha + 1, -(q + 2)^2/2), \\ S_2 &= {}_2\Lambda_{1,1}(1, -1) : (\alpha, -q^2/2) \mapsto (\alpha + 1, -(q - 2)^2/2), \\ S_3 &= {}_2\Lambda_{-1,1}(-1, -1) : (\alpha, -q^2/2) \mapsto (\alpha - 1, -(q - 2)^2/2), \\ S_4 &= {}_2\Lambda_{-1,1}(-1, 1) : (\alpha, -q^2/2) \mapsto (\alpha - 1, -(q + 2)^2/2). \end{aligned}$$

For arbitrary $k_j \in \mathbb{N} \cup \{0\}$, composing these transformations, we get

$$\begin{aligned} S_{(k_1,k_2,k_3,k_4)} &: (\alpha, -q^2/2) \\ &\mapsto \big(\alpha + k_1 + k_2 - k_3 - k_4,\ -\big(q + 2(k_1 - k_2 - k_3 + k_4)\big)^2/2\big). \end{aligned}$$

Hence, for every pair (n_1, n_2) such that $n_1+n_2 \in 2\mathbb{Z}$, and for $q = \sqrt{-2\beta}$, there exist $k_1, k_2, k_3, k_4 \in \mathbb{N}\cup\{0\}$ satisfying $n_1 = k_1+k_2-k_3-k_4$, $n_2 = k_1-k_2-k_3+k_4$, namely $n_1+n_2 = 2(k_1-k_3)$, $n_1-n_2 = 2(k_2-k_4)$. This implies that $S_{(k_1,k_2,k_3,k_4)}(\alpha,\beta) = (\widetilde{\alpha},\widetilde{\beta})$ coincides with (25.13). Furthermore, for $q = \sqrt{-2\beta}$,

$$_1\Lambda_{\mu,1} : (\alpha, -q^2/2) \mapsto \Big(-\frac{\alpha}{2}+\frac{3}{4}\mu q+\frac{\mu}{2},\ -\frac{1}{2}(\alpha\mu+q/2+1)^2\Big),$$

and hence, applying $S_{(k_1,k_2,k_3,k_4)}$ to $(-\alpha/2+(3/4)\mu q+\mu/2, -Q^2/2)$ with $Q = \alpha\mu+q/2+1$, we obtain $(\widetilde{\alpha},\widetilde{\beta})$ given by (25.14). Thus the second part of the theorem is proved.

As to the first part, namely the closedness of parameters, this may be verified by direct computation:

$$\begin{aligned}(1)\quad {}_1\Lambda_{\mu_1,\nu_1} &: \Big(\alpha+n_1, -\frac{1}{2}(2n_2+\sqrt{-2\beta})^2\Big)\\ &\mapsto \Big(\frac{1}{2}(\mu^*-\alpha)+\frac{3}{4}\mu^*\sqrt{-2\beta}+N_1,\ -\frac{1}{2}(\alpha\mu^*+1+\frac{1}{2}\sqrt{-2\beta}+2N_2)^2\Big),\end{aligned}$$

where $\mu^* = \mu_1\nu_1$,

$$\begin{aligned}N_1 &= \frac{-n_1+3\mu_1\nu_1 n_2+\mu_1(1-\nu_1)}{2} \in \mathbb{Z},\\ N_2 &= \frac{\mu_1\nu_1 n_1+n_2+(\nu_1-1)}{2} \in \mathbb{Z},\\ N_1+N_2 &= \frac{(\mu_1\nu_1-1)(n_1-n_2)+(\nu_1-1)(1-\mu_1)}{2}+2\mu_1\nu_1 n_2 \in 2\mathbb{Z}.\end{aligned}$$

$$(2)\quad {}_1\Lambda_{\mu_1,\nu_1} : \Big(\frac{1}{2}(\mu-\alpha)+\frac{3}{4}\mu\sqrt{-2\beta}+n_1,\ -\frac{1}{2}\Big(\mu\alpha+1+\frac{1}{2}\sqrt{-2\beta}+2n_2\Big)^2\Big)$$

$$\mapsto \begin{cases} \Big(\alpha+N_1', -\frac{1}{2}(\sqrt{-2\beta}+2N_2')^2\Big), & \text{if } \mu_1\nu_1 = \mu,\\ \Big(\frac{1}{2}(-\mu-\alpha)-\frac{3}{4}\mu\sqrt{-2\beta}+N_1'',\ -\frac{1}{2}(-\mu\alpha+1+\frac{1}{2}\sqrt{-2\beta}+2N_2'')^2\Big), \\ \text{if } \mu_1\nu_1 = -\mu, \end{cases}$$

where

$$\begin{aligned}N_1' &= \frac{\mu+\mu_1-n_1+3\mu n_2}{2} \in \mathbb{Z},\\ N_2' &= \frac{1+\mu\mu_1+n_1\mu+n_2}{2} \in \mathbb{Z},\\ N_1'+N_2' &= \frac{(1-\mu)(n_2-n_1)+(\mu_1+1)(\mu+1)}{2}+2\mu n_2 \in 2\mathbb{Z},\end{aligned}$$

$$N_1'' = \frac{\mu_1 - \mu - n_1 - 3\mu n_2}{2} \in \mathbb{Z},$$
$$N_2'' = \frac{-1 - \nu_1 + \mu n_1 - n_2}{2} \in \mathbb{Z},$$
$$N_1'' + N_2'' = \frac{(\mu_1 - 1)(\mu + 1) + (\mu - 1)(n_1 + n_2)}{2} - 2\mu n_2 \in 2\mathbb{Z}.$$

proving the theorem. □

Remark 3. To construct the solutions of (P_4) for arbitrary values of parameters (α, β), it is sufficient to construct solutions for every (α, β) in the domain

$$G := \{(\alpha, \beta) \mid 0 \le \operatorname{Re}\alpha \le 1,\ \operatorname{Re}\sqrt{-2\beta} \ge 0,\ \operatorname{Re}(\sqrt{-2\beta} + 2\alpha) \le 2\}. \quad (25.15)$$

The validity of this statement directly follows from Theorem 25.3, the relations (25.13), (25.14), the transformation (25.3) and the transformation $\widetilde{w}(z) = \lambda^{-1}w(\lambda z)$, $\lambda^2 = -1$. Thus (25.15) is a fundamental domain of the parameter space.

We now proceed to consider solutions of (P_4) which satisfy a first order algebraic differential equation as well, i.e. solutions corresponding to the Airy solutions in §21. The proof of Theorem 25.4 follows the basic idea of the proof of Theorem 21.1, and so could be presented in more details in the same way.

Theorem 25.4. *The Painlevé differential equation (P_4) admits a one-parameter family of solutions satisfying a first order algebraic differential equation $P(z, w, w') = 0$ if and only if the parameters α, β satisfy* (25.11) *or* (25.12). *All such solutions may be obtained from the solutions of the Riccati differential equation* (23.4), *under the condition $\beta = -2(1 + \alpha\mu)^2$, by repeated applications of the Bäcklund transformations and the transformation $\widetilde{w}(z) = \lambda^{-1}w(\lambda z)$, $\lambda^2 = -1$.*

Proof. The sufficiency of the conditions (25.11) and (25.12) has been obtained above. To proceed to the necessity, similarly as in the case of (P_2), it is sufficient to consider the equation (P_4) for the parameter values (α, β) in the fundamental domain G only. However, for the sake of simplicity, we apply the domain

$$G_1 = \{(\alpha, q) \mid 0 \le \operatorname{Re}\alpha \le 1,\ 0 \le \operatorname{Re}q < 2,\ 0 \le \operatorname{Re}(q + 2\alpha) \le 2\}, \quad (25.16)$$

where $q^2 = -2\beta$, instead of the domain G. In fact, instead of $(\alpha, q) = (0, 2) \in G$ we may introduce $(\alpha, q) = (1, 0) \in G_1$ without loss of generality as these points are equivalent with respect to the Bäcklund transformations (25.2)–(25.5) by taking $\mu = -1$ and $q = -2$.

Suppose now that $w(z)$ is transcendental. The following expansion for w' in a neighborhood of the poles of w may easily be verified:

$$G(z, w, \mu, h) = a_2 w^2 + a_1 w + a_0 + \sum_{j=1}^{\infty} a_{-j} w^{-j},$$

where $a_2 = \mu$, $a_1 = 2\mu z$, $a_0 = -2 - 2\alpha\mu$ and where $a_{-1} =: h$ is arbitrary. Hence, we obtain

$$\begin{aligned} P(z, w, w') &:= (w')^n + \sum_{j=1}^{n} P_j(z, w)(w')^{n-j} \\ &= \prod_{j=1}^{l_+} \prod_{k=1}^{l_-} \big(w' - G(z, w, 1, h_j)\big)\big(w' - G(z, w, -1, h_k)\big), \end{aligned} \tag{25.17}$$

where l_+ and l_- are the numbers of the factors of $P(z, w, w') = 0$ in w' with $\mu = 1$ and $\mu = -1$, respectively, and h_ν is the arbitrary coefficient a_{-1} in the corresponding expansion for w'. Expanding the right-hand side of (25.17) we find that

$$P(z, w, w') = (w')^n - \Big(\sigma_2 w^2 + \sigma_1 w + \sigma_0 + \sum_{j=1}^{\infty} \frac{\sigma_{-j}}{w^j}\Big)(w')^{n-1} + \cdots = 0, \tag{25.18}$$

where $\sigma_2 = l_+ - l_-$, $\sigma_1 = 2z(l_+ - l_-)$, $\sigma_0 = -2n + 2\alpha(l_- - l_+)$ and $\sigma_{-j} = \sum_{\nu=1}^{n} a_{-j}^{(\nu)}$, where $a_{-j}^{(\nu)}$ stands for the coefficient a_{-j} in the corresponding expansion. Consequently, the coefficients in the polynomial

$$P_1(z, w) = p_{10}(z)w^2 + p_{11}(z)w + p_{12}(z)$$

become

$$p_{10} = l_- - l_+, \quad p_{11} = 2z(l_- - l_+), \quad p_{12} = 2n - 2\alpha(l_- - l_+). \tag{25.19}$$

Next, we substitute $w = 1/u$ in the equations (P_4) and $P(z, w, w') = 0$ to obtain

$$u'' = \frac{3(u')^2}{2u} - \frac{3}{2u} - 4z - 2(z^2 - \alpha)u - \beta u^3, \tag{25.20}$$

and

$$\widetilde{P}(z, u, u') = (u')^n + (\widetilde{p}_{10}u^2 + \widetilde{p}_{11}u + \widetilde{p}_{12})(u')^{n-1} + \cdots = 0,$$

where

$$\widetilde{p}_{10} = -p_{12}, \quad \widetilde{p}_{11} = -p_{11}, \quad \widetilde{p}_{12} = -p_{10}. \tag{25.21}$$

We now form the expansion of $\widetilde{P}(z, u, u')$ corresponding to $G(z, w, \mu, h)$ around a pole of u, i.e. around a zero of w. First suppose that $\beta \neq 0$. Writing

$$u' = b_2u^2 + b_1u + b_0 + \frac{b_{-1}}{u} + \cdots$$

for (25.20), we find that $b_2 = \sigma\sqrt{-2\beta}$, where $\sigma^2 = 1$ and b_1 is arbitrary. Here we choose the value of $\sqrt{-2\beta}$ according to the choice of σ.

Let us denote by $\tilde{l}_+$ and $\tilde{l}_-$ the number of the factors of $\tilde{P}(z, u, u')$ with $\sigma = 1$ and $\sigma = -1$, respectively. It is clear that $\tilde{l}_- + \tilde{l}_+ = n$. Then, as in the case of $P(z, w, w') = 0$, we have $\tilde{p}_{10} = (\tilde{l}_- - \tilde{l}_+)q$. But from (25.21) and (25.19) we immediately get the following condition:

$$(\tilde{l}_- - \tilde{l}_+)q = 2\alpha(l_- - l_+) - 2n. \qquad (25.22)$$

Next suppose that $\beta = 0$. Then every pole of u is double, and u satisfies an equation of the form

$$u' = c_{3/2}u^{3/2} + c_1 u + c_{1/2}u^{1/2} + \cdots .$$

By the same argument, we have $\tilde{p}_{10} = 0$, which implies that (25.22) is valid in the case $\beta = 0$ as well.

We now proceed to consider (25.22) in the domain G_1. Clearly, $l_-, l_+, \tilde{l}_-, \tilde{l}_+ \in \mathbb{N} \cup \{0\}$, and since $n > 0$, α and q as well as $l_- - l_+$ and $\tilde{l}_- - \tilde{l}_+$ cannot vanish simultaneously.

(1) Suppose first that $l_- - l_+ = 0$ and $\tilde{l}_- - \tilde{l}_+ \neq 0$. Then

$$q = 2\frac{\tilde{l}_+ + \tilde{l}_-}{\tilde{l}_+ - \tilde{l}_-}. \qquad (25.23)$$

Therefore, $q \in \mathbb{R}$. But $0 \leq q < 2$ in G_1, contradicting (25.23).

(2) Assume next that $l_- - l_+ \neq 0, \tilde{l}_- - \tilde{l}_+ = 0$. Then,

$$\alpha = \frac{l_- + l_+}{l_- - l_+}. \qquad (25.24)$$

Hence $\alpha \in \mathbb{R}$ and so $0 \leq \alpha \leq 1$ in G_1. By (25.24), $l_+ = 0, l_-$ is arbitrary and $\alpha = 1$. But then we must have $(\alpha, q) = (1, 0) \in G_1$. By Theorem 24.1, $w(z)$ satisfies (23.4). This means that the equation $P(z, w, w') = 0$ coincides with the Riccati equation (23.4) with $\mu = -1, \alpha = 1, \beta = 0$.

(3) Finally suppose that $l_+ - l_- \neq 0$ and $\tilde{l}_+ - \tilde{l}_- \neq 0$. Let us now consider the line (25.22) in the (α, q)-plane, written as

$$H := \left\{(\alpha, q) \;\middle|\; \frac{\alpha}{L} + \frac{q}{Q} = 1,\ Q := \frac{2n}{\tilde{l}_+ - \tilde{l}_-},\ L := \frac{n}{l_- - l_+}\right\}. \qquad (25.25)$$

Clearly, if $H \cap G_1 \neq \emptyset$, then either (a) $0 \leq Q < 2$, or (b) $0 \leq L \leq 1$. As we already showed above, the case (a) is impossible. As for the case (b), we have $\alpha = 1, l_+ = 0$ in G_1. By Theorem 24.1, $w(z)$ satisfies (23.4). To complete the proof, let us consider the rational solutions. As will be shown in the next section, all rational solutions of (P_4) can be classified. For $(\alpha, q) \in G_1$, there exist only two rational solutions (except for the trivial solution $w = 0$) given by $w_0(z) = -2z$, $(\alpha, \beta) = (0, -2)$ and $w_1(z) = -2z/3$, $(\alpha, \beta) = (0, -2/9)$; this fact is verified by using (25.13) and (25.14). The solution $w_0(z)$ belongs to a one-parameter family of solutions of the Riccati equation (23.4). The other solution $w_1(z)$ does not satisfy (23.4), since $\beta \neq -2(1 + \alpha\mu)^2$. Thus, we have proved Theorem 25.4. □

Remark 4. As shown above, any transcendental solution $w(z)$ of (P_4) such that $w(z) \in \mathcal{A}_1(\mathbb{C}(z))$ belongs to a one-parameter family of solutions given in Theorem 25.4.

Remark 5. Let us again denote by ν, $\nu^2 = 1$, the choice of the value $\sqrt{-2\beta}$ in a T-transformation (25.3). Then T generates four solutions $w_j = w_j(z, \alpha_j, \beta_j)$, $j = 1, 2, 3, 4$, according to the choice of (μ, ν). These solutions may be written in the form

$$\begin{aligned} w_1 &= \big(w' - \sqrt{-2\beta} - 2zw - w^2\big)/2w, & (\mu, \nu) &= (1, 1), \\ w_2 &= -\big(w' + \sqrt{-2\beta} + 2zw + w^2\big)/2w, & (\mu, \nu) &= (-1, -1), \\ w_3 &= \big(w' + \sqrt{-2\beta} - 2zw - w^2\big)/2w, & (\mu, \nu) &= (1, -1), \\ w_4 &= -\big(w' - \sqrt{-2\beta} + 2zw + w^2\big)/2w, & (\mu, \nu) &= (-1, 1). \end{aligned}$$

Eliminating w' from the above expressions yields the following algebraic relations between the solutions $w, w_1, \dots, w_4$:

$$\begin{aligned} &w_1 + w_4 = w_2 + w_3 = -2z - w, \\ &w_1 + w_2 + \big(\sqrt{-2\beta} + 2zw + w^2\big)/w = 0, \\ &w_3 = w_1 + (\sqrt{-2\beta})/w. \end{aligned}$$

Closing this section, we obtain a nonlinear superposition formula which links solutions on two successive levels. Let $T_{\mu,\nu}$ be a T-transformation (25.3) with a fixed choice of μ, ν. One application of $T_{\mu,\nu}$ gives the four solutions $w_j = w_j(z, \alpha_j, \beta_j)$, $j = 1, 2, 3, 4$ above. Applying T twice, $T_{\mu_1,\nu_1} \circ T_{\mu_2,\nu_2}$ generates 16 solutions $w_{i,\,j} = w_{i,\,j}(z, \alpha_{i,\,j}, \beta_{i,\,j})$, $i, j \in \{1, 2, 3, 4\}$. Note that $T_{1,-1} \circ T_{-1,-1} = T_{1,-1} \circ T_{-1,1} = T_{-1,-1} \circ T_{1,1} = T_{-1,-1} \circ T_{1,-1} = I$, where I denotes the identity transformation. The relation between the solutions w, w_j and $w_{i,j}$ is given by

$$\frac{(1 + \nu_1)(2 + \nu_2 q + 2\alpha\mu_2)}{2w_j} + (2\mu_2 w + (\mu_1 + \mu_2)(2z + w_j) + 2\mu_1 w_{i,j}) = 0.$$

Using the trivial symmetry,

$$S_0 : w(z, 0, \beta) \mapsto \mu^{-1} w(\mu z, 0, \beta), \quad \mu^4 = 1$$

and T-transformations, we may obtain auto-Bäcklund transformations of the form $(T^{-1})^k \circ S_0 \circ T^k$, where $k \in \mathbb{Z}$ denotes the kth step of the application of the Bäcklund transformation, where the parameters are subjected to the condition that either $\alpha \in \mathbb{Z}$ or $\beta = -2\big(\alpha + (2n + 1)\big)^2/9$, $n \in \mathbb{Z}$.

Remark 6. Other forms of Bäcklund transformations exist for (P_4) as well as for other Painlevé equations, see e.g. Murata [1], Kitaev [1], Boiti and Pempinelli [1]. Such transformations may be obtained by using the isomonodromy deformation technique, see e.g. Jimbo and Miwa [1], [2], Jimbo, Miwa and Ueno [1], Fokas and Its [1], Flaschka and Newell [1] and Its and Novokshenov [1]. Moreover, several Bäcklund transformations for Painlevé equations have been obtained by using the Schlesinger transformation of linear differential equations. In fact, Fokas, Mugan and Ablowitz [1], see also Bassom, Clarkson and Hicks [1], applied this method for (P_4). In Bassom, Clarkson and Hicks [1] one can find a survey of various Bäcklund transformations for (P_4), relations between them and the solution hierarchies.

§26 Rational solutions of (P_4)

In this section, we aim to characterize the pairs of parameter values (α, β) in (P_4) for which rational solutions may exist. For original articles concerning this topic, see Lukashevich [1], Gromak [10] and Murata [1]. Concerning a detailed analysis of rational solutions of (P_4), omitted here, we refer to Murata [1] and Bassom, Clarkson and Hicks [1].

To start with, it is a trivial observation that any solution $w(z)$ of (P_4) is rational if and only if $z = \infty$ is either regular or a pole for $w(z)$, since $w(z)$ is meromorphic by §4. Denoting $t := z^{-1}$, (P_4) may be written as

$$w'' = \frac{(w')^2}{2w} - 2\frac{w'}{t} + \frac{3}{2}\frac{w^3}{t^4} + 4\frac{w^2}{t^5} + 2(1 - \alpha t^2)\frac{w}{t^6} + \frac{\beta}{t^4 w}. \tag{26.1}$$

We may now write (26.1) in the form

$$2(t(tw')' + tw')w - (tw')^2 = 3t^{-2}w^4 + 8t^{-3}w^3 + 4(1 - \alpha t^2)t^{-4}w^2 + 2\beta t^{-2}. \tag{26.2}$$

Assume first that $t = 0$ is a pole of $v(t) = w(t^{-1})$. As one can easily check, $v(t)$ has the following expansion in a neighborhood $t = 0$:

$$v(t) = a_{-1}t^{-1} + \sum_{j=0}^{\infty} a_j t^j, \quad a_{-1} \neq 0.$$

Since $tv' = -a_{-1}t^{-1} + \sum_{j\geq 0} j a_j t^j$, $t(tv')' = a_{-1}t^{-1} + \sum_{j\geq 0} j^2 a_j t^j$, we have

$$\begin{aligned} &2(t(tv')' + tv')v - (tv')^2 \\ &\quad = -a_{-1}^2 t^{-2} + \sum_{j\geq 0}\Big(2j(j+2)a_{-1}a_j + \sum_{k=0}^{j-1} k(3k+3-j)a_k a_{j-1-k}\Big)t^{j-1}. \end{aligned}$$

Furthermore,

$$v^2 = a_{-1}^2 t^{-2} + \sum_{j\ge 0}\Big(2a_{-1}a_j + \sum_{\substack{k_1+k_2=j-1\\ k_m\le j-1}} a_{k_1}a_{k_2}\Big)t^{j-1},$$

$$v^3 = a_{-1}^3 t^{-3} + \sum_{j\ge 0}\Big(3a_{-1}^2 a_j + \sum_{\substack{k_1+k_2+k_3=j-2\\ k_m\le j-1}} a_{k_1}a_{k_2}a_{k_3}\Big)t^{j-2},$$

$$v^4 = a_{-1}^4 t^{-4} + \sum_{j\ge 0}\Big(4a_{-1}^3 a_j + \sum_{\substack{k_1+k_2+k_3+k_4=j-3\\ k_m\le j-1}} a_{k_1}a_{k_2}a_{k_3}a_{k_4}\Big)t^{j-3}.$$

Substituting $w = v(t)$ into (26.2) and using the expressions above, we obtain

$$\begin{cases} 3a_{-1}^2 + 8a_{-1} + 4 = 0, \\ (3a_{-1}^2 + 6a_{-1} + 2)a_{-1}a_j = P_j(a_{-1}, a_1, \ldots, a_{j-1}, \alpha, \beta), \quad j \ge 0. \end{cases}$$

This implies that $a_{-1} = -2$ or $a_{-1} = -2/3$, and that, for both a_{-1}, the coefficients a_j are determined uniquely. Since (26.1) remains invariant under the substitution $(v, t) \to (-v, -t)$, the formal series $-v(-t)$ also satisfies (26.1), and hence $v(t) = -v(-t)$, implying that $a_{2j} = 0$, for $j \in \mathbb{N} \cup \{0\}$. By a simple computation, a_1 and a_3 are given by

$$\begin{cases} (3a_{-1}^2 + 6a_{-1} + 2)a_{-1}a_1 = \alpha a_{-1}^2, \\ 4(3a_{-1}^2 + 6a_{-1} + 2)a_{-1}a_3 \\ \quad = 8\alpha a_{-1}a_1 - 2\beta - a_{-1}^2 - 18a_{-1}^2 a_1^2 - 24a_{-1}a_1^2 - 4a_1^2. \end{cases} \tag{26.3}$$

For example, we have (1) $a_1 = -\alpha$ if $a_{-1} = -2$, and (2) $a_1 = \alpha$ if $a_{-1} = -2/3$.

Assume next that $t = 0$ is a holomorphic point of $v(t) = w(t^{-1})$. It is easy to see that $v(t)$ admits an expansion of the form

$$v(t) = \sum_{j=1}^{\infty} a_j t^j.$$

Observing that

$$tv' = \sum_{j\ge 1} j a_j t^j, \quad v^2 = a_1^2 t^2 + \sum_{j\ge 2}\Big(2a_1 a_j + \sum_{\substack{k_1+k_2=j+1\\ k_m\le j-1}} a_{k_1}a_{k_2}\Big)t^{j+1}$$

and so on, we conclude that $4a_1^2 + 2\beta = 0$, $a_1 a_j = \widetilde{P}_j(a_1, \ldots, a_{j-1}, \alpha, \beta)$, $j \ge 2$. By the same argument as to above, we can verify that $a_{2j} = 0$, $j \in \mathbb{N} \cup \{0\}$. The coefficients a_1, a_3 are then given by

$$\beta + 2a_1^2 = 0, \quad 2a_3 = \alpha a_1 - 2a_1^2. \tag{26.4}$$

Since the solutions of the equation (P_4) have simple poles only with residues $= \pm 1$, rational solutions for which $z = \infty$ is a pole may exist only, when $\alpha \in \mathbb{Z}$ as one can easily see by (26.3). Similarly, rational solutions for which $z = \infty$ is a holomorphic point may exist only, when $\sqrt{-\beta/2} \in \mathbb{Z}$.

Therefore, if a rational solution of (P_4) exists, then it must be of the form

$$w(z) = \lambda z + \frac{\xi(z)}{\eta(z)}, \tag{26.5}$$

where $\lambda \in \{0, -2, -2/3\}$, and

$$\xi(z) = \sum_{j=0}^{m-1} \xi_j z^j, \quad \eta(z) = \sum_{j=0}^{m} \eta_j z^j, \tag{26.6}$$

where $\eta_m = 1$, $\eta_{m-1} = \eta_{m-3} = \cdots = 0$, $\xi_{m-2} = \xi_{m-4} = \cdots = 0$. If $\xi(z) \equiv 0$, then (P_4) admits the following rational solutions:

$$\begin{cases} w = 0, \ \text{if } \beta = 0, \\ w = -2z, \ \text{if } \alpha = 0 \text{ and } \beta = -2, \\ w = -\dfrac{2}{3}z, \ \text{if } \alpha = 0 \text{ and } \beta = -\dfrac{2}{9}. \end{cases} \tag{26.7}$$

Recall the pair (23.3) of differential equations equivalent to (P_4). As we seek for rational solutions of (P_4), we may take

$$v(z) = P(z)\exp(g(z)), \quad u(z) = Q(z)\exp(g(z)), \tag{26.8}$$

where $P(z)$, $Q(z)$ are polynomials of the degrees n and m, respectively, and $g(z)$ is an entire function. Substituting (26.8) into (23.3) we obtain

$$\begin{cases} QQ'' - (Q')^2 + g''Q^2 + 2zPQ + P^2 = 0, \\ 2PQ^2P'' + 2P^2Q^2g'' + P^2(Q')^2 - 2PQP'Q' - (P')^2Q^2 \\ = P^4 + 4zP^3Q + 4(z^2 - \alpha)P^2Q^2 + 2\beta Q^4. \end{cases} \tag{26.9}$$

Similarly as to the procedure applied for (P_2) in §20, it is easy to prove the following

Theorem 26.1. *The fourth Painlevé equation* (P_4) *admits a rational solution if and only if the pair* (26.9) *of differential equations has a polynomial solution* $(P(z), Q(z))$, $Q(z) \not\equiv 0$, *under a suitable choice of a polynomial* $g(z)$ *obtained in accordance with the construction of* (23.3).

Let $w(z)$ be a rational solution of (P_4). Let $z_1, \ldots, z_m$ denote the poles of $w(z)$. Since $w(z)(w(z) + 2z) = (z - z_j)^{-2} + O(1)$ around each pole by (23.2), we have

$$w(z)(w(z) + 2z) = \sum_{j=1}^{m} \frac{1}{(z - z_j)^2} + \phi(z),$$

where $\phi(z) = -g''(z)$ is a polynomial. Substituting the series expansion $w(z) = \lambda z + a_1 z^{-1} + a_3 z^{-3} + \cdots$ around $z = \infty$, and comparing (26.2) and (26.5), we have

$$\lambda(\lambda+2)z^2+2(\lambda+1)a_1+(a_1^2+2(\lambda+1)a_3)z^{-2}+O(z^{-3}) = \phi(z)+mz^{-2}+O(z^{-3}).$$

This implies that $m = a_1^2 + 2(\lambda + 1)a_3$. If $\lambda = 0$, then by (26.3), $a_1 = \pm\sqrt{-\beta/2}$ and $a_3 = \pm\frac{\alpha}{2}\sqrt{-\beta/2} + \frac{\beta}{2}$. Therefore, in this case,

$$m = \pm\alpha\sqrt{-\beta/2} + \frac{\beta}{2}. \tag{26.10}$$

Moreover, if $\lambda = -2$, respectively $\lambda = -\frac{2}{3}$, then we apply $a_1 = -\alpha$, $a_3 = \frac{3}{4}\alpha^2 + \frac{1}{8}\beta + \frac{1}{4}$, respectively $a_1 = \alpha$, $a_3 = -\frac{3}{4}\alpha^2 - \frac{9}{8}\beta - \frac{1}{4}$, to obtain

$$m = -\frac{\alpha^2}{2} - \frac{\beta}{4} - \frac{1}{2}, \tag{26.11}$$

respectively

$$m = \frac{\alpha^2}{2} - \frac{3}{4}\beta - \frac{1}{6}. \tag{26.12}$$

To construct the rational solutions of (P_4) explicitly, the Bäcklund transformations may be applied. To this end, take first the solution $w(z, 0, -2) = -2z$ as the initial solution. We then obtain solutions as in Table 26.1 and Table 26.2 below with parameter values

$$\alpha = n_1, \quad \beta = -2(1 + 2n_2 - n_1)^2, \quad n_1, n_2 \in \mathbb{Z}. \tag{26.13}$$

These relations may be obtained by using Theorem 25.3; both types of Λ-transformations lead to the same result (26.13).

Note that, by §23, the rational solutions listed in Tables 26.1 and 26.2 contain solutions of the form

$$w = -\sqrt{\mu}\frac{H_n'(t)}{H_n(t)}, \quad \alpha = -\mu(1+n), \quad \beta = -2n^2, \quad z = \sqrt{\mu}t, \quad \mu^2 = 1,$$

$$w = -2z + i\sqrt{\mu}\frac{H_n'(t)}{H_n(t)}, \quad \alpha = n\mu, \quad \beta = -2(1+n)^2, \quad z = i\sqrt{\mu}t, \quad \mu^2 = 1,$$

where $H_n(t)$ is the Hermite polynomial defined in (23.7).

To complete the construction, let now $w(z, 0, -\frac{2}{9}) = -\frac{2}{3}z$ be the initial solution. Then we obtain solutions as in Table 26.3 with the parameter values

$$\alpha = n_1, \quad \beta = -\frac{2}{9}(6n_2 - 3n_1 + 1)^2, \quad n_1, n_2 \in \mathbb{Z}. \tag{26.14}$$

Table 26.1. The first rational solutions of (P_4) regular at $z = \infty$.

α	β	$w(z, \alpha, \beta)$
± 2	-2	$\pm \frac{1}{z}$
± 3	-8	$\pm \frac{4z}{2z^2 \pm 1}$
± 4	-2	$\pm \frac{2z^2 \pm 1}{z(2z^2 \mp 1)}$
± 5	-8	$\pm \frac{4z(2z^2 \mp 1)(2z^2 \pm 3)}{(2z^2 \pm 1)(4z^4 + 3)}$
± 5	-32	$\pm \frac{8z(2z^2 \pm 3)}{4z^4 \pm 12z^2 + 3}$

Table 26.2. The first rational solutions of (P_4) of the form $w(z, \alpha, \beta) = -2z + P(z)/Q(z)$.

α	β	$2z + w(z, \alpha, \beta)$
0	-2	0
0	-18	$-\frac{8z}{(2z^2 + 1)(2z^2 - 1)}$
± 1	-8	$\mp \frac{1}{z}$
± 2	-18	$\mp \frac{4z}{2z^2 \pm 1}$
± 3	-32	$\mp \frac{3(2z^2 \pm 1)}{z(2z^2 \pm 3)}$

Table 26.3. The first rational solutions of (P_4) of the form $w(z, \alpha, \beta) = -2z/3 + P(z)/Q(z)$.

α	β	$2z/3 + w(z, \alpha, \beta)$
0	$-\frac{2}{9}$	0
± 1	$-\frac{8}{9}$	$\pm\frac{1}{z}$
± 1	$-\frac{32}{9}$	$\pm\frac{2z^2 \pm 3}{z(2z^2 \mp 3)}$
± 2	$-\frac{2}{9}$	$\pm\frac{4z}{2z^2 \pm 3}$
± 3	$-\frac{32}{9}$	$\pm\frac{3(8z^6 \mp 20z^4 + 30z^2 \pm 45)}{z(2z^2 \mp 3)(4z^4 - 45)}$

We complete this section by collecting the above considerations in the following

Theorem 26.2. *The fourth Painlevé equation* (P_4) *admits a rational solution* $w(z)$ *if and only if the parameter values* (α, β) *are as in* (26.13) *or* (26.14). *For any of these parameter values, there exists a unique rational solution.*

Proof. Indeed, the sufficiency of these conditions follows from the fact that under the parameter values (26.13) and (26.14) with arbitrary integers n_1, n_2, the rational solutions can be constructed by means of Bäcklund transformations (25.3) starting from the nontrivial rational solutions given by (26.7). Assuming the existence of a rational solution under the values of parameters different from (26.13) and (26.14), we immediately obtain the existence of a rational solution which is different from (26.7) such that the parameters are in the fundamental domain (25.15). However, this contradicts the second necessary condition of the existence of the rational solutions. In fact, m, determined by (26.10), (26.11) and (26.12), is a non-negative integer in the fundamental domain if and only if we have one of the solutions (26.7). The uniqueness of the rational solutions under parameter values (26.13), (26.14) follows from the uniqueness of the rational solutions (26.7) in the fundamental domain. □

The next figure depicts all real pairs $(\alpha, \sqrt{-2\beta})$ described in Theorem 25.2 and Theorem 26.2. In this figure, black horizontal lines, resp. red lines, represent the Weber-Hermite solutions (25.12), resp. (25.11). The crossing points of red lines represent the real pairs $(\alpha, \sqrt{-2\beta})$ in (26.13), while the blue points represent (26.14). The gray area describes the fundamental domain G in (25.15).

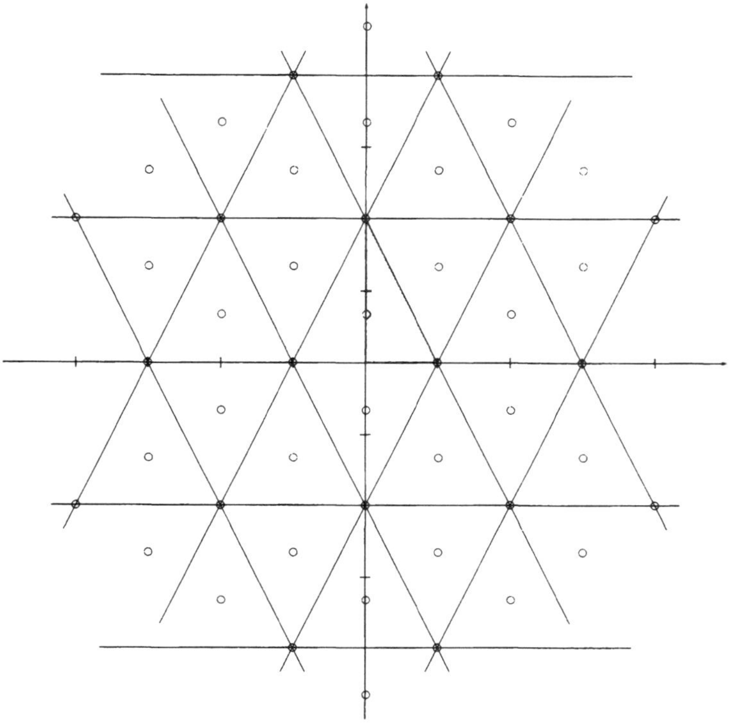

Figure 26.1.

§27 The complementary error function hierarchy

The Riccati differential equation (23.4) may be used to generate, by means of the Bäcklund transformations, in addition to the rational solutions in §26, one-parameter families of transcendental solutions of (P_4) as well. As an example, following Bassom, Clarkson and Hicks [1], we consider some families of one-parameter solutions of (P_4), generated from the seed solution $w(z, 1, 0)$ of the Riccati equation (23.4) with $\mu = -1$, i.e.

$$w' = -w^2 - 2zw. \tag{27.1}$$

In this case, the Weber–Hermite equation (23.5) may be integrated to yield

$$u(z) = B - A \operatorname{erf} c(z), \tag{27.2}$$

where A and B are arbitrary constants and $\operatorname{erf} c(z)$ is the complementary error function given by

$$\operatorname{erf} c(z) = \frac{2}{\sqrt{\pi}} \int_z^\infty \exp(-t^2)\, dt. \tag{27.3}$$

Hence we obtain the exact solution of (P_4),

$$w(z,1,0) = \frac{2A\exp(-z^2)}{\sqrt{\pi}(1 - A\,\mathrm{erf}\,c(z))}, \tag{27.4}$$

where we have set $B = 1$, since the ratio A/B of the constants only is relevant; the case $B = 0$ generates no solutions of importance and so we do not pursue this option here. Applying now the transformation $\widetilde{w}(z) = \lambda^{-1}w(\lambda z)$, $\lambda^2 = -1$, from §23, to $w(z,1,0)$, assumed to be non-vanishing, we obtain $w(z,-1,0)$. The solutions $w(z,1,0)$ and $w(z,-1,0)$ satisfy the equations

$$w' + 2zw + w^2 = 0, \quad w' - 2zw - w^2 = 0, \tag{27.5}$$

respectively. A hierarchy of solutions of (P_4) can now be derived from (27.4) by the Bäcklund transformations. Each of the solutions is expressed in terms of z, $\psi(z)$ and $\psi'(z)$, where $\psi(z) := 1 - A\,\mathrm{erf}\,c(z)$. The collection of solutions of (P_4) obtained in this way is called the complementary error function hierarchy. By setting $\alpha = 1$ and $\beta = 0$ in Theorem 25.3 we see that the solutions in the complementary error function hierarchy will have parameter values given by

$$\alpha_n = m_1 + 1, \quad \beta_n = -2m_2^2$$

provided that $m_1 + m_2 = 2m_3$ and m_1, m_2, m_3 are integers.

In Tables 27.1 and 27.2 we present the first solutions in this complementary error function hierarchy with $\beta \neq 0$. Observe that these solutions will be expressed as rational functions of z and $\Psi(z) \equiv \psi'(z)/\psi(z)$. Moreover, they are listed either being of the form $P(z,\Psi(z))/Q(z,\Psi(z))$ or of the form $-2z + P(z,\Psi(z))/Q(z,\Psi(z))$.

Remark 1. If $A = 0$ above, then $\Psi \equiv 0$ and complementary error function solutions reduce to rational solutions. Not all rational solutions arise as special limits of complementary error function solutions in this way.

Remark 2. We are not going to examine here another hierarchy of (P_4) related to complementary error functions the solutions of which are complex valued for real z. To derive these solutions we apply the transformation $\widetilde{w}(z) = \lambda^{-1}w(\lambda z)$ for $\lambda^2 = -1$ to the solutions of complementary error function hierarchy. For example, from the solution (27.4) the transformation

$$z \mapsto iz, \quad w \mapsto -iw, \quad \alpha \mapsto -\alpha$$

results in

$$w(z,-1,0) = -\frac{2iA\exp(z^2)}{\sqrt{\pi}(1 - A\,\mathrm{erf}\,c(iz))} = -\frac{\Phi'(z)}{\Phi(z)},$$

where $\Phi(z) = 1 - A\,\mathrm{erf}\,c(iz)$.

Table 27.1. Complementary error function solutions of the form $P(z, \Psi)/Q(z, \Psi)$.

α	β	$w(z, \alpha, \beta)$
-2	-2	$\dfrac{\Psi^2 + 2z\Psi - 2}{\Psi + 2z}$
2	-2	$\dfrac{2}{\Psi + 2z}$
3	-8	$\dfrac{2(\Psi + 2z)}{z\Psi + (2z^2 + 1)}$
4	-18	$\dfrac{3\big(z\Psi + (2z^2 + 1)\big)}{(z^2 + 1)\Psi + (2z^3 + 3z)}$
-4	-2	$\dfrac{-2\big(\Psi^2 + 3z\Psi + (2z^2 - 1)\big)}{(\Psi + 2z)(z\Psi + (2z^2 + 1))}$

Table 27.2. Complementary error function solutions of the form $-2z + P(z, \Psi)/Q(z, \Psi)$.

α	β	$2z + w(z, \alpha, \beta)$
0	-2	$-\Psi$
1	-8	$\dfrac{-2}{\Psi + 2z}$
2	-18	$\dfrac{-2(\Psi + 2z)}{z\Psi + (2z^2 + 1)}$
3	-32	$\dfrac{-3\big(z\Psi + (2z^2 + 1)\big)}{(z^2 + 1)\Psi + (2z^2 + 3)z}$
-1	-8	$\dfrac{2\big(z\Psi^3 + 2(2z^2 + 1)\Psi^2 + 4z(z^2 + 1)\Psi - 2\big)}{(\Psi + 2z)(\Psi^2 + 2z\Psi - 2)}$

§28 A half-integer hierarchy of solutions

In this section, we consider a one-parameter family of solutions of (P_4) associated with certain half-integer values of the parameters α, β, see Bassom, Clarkson and Hicks [1] and Gromak [10] for more details.

Choosing $\alpha = -\mu/2$ in (23.4) implies that $\beta = -2(1+\alpha\mu)^2 = -1/2$. The general solution of (23.4) with $\alpha = -\mu/2$ may be obtained from

$$\eta(\xi) = A_1 D_{-\frac{1}{2}}(\xi) + A_2 D_{-\frac{1}{2}}(-\xi),$$

where A_1 and A_2 are arbitrary constants and $D_\nu(\xi)$ is the parabolic cylinder function, see (23.8) and (23.9). The general solution of the Weber–Hermite differential equation (23.5) now reads as

$$u(z,\mu) = \exp\left(\frac{\mu}{2}z^2\right)\left(A_1 D_{-\frac{1}{2}}(\sqrt{2}z) + A_2 D_{-\frac{1}{2}}(-\sqrt{2}z)\right).$$

Thus we obtain

$$w\left(z, -\frac{1}{2}, -\frac{1}{2}\right) = -\frac{u'(z,1)}{u(z,1)} = -z - U(z), \tag{28.1}$$

where

$$U(z) = \frac{\sqrt{2}\left[A_1 D'_{-\frac{1}{2}}(\sqrt{2}z) - A_2 D'_{-\frac{1}{2}}(-\sqrt{2}z)\right]}{A_1 D_{-\frac{1}{2}}(\sqrt{2}z) + A_2 D_{-\frac{1}{2}}(-\sqrt{2}z)}$$

and

$$w\left(z, \frac{1}{2}, -\frac{1}{2}\right) = \frac{u'(z,-1)}{u(z,-1)} = -z + U(z). \tag{28.2}$$

We note that the solutions $w(z,-1/2,-1/2)$ and $w(z,1/2,-1/2)$ satisfy the equations

$$w' = 2zw + w^2 - 1, \quad w' = -2zw - w^2 - 1,$$

respectively. The solutions (28.1) and (28.2) may now be used to generate the required hierarchy of solutions. Substituting $\alpha = \pm 1/2$ and $\beta = -1/2$ into Theorem 25.3 leads to the conclusion that the solutions in the hierarchy will have the parameters α_n and β_n given by

$$\alpha_n = n_1 + \frac{1}{2}, \quad \beta_n = -\frac{1}{2}(2n_2+1)^2,$$

where n_1 and n_2 are any integers, or

$$\alpha_n = m_1, \quad \beta_n = -2m_2^2,$$

where m_1 and m_2 are integers satisfying $(m_1 + m_2)/2 \in \mathbb{Z}$. Thus, α and β can take both half-integer and integer values. The hierarchy of solutions generated by this procedure will be usually referred to as the half-integer hierarchy. The first solutions in this hierarchy are given in Table 28.1 below, expressed in terms of $U(z) := \sqrt{2}[A_1 D'_{-1/2}(\sqrt{2}z) - A_2 D'_{-1/2}(-\sqrt{2}z)]/[A_1 D_{-1/2}(\sqrt{2}z) + A_2 D_{-1/2}(-\sqrt{2}z)]$.

Table 28.1. The first solutions in the half-integer hierarchy.

α	β	$w(z, \alpha, \beta)$
$\frac{1}{2}$	$-\frac{1}{2}$	$U - z$
$-\frac{1}{2}$	$-\frac{1}{2}$	$-U - z$
1	-2	$\frac{-U^2 + z^2 - 1}{U - z}$
$\frac{3}{2}$	$-\frac{9}{2}$	$\frac{3U^2 - 4zU + z^2 + 1}{(U - z)(U^2 - z^2 + 1)}$
2	-8	$-\frac{4(U - z)\big(zU^3 - (z^2 - 2)U^2 - (z^3 + z)U + z^4 - z^2 + 1\big)}{(U^2 - z^2 + 1)(3U^2 - 4zU + z^2 + 1)}$

Chapter 7

The third Painlevé equation (P_3)

The third Painlevé equation (P_3) is different from the first, second and fourth Painlevé equations (P_1), (P_2) and (P_4), treated in the preceding chapters, in the sense that (P_3) may admit non-meromorphic solutions. However, as described in §3, and proved by Hinkkanen and Laine [2] in detail, a simple transformation, see (3.1), carries (P_3) over to a modified form, all solutions of which are meromorphic in the complex plane. Therefore, by Theorem 3.1, and an application of the reversed transformation (3.1), all local solutions of (P_3) around $z \neq 0$ permit an unrestricted analytic (meromorphic) continuation into $\mathbb{C}\backslash\{0\}$. A similar conclusion applies for (P_5) as well, by Theorem 5.1. However, it should be kept in mind that a solution of (P_3), resp. (P_5), may admit a logarithmic branch point at $z = 0$. To describe algebraic solutions in this and the subsequent chapters, we shall use root expressions for specific solutions of the Painlevé equation in question.

The contents of this chapter largely follows the pattern of the previous Chapters 4 through 6. A new aspect will appear by some analysis devoted to considering about conditions on the coefficients of (P_3) which imply the existence of a solution $w(z)$ meromorphic, resp. algebraic, at $z = 0$.

§29 Preliminary remarks

The third Painlevé equation

$$w'' = \frac{(w')^2}{w} - \frac{w'}{z} + \frac{1}{z}(\alpha w^2 + \beta) + \gamma w^3 + \frac{\delta}{w} \qquad (P_3)$$

with arbitrary complex parameters $\alpha, \beta, \gamma, \delta$ cannot be integrated, in general, in terms of classical transcendental functions. However, under suitable conditions imposed upon the parameters, the equation (P_3) may be integrated in terms of elementary functions, or (P_3) may admit solutions which are elementary functions of certain classical transcendental functions.

(1) To present some of such situations described above, we first apply the transformation $z = \exp(\zeta)$ for (P_3). Rewriting, after the transformation, z again in place of ζ, we obtain

$$w'' = \frac{(w')^2}{w} + e^z(\alpha w^2 + \beta) + e^{2z}\left(\gamma w^3 + \frac{\delta}{w}\right). \qquad (29.1)$$

It is now elementary to verify that the following identities are valid for all solutions of (29.1):

$$\begin{aligned}&\frac{d}{dz}\left(\left(\frac{w'}{w}\right)^2+\left(\frac{\delta}{w^2}-\gamma w^2\right)e^{2z}+2\left(\frac{\beta}{w}-\alpha w\right)e^z\right)\\&=2\left(\frac{\delta}{w^2}-\gamma w^2\right)e^{2z}+2\left(\frac{\beta}{w}-\alpha w\right)e^z\end{aligned}\tag{29.2}$$

and

$$\left(\frac{w'}{w}\right)'=\left(\frac{\delta}{w^2}+\gamma w^2\right)e^{2z}+\left(\frac{\beta}{w}+\alpha w\right)e^z.\tag{29.3}$$

Adding now (29.3) twice to (29.2) results in

$$\frac{d}{dz}\left(\left(\frac{w'}{w}\right)^2+2\frac{w'}{w}+G(z,w)\right)=4(\delta e^{2z}+\beta e^z w)\frac{1}{w^2},\tag{29.4}$$

where

$$G(z,w):=\left(\frac{\delta}{w^2}-\gamma w^2\right)e^{2z}+2\left(\frac{\beta}{w}-\alpha w\right)e^z.$$

By the corresponding subtraction,

$$\frac{d}{dz}\left(\left(\frac{w'}{w}\right)^2-2\frac{w'}{w}+G(z,w)\right)=-4(\gamma e^{2z}w^2+\alpha e^z w).\tag{29.5}$$

If now $\alpha=\beta=\gamma=\delta=0$, then it immediately follows from (29.4) and (29.5), recalling the transformation $z=\exp(\zeta)$ we made for (P_3), that all solutions of (P_3) are power functions $w(z)=C_1 z^{C_2}$ with $C_1, C_2\in\mathbb{C}$.

Let us now consider the cases $\beta=\delta=0$, resp. $\alpha=\gamma=0$, in (29.4), resp. in (29.5). Then we immediately get, respectively,

$$\left(\frac{w'}{w}\right)^2+2\frac{w'}{w}-2\alpha e^z w-\gamma e^{2z}w^2=C_1,\tag{29.6}$$

$$\left(\frac{w'}{w}\right)^2-2\frac{w'}{w}+\delta e^{2z}\frac{1}{w^2}+2\beta e^z\frac{1}{w}=C_2.\tag{29.7}$$

The equations (29.6) and (29.7) may be integrated in terms of elementary functions. For example, transforming (29.6) by $1/v=e^z w$, we obtain

$$(v')^2=2\alpha v+\gamma+(1+C_1)v^2.\tag{29.8}$$

Correspondingly, for (29.7) we get

$$(v')^2=-2\beta v-\delta+(1+C_2)v^2$$

by means of the transformation $v = e^{-z}w$. These two equations may now be explicitly integrated. For instance, for (29.8) we obtain the following general solution, provided $1 + C_1 \neq 0$:

$$v(z) = -\frac{\alpha}{1+C_1} + \frac{(1+C_1)e^{2(1+C_1)^{1/2}(c+\mu z)} + 4(\alpha^2 - \gamma - \gamma C_1)}{4(1+C_1)^{3/2}e^{(1+C_1)^{1/2}(c+\mu z)}},$$

where $\mu^2 = 1$ and c is an arbitrary constant. On the other hand, if $1 + C_1 = 0$ and $\alpha \neq 0$, then (29.8) has the general solution

$$v(z) = \frac{1}{2}(\alpha c^2 - \frac{\gamma}{\alpha} - 2\mu\alpha cz + \alpha z^2),$$

while for $1 + C_1 = 0$ and $\alpha = 0$, we obtain

$$v(z) = c + \mu\sqrt{\gamma}z.$$

(2) Similarly as in the previous equations (P_2) and (P_4), we are interested about seeking for such combinations of parameters that a solution $w(z)$ of (P_3) simultaneously satisfies a Riccati differential equation. To this end, suppose that all solutions of a Riccati differential equation

$$w' = a(z)w^2 + b(z)w + c(z), \quad a(z) \not\equiv 0, \tag{29.9}$$

with meromorphic coefficients are solutions of (P_3) at the same time. Substituting (29.9) into (P_3), it is an elementary computation to see that necessary and sufficient conditions for the existence of such a solution family are

$$a^2 = \gamma \neq 0, \quad b = \frac{\alpha - a}{az}, \quad c^2 = -\delta, \quad a\beta + c(\alpha - 2a) = 0. \tag{29.10}$$

Hence, (29.9) must be of the form

$$w' = aw^2 + \frac{\alpha - a}{az}w + c. \tag{29.11}$$

It is now an easy computation to check that the linear differential equation

$$\frac{d^2u(\tau)}{d\tau^2} + \frac{a-\alpha}{a\tau}\frac{du(\tau)}{d\tau} + u(\tau) = 0 \tag{29.12}$$

reduces back to (29.11) by the transformation $w(z) = -\frac{1}{au(z)}\frac{du(z)}{dz}$, $z = \lambda\tau$, $\lambda^2 = (ac)^{-1}$, provided $c \neq 0$. As is well-known from the theory of special functions, or may be verified by a suitable mathematical software, the general solution of (29.12) is

$$u(\tau) = \tau^{\alpha/(2a)}(C_1 J_{\alpha/(2a)}(\tau) + C_2 J_{-\alpha/(2a)}(\tau)), \tag{29.13}$$

where $J_\nu(\tau)$ is a Bessel function, whenever $\alpha/(2a) \notin \mathbb{Z}$, while if $\alpha/(2a) \in \mathbb{Z}$, then $J_{-\alpha/(2a)}$ in (29.13) has to be replaced by a cylinder function of the second kind $Y_{\alpha/(2a)}$. Clearly, the solutions of (29.12) generate all solutions of (29.11), hence a family of solutions of (P_3). If $c = 0$ and $a \neq 0$ in (29.11), then (29.11) may be directly solved in the form $w(z) = z^{(\alpha-a)/a}(C - a^2 z^{\alpha/a}/\alpha)^{-1}$.

In some particular cases, the solutions of (29.12) and hence those of (P_3) may be elementary functions. As an example, if $\alpha = a(1+2n)$, $n \in \mathbb{N} \cup \{0\}$, then by the well-known expressions for Bessel functions,

$$u(\tau) = \tau^{1+2n}\left(C_1 \frac{d^n}{(\tau d\tau)^n}\frac{\cos\tau}{\tau} + C_2 \frac{d^n}{(\tau d\tau)^n}\frac{\sin\tau}{\tau}\right).$$

Finally, we remark that the possible rational solutions for (29.11) and (P_3) will be considered in §35 below.

(3) As in the case of all previous Painlevé equations as well, it is sometimes convenient to consider a pair of differential equations of first order, equivalent to (P_3). To this end, we consider a pair of first order differential equations of the form

$$\begin{cases} w' = a_0(z) + a_1(z)w + a_2(z)w^2 + a_3(z)w^2 v, \\ v' = b_0(z) + b_1(z)w + b_2(z)v + b_3(z)wv + b_4(z)wv^2, \end{cases} \tag{29.14}$$

where v is an auxiliary function and w has to satisfy (P_3). Substituting (29.14) in (P_3), and assuming that the power combinations $w^j v^k$ are linearly independent, we obtain the following equations for determining the coefficients $a_j(z)$, $b_j(z)$:

$$\begin{cases} a_0^2 = -\delta, \quad a_1' = -\frac{1}{z}a_1 \\ a_2' + a_1 a_2 + b_0 a_3 = \frac{1}{z}(\alpha - a_2), \\ a_3' + a_1 a_3 + b_2 a_3 = -\frac{1}{z}a_3, \\ a_0' - a_0 a_1 = \frac{1}{z}(\beta - a_0), \\ a_2^2 + b_1 a_3 = \gamma, \quad b_3 + 2a_2 = 0, \quad b_4 = -a_3. \end{cases} \tag{29.15}$$

Looking at (29.15), observe that we may assume $a_3 \neq 0$. In fact, if $a_3 = 0$, the first equation in (29.14) reduces back to (29.9) above. If $\delta = 0$, then (29.15) implies that $\beta = 0$ as well, and so we are back in (29.6). Assuming that $\delta \neq 0$, $a_2 = 0$ and $a_3 = 1$, as we may do without loosing generality, we obtain from (29.15) the following pair of differential equations:

$$\begin{cases} zw' = (-\delta)^{1/2}z + hw + zw^2 v, \\ zv' = \alpha + \gamma zw - (1+h)v - zwv^2, \end{cases} \tag{29.16}$$

where $(h-1)(-\delta)^{1/2} = -\beta$. Before proceeding, we remark that by $w = 1/W$, (P_3) transforms into the same equation with $(-\beta, -\alpha, -\delta, -\gamma)$ replacing $(\alpha, \beta, \delta, \gamma)$, see (29.18) below. Then, in (29.16) with respect to (W, v), we may choose the coefficients $(-\beta, -\alpha, -\delta, -\gamma^{1/2})$ to replace $(\alpha, \beta, \gamma, (-\delta)^{1/2})$. Putting $w = 1/W$, $w_1 = -v$, we obtain the following pair of differential equations:

$$\begin{cases} zw' = (\alpha/\sqrt{\gamma} - 1)w + zw_1 + \sqrt{\gamma}zw^2, \\ zww_1' = \beta w + (\alpha/\sqrt{\gamma} - 2)ww_1 + zw_1^2 + \delta z, \end{cases} \tag{29.17}$$

which is equivalent to (P_3), provided $\gamma \neq 0$.

(4) We next consider a transformation property of (P_3), without going into details. Denote by $\phi(z, \alpha, \beta, \gamma, \delta)$ a solution of (P_3) and consider transformations of the following type:

$$T : w \mapsto \sigma_1 w^{k_1}, \quad z \mapsto \sigma_2 z^{k_2},$$

where $\sigma_j, k_j \neq 0$. Then the transformations

$$\begin{aligned} T_1(\sigma_1, \sigma_2) : \phi \mapsto \tilde{\phi}(z, \alpha\sigma_1\sigma_2, \beta\sigma_1^{-1}\sigma_2, \gamma\sigma_1^2\sigma_2^2, \delta\sigma_1^{-2}\sigma_2^2) \\ := \sigma_1^{-1}\phi(\sigma_2 z, \alpha, \beta, \gamma, \delta), \\ T_2 : \phi \mapsto \tilde{\phi}(z, -\beta, -\alpha, -\delta, -\gamma) := \big(\phi(z, \alpha, \beta, \gamma, \delta)\big)^{-1}, \\ T_3 : \phi(z, \alpha, \beta, 0, 0) \mapsto \tilde{\phi}(z, 0, 0, 2\alpha, 2\beta) := \phi(z^2, \alpha, \beta, 0, 0)^{1/2} \end{aligned} \tag{29.18}$$

may be used to generate the solutions of (P_3). Leaving aside the integrable cases $\alpha = \beta = \gamma = \delta = 0, \alpha = \gamma = 0, \beta = \delta = 0$, we may restrict ourselves to considering the four cases

$$\begin{aligned} &\text{(a)}\ \ \gamma\delta \neq 0, \\ &\text{(b)}\ \ \gamma = 0, \quad \alpha\delta \neq 0, \\ &\text{(c)}\ \ \delta = 0, \quad \beta\gamma \neq 0, \\ &\text{(d)}\ \ \delta = \gamma = 0, \quad \alpha\beta \neq 0. \end{aligned} \tag{29.19}$$

By the transformations T_3 and T_2, the cases (29.19(d)) and (29.19(c)) reduce back to (29.19(a)) and (29.19(b)), respectively. Therefore, it suffices to consider these two cases only. Moreover, as it is easy to see, we may assume that $\gamma = 1, \delta = -1$ in (29.19(a)) and $\alpha = 1, \delta = -1$ in (29.19(b)) without loss of generality.

(5) Finally, observe that the third Painlevé equation (P_3) is equivalent to the following nonlinear chain, see Adler [1]:

$$\begin{cases} f_1' = \varepsilon_2^2 f_2 + f_1^2 g_2, \\ f_2' = \varepsilon_1^2 f_1 + f_2^2 g_1, \\ -g_1' = \varepsilon_2^2 g_2 + g_1^2 f_2, \\ -g_2' = \varepsilon_1^2 g_1 + g_2^2 f_1, \end{cases} \tag{29.20}$$

where $\varepsilon_1' = \alpha_1\varepsilon_1$, $\varepsilon_2' = \alpha_2\varepsilon_2$ and $\alpha_1 + \alpha_2 = 1$.

Substituting $v(z_1) = \varepsilon_1(z) f_1(z)/(\varepsilon_2 f_2(z))$, $\phi(z_1) = f_1(z)g_2(z) - f_2(z)g_1(z)$, $z_1 = \varepsilon_1\varepsilon_2$ and eliminating ϕ from the equations

$$v' = 1 - v^2 + (\alpha_1 - \alpha_2 + \phi)v/z_1, \quad \phi' = (\phi - s)v + (\phi + s)/v,$$

where $s = f_1(z)g_2(z) + f_2(z)g_1(z)$ is the first integral of (29.20), we obtain that the function $v(z_1)$ that satisfies (P_3) with parameters $\alpha = -s - 2\alpha_1$, $\beta = s + 2\alpha_2$, $\gamma = -\delta = 1$.

§30 Behavior of solutions around $z = 0$ and $z = \infty$

This section is now devoted to finding out conditions on the coefficients of (P_3) which imply the existence of a solution meromorphic, resp. algebraic, in a neighborhood of $z = 0$. We also include similar considerations at $z = \infty$. Throughout of this section, we omit the cases $\beta = \delta = 0$ and $\alpha = \gamma = 0$, which have been treated in §29.

(1) We first proceed to finding necessary and sufficient conditions for the existence of a solution $w(z)$ of (P_3) with a pole at $z = 0$. To this end, let us assume that a formal solution $w(z)$ of (P_3) in a neighborhood of $z = 0$ is of the form

$$w(z) = \sum_{j=-l}^{\infty} a_j z^j, \quad a_{-l} \neq 0. \tag{30.1}$$

Writing (P_3) in the form

$$z(zw')'w - (zw')^2 = \alpha z w^3 + \beta z w + \gamma z^2 w^4 + \delta z^2.$$

and substituting (30.1) into (P_3), we easily deduce that the cases $\alpha = 0$, $\gamma \neq 0$ and $\alpha \neq 0$, $\gamma = 0$ result in an immediate contradiction. Therefore, a necessary condition for a pole at $z = 0$ is that $\alpha\gamma \neq 0$. In this case, by an elementary pole multiplicity comparison, $l = 1$. By the expressions

$$z(zw')'w - (zw')^2 = \sum_{j\geq 0}\Big((j+1)^2 a_j a_{-1} + \sum_{k=0}^{j-1} k(j-1)a_k a_{j-1-k}\Big)z^{j-1},$$

$$w^p = a_{-1}^p z^{-p} + \sum_{j\geq 0}\Big(p a_{-1}^{p-1} a_j + \sum_{\substack{k_1+\cdots+k_p=j-p+1\\ k_m\leq j-1}} a_{k_1}\ldots a_{k_p}\Big)z^{j-p+1},$$

$p = 2, 3, 4,$

the system of equations for determining the coefficients a_j in (30.1) takes the form:

$$\begin{cases} \gamma a_{-1} + \alpha = 0, \\ (1 - 3\alpha a_{-1} - 4\gamma a_{-1}^2)a_0 = 0, \\ (4 - 3\alpha a_{-1} - 4\gamma a_{-1}^2)a_1 = 3\alpha a_0^2 + 6\gamma a_0^2 a_{-1} + \beta, \\ \left[(n+1)^2 - 3\alpha a_{-1} - 4\gamma a_{-1}^2\right]a_{-1}a_n \\ = P_n(a_{-1}, a_0, \dots, a_{n-1}, \alpha, \beta, \gamma, \delta), \quad n \geq 2, \end{cases} \tag{30.2}$$

where P_n is a polynomial in the variables described. Solving a_{-1} from the first equation of (30.2), the coefficients a_n can be obtained, recursively, from the following set of equations:

$$\begin{aligned} \left[(n+1)^2\gamma - \alpha^2\right]a_n &= -\frac{\gamma^2}{\alpha} P_n(a_{-1}, a_0, \dots, a_{n-1}, \alpha, \beta, \gamma, \delta) \\ &=: \sigma_n(a_{-1}, a_0, \dots, a_{n-1}, \alpha, \beta, \gamma, \delta). \end{aligned} \tag{30.3}$$

Therefore, necessary conditions for a pole at $z = 0$ are that either

$$k^2\gamma - \alpha^2 \neq 0, \quad k \in \mathbb{N}, \tag{30.4}$$

or

$$\sigma_{k-1} = 0, \quad \text{if } k^2\gamma - \alpha^2 = 0, \tag{30.5}$$

for some $k \in \mathbb{N}$. Moreover, we infer from (30.2) that either $a_{2k} = 0$, $k \in \mathbb{Z}$, and a_{2k+1} is uniquely determined, when $\alpha^2 \neq (2k+2)^2\gamma$, or a_{2k} is arbitrary, when $\alpha^2 = (2k+1)^2\gamma$. Therefore, the coefficients a_n of the expansion (30.1) are uniquely defined in the case of (30.4) and depend on one arbitrary coefficient a_{k-1} in the case (30.5).

We now proceed to show that the conditions (30.4) and (30.5) are, in fact, sufficient. In (29.17), we make the change of variables:

$$w = u^{-1}, \quad u = zU, \quad \text{and} \quad z^2 w_1 = V. \tag{30.6}$$

Then we have

$$\begin{cases} zU' = -\gamma^{1/2} - \alpha\gamma^{-1/2}U - U^2V, \\ zV' = \beta z^2 + \alpha\gamma^{-1/2}V + \delta z^4 U + UV^2. \end{cases} \tag{30.7}$$

For the formal series of $w(z)$ constructed above, the corresponding formal series $U(z)$ obtained from (30.6) contains no terms of negative power and satisfies $U(0) = u_0 := -\gamma/\alpha \neq 0$. Furthermore, it is easy to see that the function $V(z)$ defined by the first equation of (30.7) has the same property. Denote now $V(0) =: v_0$. Then, the pair of functions $(\phi(z), \psi(z))$ given by $\phi(z) := U(z) - u_0$, $\psi(z) := V(z) - v_0$ satisfies a system of the form

$$\begin{cases} z\phi' = F(\phi, \psi, z), \\ z\psi' = G(\phi, \psi, z), \end{cases}$$

where $F(\phi, \psi, z)$, $G(\phi, \psi, z)$ are polynomials of ϕ, ψ, z satisfying $F(0,0,0) = G(0,0,0)$. By Theorem A.12, we conclude the convergence of $w(z)$. Therefore, having deleted the integrable cases treated in §29, we have proved

Theorem 30.1. *The third Painlevé equation* (P_3) *admits solutions meromorphic in the whole complex plane and with a pole at* $z=0$ *if and only if* $\alpha\gamma \neq 0$ *and the conditions* (30.4) *or* (30.5) *are fulfilled. If* (30.4) *is satisfied, exactly one such solution exists. In the case of* (30.5), *there exists a one-parameter family of solutions of the required type.*

(2) We next consider the existence of a meromorphic solution for (P_3) analytic in a neighborhood of $z=0$. To this end, let us assume that a formal solution of (P_3) in a neighborhood of $z=0$ of the form

$$w(z) = \sum_{j=0}^{\infty} a_j z^j \tag{30.8}$$

exists. Substituting (30.8) into (P_3), and observing that

$$z(zw')'w - (zw')^2 = \sum_{j\geq 0}\Big(j^2 a_j a_0 + \sum_{k=1}^{j-1} k(2k-j)a_k a_{j-k}\Big)z^j,$$

we get the following system of equations for determining the coefficients a_j:

$$\begin{cases} a_0(a_1 - \alpha a_0^2 - \beta) = 0, \\ 4a_0a_2 = 3\alpha a_0^2 a_1 + \gamma a_0^4 + \beta a_1 + \delta, \\ 9a_0a_3 = -a_1a_2 + 3\alpha(a_0^2a_2 + a_0a_1^2) + 4\gamma a_0^3 a_1 + \beta a_2, \\ n^2 a_0 a_n = P_n(a_0, a_1, \dots, a_{n-1}, \alpha, \beta, \gamma, \delta), \quad n \geq 2. \end{cases} \tag{30.9}$$

By the first equation of (30.9), we have to consider the two cases $a_0 = 0$ and $a_0 \neq 0$ separately.

First consider the case $a_0 = 0$. By the transformation T_2 in (29.18), $w(z)$ transforms to a solution with a pole at $z=0$. As shown above, this case occurs when $\beta\delta \neq 0$ or when $\beta = \delta = 0$. Hence, by Theorem 30.1, the convergence of (30.8) follows.

We may now assume that $a_0 \neq 0$. Then all coefficients a_j, $j \geq 1$, may be uniquely expressed in terms of $\alpha, \beta, \gamma, \delta, a_0$ by using (30.9). Convergence of the series (30.8) can be proved by reducing (P_3) to a Briot–Bouquet system. Putting $w = a_0 + u$, $v = (v_1 - h/a_0)/z$, $h = (\sqrt{-\delta} - \beta)/\sqrt{-\delta}$, $\delta \neq 0$ in the pair (29.16) of differential equations equivalent to (P_3), we get the following pair of differential equations:

$$\begin{cases} zu' = -hu + a_0^2 v_1 + \sqrt{-\delta}z + 2a_0uv_1 + v_1u^2 - (h/a_0)u^2, \\ zv_1' = -(h^2/a_0^2)u + hv_1 + \alpha z + \gamma z^2(a_0 + u) - a_0 v_1^2 \\ \qquad - v_1^2 u + 2huv_1/a_0 \end{cases} \tag{30.10}$$

with eigenvalues $\lambda_{1,2} = 0$. Therefore, (30.10) has a unique solution analytic in a neighborhood of $z = 0$ such that $u \to 0$, $v \to 0$, as $z \to 0$. Consider the formal series $v(z)$ defined by the first equation of (29.16) and the formal series of $w(z)$. Then it is easy to check that the corresponding series $v_1(z) = h/a_0 + zv(z)$ satisfies $v_1(0) = 0$. Since (30.10) admits a formal series solution $(u(z), v_1(z))$ vanishing at $z = 0$, the convergence of $w(z)$ now follows, provided $\delta \neq 0$. If $\delta = 0$, $\gamma \neq 0$, then by applying the transformation T_2 from (29.18) into (P_3) this case is reduced to $\delta \neq 0$, $\gamma = 0$. Finally, the case $\delta = \gamma = 0$, $\alpha\beta \neq 0$ may be reduced to $\delta\gamma \neq 0$ by using the transformation T_3 from (29.18). Then, (30.8) is changed into $\widetilde{w} = \sum_{j=0}^{\infty} b_j z^{j/2}$, $b_0 \neq 0$, which satisfies (P_3) with $(\widetilde{\alpha}, \widetilde{\beta}, \widetilde{\gamma}, \widetilde{\delta}) = (0, 0, 2\alpha, 2\beta)$. The system (29.16) is now given by

$$\begin{cases} z\widetilde{w}' = (-\widetilde{\delta})^{1/2} z + \widetilde{w} + z\widetilde{w}^2 v, \\ z\widetilde{v}' = \widetilde{\gamma} z\widetilde{w} - 2v - z\widetilde{w}v^2. \end{cases}$$

Making the change of variables $\widetilde{w} = b_0 + u$, $zv + b_0^{-1} = v_1$, $z^{1/2} = t$, we obtain a Briot–Bouquet system with respect to (u, v_1). This admits a formal series solution $(u(t), v_1(t))$ such that $u(0) = v_1(0) = 0$. Thus the convergence is verified.

Thus, except for the special cases treated in §29, the series (30.8) is either reduces to a polar series (30.1) by means of T_2-transformation from (29.18), implying the convergence by Theorem 30.1, or (30.8) has an arbitrary coefficient and according to (30.10) the convergence follows.

(3) We now proceed to consider the behavior of solutions of (P_3) around $z = \infty$. Putting $zt = 1$ in (P_3), and observing that $z(d/dz) = -t(d/dt)$, we get

$$t(tw')'w - (tw')^2 = \frac{1}{t}(\alpha w^3 + \beta w) + \frac{1}{t^2}(\gamma w^4 + \delta). \tag{30.11}$$

We leave it as an exercise for the reader to show that $t = 0$ cannot be a pole for the solutions of (30.11), except for the solvable cases corresponding to the special cases treated in §29. However, formal solutions of the form

$$w(t) = \sum_{j=0}^{\infty} a_j t^j \tag{30.12}$$

may exist in a neighborhood of $t = 0$. Substituting the series (30.12) into (30.11), we obtain the following system of equations for recursive determination of the coefficients:

$$\begin{cases} \gamma a_0^4 + \delta = 0, \\ 4\gamma a_0^3 a_1 + \alpha a_0^3 + \beta a_0 = 0, \\ 4\gamma a_0^3 a_2 + 3\alpha a_0^2 a_1 + 6\gamma a_0^2 a_1^2 + \beta a_1 = 0, \\ -\left(a_0(a_1 - 3\alpha a_1^2 - 4\gamma a_1^3)\right) + \beta a_2 + 3a_0^2(\alpha + 4\gamma a_1)a_2 \\ \qquad + 4\gamma a_0^3 a_3 = 0, \\ 4\gamma a_0^3 a_n + P_n(a_0, a_1, \ldots, a_{n-1}, \alpha, \beta, \gamma, \delta) = 0, \quad n \geq 4. \end{cases} \tag{30.13}$$

It now follows from (30.13) that if $\gamma\delta \neq 0$, then all a_j, $j \geq 1$, may be uniquely expressed in terms of $a_0, \alpha, \beta, \gamma, \delta$. If $\gamma = 0$, then $\delta = 0$ as well by the first equation of (30.13). This case may be reduced to $\gamma\delta \neq 0$ by using the transformation T_3 from (29.18). Indeed, (30.12) transforms into $\widetilde{w}(t) = \sum_{j=0}^{\infty} \widetilde{a}_j t^{j/2}$ such that $\widetilde{w}(1/z)$ formally satisfies (P_3) with $(\widetilde{\alpha}, \widetilde{\beta}, \widetilde{\gamma}, \widetilde{\delta}) = (0, 0, 2\alpha, 2\beta)$. As will be shown later in this section, for a Puiseux series around $z = \infty$, the branching multiplicity is either 1 or 3. Hence, $\widetilde{a}_{2k-1} = 0$ for all $k \in \mathbb{N}$. Thus this case is reduced to that of $\gamma\delta \neq 0$. If now $\delta = 0$, then $\gamma a_0 = 0$ by the first equation of (30.13) again. Therefore, we may now assume that $a_0 = 0$. Substituting now $w(t) = a_m t^m + \cdots$, $a_m \neq 0$, $m \geq 1$ into (30.11) and comparing the terms of the lowest degree, we obtain $\beta = 0$, which is a special case in §29.

Clearly, the series (30.12) converges if it is an expansion of a rational solution in a neighborhood of $z = \infty$. Necessary and sufficient conditions of the existence of rational solutions of (P_3), hence also sufficient conditions for the convergence of (30.12) will be treated in §35 below.

(4) In a similar way, necessary and sufficient conditions for the existence of solutions of (P_3) for which $z = 0$ and $z = \infty$ are algebraic branch points may be found out. Actually, we are going to show below that the order of the branching of a solution at these points has to be equal to three. If $z = 0$ and $z = \infty$ are algebraic points with branching multiplicity $s \in \mathbb{N}$, then the solution $w(z)$ is expressible in the form $w(z) = R(z^{1/s})$ for some rational function $R(\tau)$. In fact, since $w(z)$ is at most meromorphic around each point $z_0 \neq 0, \infty$, $w(\tau^s) = R(\tau)$ is a rational function of $\tau \in \widehat{\mathbb{C}} = \mathbb{C} \cup \{\infty\}$, and hence, by the substitution $\tau = z^{1/s}$, the expression follows. If $s = 1$, then an algebraic solution degenerates into a rational function. To obtain the required conditions for the coefficients, denote $z = \tau^s$, where $s \in \mathbb{N}$. Observing that $z(d/dz) = s^{-1}\tau(d/d\tau)$, the third Painlevé equation (P_3) takes the form

$$\tau(\tau w')' w - (\tau w')^2 = s^2\big(\tau^s(\alpha w^3 + \beta w) + \tau^{2s}(\gamma w^4 + \delta)\big). \tag{30.14}$$

Let now

$$w(\tau) = P(\tau)/Q(\tau) \tag{30.15}$$

be a rational solution of the equation (30.14), where $P(\tau)$ and $Q(\tau)$ are irreducible polynomials of degrees p and q, respectively. If $s > 1$, then (30.15) defines an algebraic solution of the equation (P_3). Substituting (30.15) as a power series around $\tau = \infty$ into (30.14) and comparing the coefficients of the highest powers of τ, we obtain

$$\begin{aligned} &\text{(a)} \quad p = q, \quad \text{if } \gamma\delta \neq 0; \\ &\text{(b)} \quad p = q + s/3, \quad \text{if } \alpha\delta \neq 0,\ \gamma = 0. \end{aligned} \tag{30.16}$$

Observe that (30.16) is related to the cases (a) and (b) in (29.19). The case (c) in (29.19), i.e. $\delta = 0$, $\beta\gamma \neq 0$, reduces to (30.16(b)) by means of the transformation T_2 in (29.18), $p = q + s/3$. Finally, in the case (d) in (29.19), i.e. $\gamma = \delta = 0$, $\alpha\beta \neq 0$, we have $p = q$ since (d) reduces to (30.16(a)) by means of the transformation T_3 from (29.18).

We next consider the equation (30.14) at infinity by putting $\tau = 1/t$. Then we obtain

$$t(tw')'w - (tw')^2 = s^2\big(t^{-s}(\alpha w^3 + \beta w) + t^{-2s}(\gamma w^4 + \delta)\big). \tag{30.17}$$

Consider first the case (30.16(a)), i.e. $\gamma\delta \neq 0$. Then the solution (30.15) is analytic in a neighborhood of $t = 0$. We next show that the solution (30.15) is a function of t^s which means that it is a rational function of z. To this end, we compute the expansion of (30.15) in a neighborhood of $\tau = \infty$, i.e. of $t = 0$, and show that in a neighborhood of $t = 0$ the solution (30.15) can be expressed in the form

$$w(t) = \sum_{j=0}^{\infty} b_{sj} t^{sj}. \tag{30.18}$$

This must coincide with the expansion (30.12). In fact, let the expansion of (30.15) be of the following form

$$w(t) = \sum_{k=0}^{\infty} b_k t^k. \tag{30.19}$$

Substituting (30.19) into (30.17), we easily see that $\gamma b_0^4 + \delta = 0$ and that the coefficients b_k, $k \geq 1$, are determined uniquely up to the choice of b_0. Recall the formal solution $\phi(z) = \sum_{j=0}^{\infty} a_j z^{-j}$ of (P_3) around $z = \infty$, see (30.12). Note that $\gamma a_0^4 + \delta = 0$ by the first equation of (30.13), and that a_j, $j \geq 1$, are determined uniquely up to a_0. Since $w(t) = \phi(t^{-s}) = \sum_{j=0}^{\infty} a_j t^{sj}$ also satisfies (30.17), this must coincide with (30.19). Hence,

$$b_k = \begin{cases} a_j, & \text{if } k = sj,\ j \in \mathbb{N} \cup \{0\} \\ 0, & \text{otherwise,} \end{cases}$$

provided that the choice of the root a_0 is made suitably. Consequently, the solution (30.15) has the expansion (30.18) in a neighborhood of the infinity $t = 0$. Hence, (30.15) is a rational function of $z = \tau^s$.

It remains to consider the case (30.16(b)), i.e., $\gamma = 0$, $\alpha\delta \neq 0$. In this case $s = 3r$, $r \in \mathbb{N}$. Then $t = 0$ is a pole of multiplicity r of the solutions. As in the previous case, we proceed to show that the only essential case is $s = 3$. We now write (30.15) in the form

$$w(t) = \sum_{k=-r}^{\infty} c_k t^k, \quad c_{-r} \neq 0. \tag{30.20}$$

Substituting this into (30.17) with $s = 3r$, we see that $\alpha c_{-r}^3 + \delta = 0$, and that c_k, $k > -r$ are determined uniquely up to c_{-r}. Make now the change of variables $t^r = \rho$.

Then $t\frac{d}{dt} = r\rho\frac{d}{d\rho}$. Hence (30.17) transforms into

$$\rho(\rho w')'w - (\rho w')^2 = 3^2\big(\rho^{-3}(\alpha w^3 + \beta w) + \rho^{-6}\delta\big). \tag{30.21}$$

This equation admits a solution $w(\rho) = \psi(\rho) = \sum_{j=-1}^{\infty} d_j\rho^j$. Here, $\alpha d_{-1}^3 + \delta = 0$, and d_j, $j \geq 0$, are determined uniquely. It is easy to see that $\psi(t^r)$ satisfies (30.17). This implies that $\psi(t^r)$ coincides with (30.20):

$$c_k = \begin{cases} d_j, & \text{if } k = rj,\ j \in \mathbb{N} \cup \{-1, 0\} \\ 0, & \text{otherwise,} \end{cases}$$

provided that the root d_{-1} is chosen suitably. Consequently, the solution (30.15) has the expansion (30.20) in a neighborhood of $t = 0$ when $\gamma = 0$, $\alpha\delta \neq 0$. This means that (30.15) is a rational function of $z^{1/3} = \tau^r$.

Hence, we are now ready to formulate the following

Theorem 30.2. *The equation (P_3) with $\gamma\delta \neq 0$ admits no algebraically branched solutions at $z = 0$ and $z = \infty$. On the other hand, the equation (P_3) with $\gamma = 0$, $\alpha\delta \neq 0$ may admit solutions algebraically branched at $z = 0$ and $z = \infty$. In this case, the branching multiplicity of such solutions must be equal to* 3.

§31 Poles of third Painlevé transcendents

Similarly as in the preceding section, we omit the special case $\gamma = \alpha = 0$, treated in §29, from our considerations in this section. We first consider the modified third Painlevé equation $(\widetilde{P}_3)$, writing it in the form

$$ww'' = (w')^2 + \alpha w^3 + \gamma w^4 + \beta e^z w + \delta e^{2z}. \tag{$\widetilde{P}_3$}$$

Suppose first that $\gamma = 0$ and $\alpha \neq 0$, and let z_0 be a pole of $w(z)$. Substituting the Laurent expansion at z_0 into $(\widetilde{P}_3)$, we immediately see that the pole has to be double and the expansion is of the form

$$w(z) = \frac{2/\alpha}{(z - z_0)^2} + \Phi(z),$$

where $\Phi(z)$ is analytic in a neighborhood of z_0. As an immediate consequence of the Clunie lemma, Lemma B.11, $w(z)$ has infinitely many poles in $\mathbb{C}$, at least if $w(z)$ satisfies $r/T(r, w) = o(1)$. If then $\gamma \neq 0$, and z_0 is a pole of $w(z)$, we use the same reasoning to infer that the pole must be simple with residue equal to $1/\sqrt{\gamma}$. Moreover, by the Clunie reasoning, the number of poles must be infinite in this case as well.

As there are two possible residue values in the case of $\gamma \neq 0$, the question again arises whether there are infinitely many poles of both residue type. For simplicity, we first restrict ourselves to considering the original third equation (P_3), assuming that $w(z)$ is a meromorphic solution in the whole complex plane.

So, let us assume that $z_0 \neq 0$ is a pole of a meromorphic solution of (P_3). Substituting the series

$$w = \sum_{j=-k}^{\infty} a_j (z - z_0)^j \tag{31.1}$$

into (P_3) and comparing the coefficients, we find that $k = 1$ if $\gamma \neq 0$ and $k = 2$ if $\gamma = 0$ and $\alpha \neq 0$. In the latter case, we infer, by substituting (31.1) in (P_3), that $a_{-2} = 2z_0/\alpha$, $a_{-1} = 0$, a_0 is arbitrary, and all coefficients a_j, $j \geq 2$, may be uniquely determined in terms of α, β, γ, δ, a_0 and z_0. Applying Lemma B.11 to the equation (P_3), we see at once that $w(z)$ admits infinitely many poles. Therefore, we may assume in this section, from now on, that $\gamma \neq 0$. Then we have the following system for determining the coefficients in (31.1):

$$\begin{cases} z_0(1 - \gamma a_{-1}^2) = 0, \\ 2a_0 z_0 = \alpha a_{-1}^2 + 4z_0 \gamma a_0 a_{-1}^2 + \gamma a_{-1}^3, \\ a_0 + 4z_0 a_1 = 3\alpha a_0 a_{-1} + \gamma z_0 (4a_1 a_{-1}^2 + 6a_0^2 a_{-1}) + 4\gamma a_0 a_{-1}^2, \\ z_0(n^2 + n - 2)a_n = P_n(a_{-1}, a_0, \dots, a_{n-1}), \quad n \geq 2. \end{cases} \tag{31.2}$$

By (31.2), a routine computation results in

$$a_{-1} = \frac{1}{\sqrt{\gamma}}, \qquad a_0 = -\frac{1}{2z_0}\left(\frac{\alpha}{\gamma} + \frac{1}{\sqrt{\gamma}}\right). \tag{31.3}$$

Moreover, the coefficient a_1 is arbitrary and all coefficients a_j with $j \geq 2$ are uniquely determined in terms of α, β, γ, δ, a_1 and z_0.

Recall now the pair (29.17) of differential equations equivalent to (P_3), and observe that the values of $\sqrt{\gamma}$ may be different in (29.17) and in (31.3). Therefore, to prevent misunderstanding, we denote their signs, respectively, with μ and ν, $\mu^2 = \nu^2 = 1$. If now w_1 is eliminated from (29.17), then we get the equation (P_3). Suppose now that

$$zw_1' - \beta - (\alpha\mu/\sqrt{\gamma} - 2)w_1 \not\equiv 0. \tag{31.4}$$

Then, eliminating w from (29.17), we get an algebraic differential equation for w_1 as follows:

$$\begin{aligned} \mu\gamma\sqrt{\gamma}z^2 w_1^4 &= -2\beta\gamma\delta + \mu\sqrt{\gamma}\delta(\alpha\beta - \gamma\delta z^2) \\ &\quad - w_1\big(\beta^2\gamma - \delta\alpha^2 + 4\gamma\delta - 4\mu\alpha\sqrt{\gamma}\delta - \gamma z^2(w_1')^2\big) - \gamma\delta z(w_1' + zw_1'') \\ &\quad - w_1^2\big(-2\beta\gamma + \mu\sqrt{\gamma}(\alpha\beta + 2\gamma\delta z^2) + \gamma z(w_1' + zw_1'')\big), \end{aligned} \tag{31.5}$$

with just one term $\mu\gamma\sqrt{\gamma}z^2 w_1^4$ of maximal total order. The relation between the solutions of (P_3) and those of (31.5) under the condition (31.4) now reads as follows:

$$\begin{cases} w_1 = w' - \mu\sqrt{\gamma}w^2 + \left(1 - \dfrac{\alpha\mu}{\sqrt{\gamma}}\right)\dfrac{w}{z}, \\ w = (\delta z + zw_1^2)\left(zw_1' + \left(2 - \dfrac{\alpha\mu}{\sqrt{\gamma}}\right)w_1 - \beta\right)^{-1}. \end{cases} \tag{31.6}$$

Let $z_0 \neq 0$ be a pole of the solution $w(z)$ of (P_3). Then the principal part of the expansion of w_1 in a neighborhood of z_0 is

$$G(w_1) := -\frac{\nu+\mu}{\sqrt{\gamma}}(z-z_0)^{-2} + \frac{\nu+\mu}{z_0\sqrt{\gamma}(z-z_0)}.$$

It is now immediate to observe that the function w_1 is analytic in a neighborhood of z_0, if z_0 is a pole of $w(z)$ with residue $\nu/\sqrt{\gamma}$, and $\mu = -\nu$, and w_1 has a double pole at z_0, if $\mu = \nu$. Applying the Clunie reasoning again to the equation (31.5), we infer that w_1 has infinitely many poles in both possible cases of μ which, in fact, may be chosen arbitrarily. Therefore, we have obtained

Theorem 31.1. *Any transcendental meromorphic solution of (P_3) has an infinite number of poles with both residues $1/\sqrt{\gamma}$ and $-1/\sqrt{\gamma}$, provided the condition* (31.4) *holds.*

If the condition (31.4) is not fulfilled, then we get $w_1^2 + \delta = 0$ from the second equation of (29.17). Therefore, w_1 reduces to a constant. This means that the first equation in (29.17) becomes a Riccati differential equation for $w(z)$, and so, by the Clunie reasoning again, $w(z)$ has infinitely many poles, all with the same residue $\nu/\sqrt{\gamma}$ of the same sign. Examples of such solutions are the solutions generated by (29.13).

Let us next consider an arbitrary solution $w(z)$ of (P_3). We denote by $w_1(z)$ the function defined by the first equation of (31.6). Consider now $W(z) := w(e^z)$, $W_1(z) := w_1(e^z)$. Clearly, $W(z)$ is a solution of (29.1). Suppose now that (31.4) is fulfilled, namely that $w(z)$ is not a solution of the Riccati differential equation

$$zw' = \mu\sqrt{\gamma}zw^2 + (\alpha\mu/\sqrt{\gamma} - 1)w + \mu'\sqrt{-\delta}z, \quad \mu^2 = (\mu')^2 = 1, \tag{31.7}$$

and that $W(z)$ satisfies

$$\liminf_{r\to\infty} r/T(r, W) = 0. \tag{31.8}$$

By (31.6),

$$e^z W_1(z) = W'(z) - \mu\sqrt{\gamma}e^z W(z)^2 + (1 - \alpha\mu/\sqrt{\gamma})W(z),$$

from which $T(r, W_1) \ll T(r, W) + r \ll \exp(\Lambda_0 r)$ follows, see Theorem 9.1. By the Clunie reasoning, from (31.5) with e^z in place of z, we obtain

$$m(r, W_1) = O(r). \tag{31.9}$$

Furthermore, under the supposition (31.4), the second equation of (31.6)

$$W(z) = e^z(\delta + W_1(z)^2)\big(W_1'(z) + (2 - \alpha\mu/\sqrt{\gamma})W_1(z) - \beta\big)^{-1}$$

yields $T(r, W) \ll T(r, W_1) + r$, implying that $\liminf_{r\to\infty} r/T(r, W_1) = 0$ by (31.8). Combining this with (31.9), we conclude that $\liminf_{r\to\infty} r/N(r, W_1) = 0$. Hence, for both μ and $-\mu$, $w_1(z)$ admits infinitely many poles. Indeed, if $w_1(z)$ has a finite number of poles only, then $N(r, W_1) = O(r)$, a contradiction. By the same argument as above, we obtain

Theorem 31.2. *Suppose that $\gamma \neq 0$. Let $w(z)$ be a solution of (P_3) such that $W(z) = w(e^z)$ satisfies* (31.8). *Then $w(z)$ has an infinite number of poles with both residues $1/\sqrt{\gamma}$ and $-1/\sqrt{\gamma}$ if and only if $w(z)$ is not a solution of* (31.7).

§32 Canonical representation of solutions

By the transformation $z = \exp\zeta$, (P_3) takes the following modified form:

$$w'' = (w')^2/w + e^{\zeta}(\alpha w^2 + \beta) + e^{2\zeta}(\gamma w^3 + \delta/w). \tag{32.1}$$

Comparing with the substitution (3.1), it follows immediately from Theorem 3.1 that all solutions of (32.1) are meromorphic functions. We now proceed to construct a canonical representation for solutions of (32.1) in the form

$$w(\zeta) = v(\zeta)/u(\zeta), \tag{32.2}$$

where $v(\zeta)$ and $u(\zeta)$ are some entire functions, to be constructed. Making use of the Laurent expansions (31.1) in a neighborhood of the poles of $w(\zeta)$, it is not difficult to check that the function

$$u(\zeta) = \exp\left\{-\int^{\zeta}\int^{t}(\gamma e^{2t}w(t)^2 + \alpha e^t w(t))\,dt\,dt\right\} \tag{32.3}$$

has a simple zero at a simple pole of $w(\zeta)$, if $\gamma \neq 0$, and a double zero at a double pole of $w(\zeta)$, if $\alpha \neq 0$ and $\gamma = 0$. Obviously, $u(\zeta)$ is analytic, whenever $w(\zeta)$ is analytic, and so $u(\zeta)$ is an entire function. Consequently, the function $v(\zeta) := w(\zeta)u(\zeta)$ is entire as well.

Differentiating now (32.3) and taking into account (32.2), we obtain a pair of differential equations

$$\begin{cases} uu'' = (u')^2 - \gamma e^{2\zeta}v^2 - \alpha e^{\zeta}uv, \\ vv'' = (v')^2 + \beta e^{\zeta}uv + \delta e^{2\zeta}u^2, \end{cases} \tag{32.4}$$

where the second equation follows by substituting (32.2) into (32.1) and observing that

$$w'' - \frac{(w')^2}{w} = w\left(\frac{w'}{w}\right)' = \frac{v}{u}\left(\left(\frac{v'}{v}\right)' - \left(\frac{u'}{u}\right)'\right).$$

By (32.3) combined with (29.5), we have

$$\frac{d}{d\zeta}\left(\left(\frac{w'}{w}\right)^2 - 2\frac{w'}{w} + G(\zeta, w)\right) = 4\frac{d}{d\zeta}\left(\frac{u'}{u}\right).$$

Substitution of $w = v/u$ yields the first integral

$$(u'v-uv')^2-2(u'v+uv')uv+(\delta u^4-\gamma v^4)e^{2\zeta}+2(\beta u^3 v-\alpha uv^3)e^{\zeta} = Cu^2v^2 \quad (32.5)$$

of (32.4) with an arbitrary complex constant C.

We may now construct solutions of (32.4) in the form of exponential series

$$v(\zeta) = \sum_{j=0}^{\infty} a_j e^{j\zeta}, \quad u(\zeta) = \sum_{j=0}^{\infty} b_j e^{j\zeta}. \quad (32.6)$$

Substituting these series into the first equation of (32.4) and comparing the coefficients, we obtain

$$\begin{cases} a_0(a_1 - \beta b_0) = 0, \\ b_0(b_1 + \alpha a_0) = 0, \\ 4a_0a_2 = b_0(\delta b_0 + \beta a_1) + a_0 b_1 \beta, \\ 4b_0b_2 = -a_0(\alpha b_1 + \gamma a_0) - a_1 b_0 \alpha, \\ 9a_0a_3 + a_1a_2 = \beta a_1 b_1 + b_0(\beta a_2 + 2\delta b_1) + b_2 a_0 \beta, \\ 9b_0b_3 + b_1b_2 = -\alpha a_1 b_1 + a_0(-2\gamma a_1 - \alpha b_2) - a_2 b_0 \alpha, \\ \cdots\cdots \\ n^2 a_0 a_n + (n-2)^2 a_1 a_{n-1} + \cdots = P_n(a_0, \ldots, a_{n-1}, b_0, \ldots, b_{n-1}), \\ n^2 b_0 b_n + (n-2)^2 b_1 b_{n-1} + \cdots = Q_n(a_0, \ldots, a_{n-1}, b_0, \ldots, b_{n-1}). \end{cases} \quad (32.7)$$

We divide our consideration of these equations in four subcases:

(1) Let us first suppose that $b_0 a_0 \neq 0$. Then the coefficients in the exponential series (32.6) may be computed from (32.7) successively: $a_1 = \beta b_0$, $b_1 = -\alpha a_0$, $a_2 = -\alpha\beta a_0/4 + (\beta^2 + \delta)b_0^2/4a_0$, $b_2 = (\alpha^2 - \gamma)a_0^2/4b_0 - \alpha\beta b_0/4$. As one can immediately see, all coefficients a_j and b_j, with $j \geq 2$, may be uniquely expressed in terms of $a_0, b_0, \alpha, \beta, \gamma$, and δ.

(2) Let us next suppose that $a_0 = 0$, $b_0 \neq 0$. By (32.7), we observe that $b_1 = 0$ and $a_1\beta + \delta b_0 = 0$. If $\beta = 0$, then necessarily $\delta = 0$, and this reduces back to a

special case treated in §29, solvable in terms of elementary functions. If $\beta \neq 0$, then $a_1 = -b_0\delta/\beta$, and all subsequent coefficients may be found uniquely. In particular, we see that a_{2k} is arbitrary in the series (32.6) for $v(\zeta)$, provided $\beta^2 + (2k-1)^2\delta = 0$.

(3) If $a_0 \neq 0$ and $b_0 = 0$, then by (32.7) again, we find successively $a_1 = 0$, $a_2 = \beta b_1/4$, $\alpha b_1 + a_0\gamma = 0$. If $\alpha = 0$, then $\gamma = 0$ as well, and we are again back in the special cases of §29. If $\alpha \neq 0$, then $b_1 = -a_0\gamma/\alpha$, and the solution depends on an arbitrary parameter a_0. Of course, all other coefficients can be found uniquely and we obtain that b_{2k} is an arbitrary coefficient in the series (32.6) for $u(\zeta)$, provided $\alpha^2 - (2k-1)^2\gamma = 0$.

(4) The case $a_0 = b_0 = 0$ may be treated in a similar way, and we leave it as an exercise for the reader.

§33 The special case $\gamma = 0, \alpha\delta \neq 0$

In this section, we may assume, for simplification of the calculations, that $\alpha = -\delta = 1$. Hence, (P_3) now takes the special form

$$w'' = \frac{(w')^2}{w} - \frac{w'}{z} + \frac{1}{z}(w^2 + \beta) - \frac{1}{w}. \tag{33.1}$$

In fact, we may fix $\alpha = -\delta = 1$ by putting $\sigma_1 = 1/(\alpha\sigma_2)$, $\sigma_2 = 1/\sqrt[4]{-\alpha^2\delta}$ in (29.18). We can rewrite (33.1) in the form of an equivalent system

$$\begin{cases} zw' = \varepsilon z + (1 - \varepsilon\beta)w + zw^2 v, \\ zv' = 1 - (2 - \varepsilon\beta)v - zwv^2, \end{cases} \tag{33.2}$$

by fixing $\varepsilon^2 = 1$, $\sqrt{-\delta} = \varepsilon$, $\gamma = 0$, $\alpha = 1$ in (29.16). From the first equation of (33.2) we infer that the equation

$$v'' = \frac{(v')^2}{v} - \frac{v'}{z} - \varepsilon v^2 + \frac{2 - \varepsilon\beta}{z^2} - \frac{1}{z^2 v} \tag{33.3}$$

is of the considered type by a simple substitution. Indeed, if we define $\widetilde{w}(z) = -\varepsilon z v(z)$, then the function $\widetilde{w}(z)$ satisfies the equation (33.1) with the parameter value $\widetilde{\beta} := \beta - 2\varepsilon$. Therefore, we have proved the following result, see Gromak [1]:

Theorem 33.1. *Let $w = w(z, \beta)$ be a solution of* (33.1). *Then the function*

$$\widetilde{w}(z, \widetilde{\beta}) := w^{-2}\big[(\varepsilon - \beta)w + z - \varepsilon z w'\big] \tag{33.4}$$

also solves an equation of the form (33.1), with the parameter value

$$\widetilde{\beta} := \beta - 2\varepsilon,$$

where $\varepsilon^2 = 1$.

Remark 1. The equation (33.1) does not admit a solution satisfying $(\varepsilon - \beta)w + z - \varepsilon z w' = 0$ simultaneously. Hence, the above transformation may be repeated unconditionally.

Remark 2. The system (33.2) has the following symmetry property: By $\widetilde{w} = -\varepsilon z v$, $\widetilde{v} = \varepsilon z^{-1} w$, (33.2) is changed into

$$\begin{cases} z\widetilde{w}' = \widetilde{\varepsilon} z + (1 - \widetilde{\varepsilon}\widetilde{\beta})\widetilde{w} + z\widetilde{w}^2\widetilde{v}, \\ z\widetilde{v}' = 1 - (2 - \widetilde{\varepsilon}\widetilde{\beta})\widetilde{v} - z\widetilde{w}\widetilde{v}^2, \end{cases}$$

with $\widetilde{\varepsilon} = -\varepsilon$, $\widetilde{\beta} = \beta - 2\varepsilon$.

If now $w(z, \beta)$ is the general solution of (33.1) for a fixed value of the parameter β, then $w(z, \widetilde{\beta})$ is the corresponding general solution as well.

Corollary 33.2. *To construct the general solution of* (33.1) *for arbitrary values of the parameter β it is sufficient to construct the general solution for all β from a fundamental domain*

$$r \leq \operatorname{Re} \beta < r + 2,$$

for an arbitrary $r \in \mathbb{R}$.

Denote now the transformation described in Theorem 33.1 by $T(\varepsilon)$, i.e., $T(\varepsilon) : w(z, \beta) \mapsto \widetilde{w}(z, \widetilde{\beta})$. In particular, we use the notations $T := T(-1)$, $T^{-1} = T(1)$. Then, by direct computation, we verify that $T^{-1} \circ T = T \circ T^{-1} = I$, where I is the identity transformation. Similar to the equations (P_2), (P_4) in §19 and §25, respectively, we may define simple auto-Bäcklund transformations of the equation (33.1):

$$S_\mu : w(z, 0) \mapsto \mu w(\mu z, 0), \quad \mu^2 = -1,$$
$$S : w(z, \beta) \to -w(-z, \beta).$$

Thus, S_μ are auto-Bäcklund transformations for $\beta = 0$, S is an auto-Bäcklund transformation for an arbitrary β and $T^k \circ S_\mu \circ T^{-k}$ are auto-Bäcklund transformations for $\beta = 2k$, $k \in \mathbb{Z}$.

Moreover, the solutions $w_\beta := w(z, \beta)$, $w_{\beta+2} := w(z, \beta+2)$, $w_{\beta+4} := w(z, \beta+4)$, obtained by repeated applications of the Bäcklund transformation T, are connected by means of the following nonlinear superposition formula, as one may verify by a routine computation:

$$w_{\beta+4} = (2z - (2 + \beta)w_{\beta+2} - w_\beta w_{\beta+2}^2)/w_{\beta+2}^2.$$

We continue this section by constructing some algebraic solutions of (P_3) with the help of Theorem 33.1. By §30, algebraic solutions may exist only if (a) $\gamma = 0$, $\alpha\delta \neq 0$ or if (b) $\delta = 0$, $\beta\gamma \neq 0$ except for the integrable cases treated in §29, and again omitted from this consideration. The case (b) reduces back to the case (a) by using the T_2-transformation, see (29.18). Hence, we may restrict ourselves to considering the case (a) only. Moreover, we may normalize the situation by assuming that $\alpha = 1$, $\delta = -1$, as stated in the beginning of this section. By §30, we know that the algebraic solutions have to be rational in $z^{1/3}$. Therefore, we make the transformation $z = \tau^3$ to start with. Then, instead of (33.1) and (33.4), we have

$$w'' = \frac{(w')^2}{w} - \frac{w'}{\tau} + 9\tau\left(w^2 + \beta - \frac{\tau^3}{w}\right) \tag{33.5}$$

and

$$\widetilde{w}(\tau, \beta - 2\varepsilon) = \frac{3(\varepsilon - \beta)w + 3\tau^3 - \varepsilon\tau w'_\tau}{3w^2}. \tag{33.6}$$

Therefore, the construction problem is reverted to finding conditions for the existence of rational solutions of (33.5) and to constructing them.

Theorem 33.3. *For the existence of rational solutions of* (33.5), *it is necessary and sufficient to have*

$$\beta = 2k, \quad k \in \mathbb{Z}. \tag{33.7}$$

Proof. Proving the sufficiency is almost trivial. In fact, it is immediate to see that (33.5) has a solution

$$w = \lambda\tau, \quad \lambda^3 = 1, \quad \beta = 0. \tag{33.8}$$

Hence, by using the transformation (33.4) inductively, we obtain rational solutions for all $\beta = 2k$, $k \in \mathbb{Z}$. The first such rational solutions are listed in the following table.

Table 33.1. The first rational solutions of (33.5) with $\gamma = 0$, $\alpha\delta \neq 0$.

β	$w(\tau, \alpha, \beta)$
0	$w = \lambda\tau, \quad \lambda^3 = 1$
-2ε	$\dfrac{2\varepsilon\lambda + 3\tau^2}{3\lambda^2\tau}$
-4ε	$\dfrac{9\lambda\tau^5 + 24\varepsilon\lambda^2\tau^3 + 20\tau}{(2\varepsilon\lambda + 3\tau^2)^2}$
-6ε	$\dfrac{243\tau^{10} + 7920\tau^4\varepsilon + 5040\tau^2\lambda + 1782\tau^8\varepsilon\lambda + 5400\tau^6\lambda^2 + 1120\varepsilon\lambda^2}{3\tau(20 + 9\tau^4\lambda + 24\tau^2\varepsilon\lambda^2)^2}$

To prove the necessity, let $w(\tau)$ be a nontrivial rational solution of (33.5) and make use of the transformation

$$y := \tau/w, \quad \tau^2 = \eta. \tag{33.9}$$

Then (33.5) transforms to a differential equation

$$4\eta yy'' = 4\eta(y')^2 - 4yy' - 9(\eta y + \beta y^3 - \eta y^4) \tag{33.10}$$

for $y = y(\eta, \beta)$, while the Bäcklund transformation (33.6) takes the form

$$\tilde{y}(\eta, \beta - 2\varepsilon) = \frac{3\eta}{(2\varepsilon - 3\beta)y + 3\eta y^2 + 2\varepsilon\eta y'}. \tag{33.11}$$

The initial solution $w(\tau, 0)$ reduces to $y(\eta, 0) = \lambda^2$.

Recall that $w(\tau)$ admits a formal series expansion around $\tau = \infty$ of the form $\tau \sum_{j\geq 0} a_j \tau^{-j}$, $a_0 \neq 0$, and that these coefficients are determined uniquely up to λ, $\lambda^3 = 1$. The form of (33.5) indicates that $w(-\tau) = -w(\tau)$, implying that $a_j = 0$ if $j \in 2\mathbb{Z}$. Hence, by (33.9), $w(\tau)$ is taken into a solution $y(\eta)$ of (33.10), which is also rational in η. Next let us consider the series expansion of $w(\tau)$ around $\tau = 0$. By comparison of powers on both sides of (33.5), concerning the term of the lowest power τ^l, two possibilities remain: (a) $l = \pm 1$ and (b) $l = 3$, $\beta \neq 0$.

Now consider the case (a). Then $y(\eta)$ is analytic at $\eta = 0$. Looking at a pole $\eta_0 \neq 0$ of $y(\eta)$, we obtain

$$y(\eta) = \frac{a_{-1}}{\eta - \eta_0} + a_0 + \cdots, \quad a_{-1}^2 = \frac{4}{9}.$$

By (33.9), in a neighborhood of the infinity, the solution $y(\eta)$ has the following expansion:

$$y = c_0 + \frac{c_1}{\eta} + \frac{c_2}{\eta^2} + \cdots, \quad c_0^3 = 1, \; c_1 = \frac{\beta}{3}.$$

Consequently, if $y(\eta)$ has l_1 poles with the residue equal to $2/3$ and l_2 poles with the residue equal to $-2/3$, where $l_1, l_2 \in \mathbb{N} \cup \{0\}$, then the necessary condition is $2l_1/3 - 2l_2/3 = \beta/3$. From this formula, in the case (a), we obtain the necessary condition $\beta = 2k$, $k \in \mathbb{Z}$.

To complete the proof, we consider the case (b). By the transformation (33.6), it is sufficient to consider the case $\beta \in G_1 = \{\beta \mid -1 \leq \mathrm{Re} < 1\}$. By T_2 of (29.18), the series $W(\tau) = w(\tau)^{-1} = \sum_{j\geq -3} c_j \tau^j$, $c_{-3} = \beta \neq 0$ satisfies

$$\tau(\tau W')'W - (\tau W')^2 = 9\tau^3(-\beta W^3 - W + \tau^3 W^4). \tag{33.12}$$

The coefficients c_j satisfy the recursive relations:

$$c_{-3} - \beta = 0, \quad f(j, \beta)c_{-3}c_j = R_j(c_l; l \leq j - 1), \quad j \geq -2,$$

where $f(j, \beta) = (j + 3)^2 + 9(3\beta c_{-3} - 4c_{-3}^2) = (j + 3)^2 - 9\beta^2$. Note that

$$R_{-2} = 0, \quad R_{-1} = c_{-2}R_{-1}^*(\beta), \quad R_0 = c_{-2}R_{-1}^*(\beta, c_{-1}).$$

On the other hand, by $\tau^3 = z$, (33.12) transforms into

$$z(zv')'v - (zv')^2 = z(-\beta v^3 - v) + z^2 v^4.$$

The formal series solution $v = v(z) = \sum_{k\geq -1} d_k z^k$ also satisfies a recursive relation

$$d_{-1} - \beta = 0, \quad \widetilde{f}(k, \beta) d_{-1} d_k = \widetilde{R}_k(d_l; l \leq k-1), \quad k \geq 0$$

with $\widetilde{f}(k, \beta) = f(3k, \beta)/9 = (k+1)^2 - \beta^2$, $\widetilde{R}_0 = 0$. Now we divide into three subcases with respect to $\beta \in G_1$: (b.1) $\beta = -1$; (b.2) $\beta = \pm 1/3, \pm 2/3$; (b.3) otherwise.

Case (b.1): In this case, $c_{-2} = c_{-1} = 0$, and $f(0, \beta) = R_0 = 0$. Since $f(j, \beta) \neq 0$ for $j \geq 1$, we have $c_j = p_j(\beta, c_0)$ for $j \geq 1$, where $p_j(\beta, c_0)$ is a polynomial. Furthermore, $d_k = d_k(\beta, d_0)$, $k \geq 1$ is determined uniquely up to d_0. Since $v(\tau^3)$ satisfies (33.12), if we choose $d_0 = c_0$, $v(\tau^3)$ coincides with $W(\tau)$, implying that $c_j = 0$ for $j \notin 3\mathbb{Z}$. Since $w(\tau)$ is rational, the solution $w(z^{1/3}) = \sum_{j\geq 3} c_{3j} z^j$ of (33.1) admits an expression $w(z^{1/3}) = \sum_{j\geq 0} a_j z^{-j}$ around $z = \infty$, which contradicts the first equation of (33.13) (or (33.16)).

Case (b.2): In this case, it is easy to see that $y(\eta)$ possesses a simple pole with residue β at $\eta = 0$ as well. Applying the same argument as in case (a), we obtain $2l_1/3 - 2l_2/3 + \beta = \beta/3$, which implies $\beta \in \mathbb{Z}$, a contradiction.

Case (b.3): Since $f(j, \beta) \neq 0$, $\widetilde{f}(k, \beta) \neq 0$ for all j, k, the coefficients c_j, d_k are determined uniquely. We can derive a contradiction by the same method as in case (b.1), completing the proof. □

Theorem 33.4. *Any rational solution of* (33.5) *has* $\beta^2/4$ *poles and* $\beta^2/4 + 1$ *zeros in the complex plane.*

Proof. Let $w = P(\tau)/Q(\tau)$ be a rational solution of (33.5), where $P(\tau)$, $Q(\tau)$ are irreducible polynomials in τ with degrees m and n, respectively. Substituting this into (33.5) and comparing terms, we have $m = n + 1$.

To find the number of poles and zeros in terms of β, we put $\tau = \exp \xi$ in (33.5). Then we get the equation

$$ww'' = (w')^2 + 9e^{3\xi}(w^3 + \beta w - e^{3\xi}). \tag{33.13}$$

Since all solutions of the equation (33.13) are meromorphic, we obtain a system

$$\begin{cases} uu'' = (u')^2 - 9e^{3\xi} uv, \\ vv'' = (v')^2 + 9\beta e^{3\xi} uv - 9e^{6\xi} u^2, \end{cases} \tag{33.14}$$

where the two functions $u(\xi)$ and $v(\xi)$ which are such that $w(\xi) = v(\xi)/u(\xi)$, correspond to (32.4). By using (32.5), it is easy to verify by direct calculation that the expression

$$(v'u - u'v)^2 - 6uv(u'v + v'u) - 9e^{6\xi} u^4 + 18e^{3\xi} uv(\beta u^2 - v^2) = Cu^2 v^2 \tag{33.15}$$

is the first integral of the system (33.14). The entire solution

$$v(\xi) = P(\xi)\exp g(\xi), \quad u(\xi) = Q(\xi)\exp g(\xi), \tag{33.16}$$

where $g(\xi)$ is an entire function to be determined, corresponds to a solution $w(\xi) = P(\xi)/Q(\xi)$ of the equation (33.13) rational in e^{ξ}. Substituting now (33.16) into (33.14) and (33.15), we get

$$QQ'' - (Q')^2 + g''Q^2 + 9e^{3\xi}PQ = 0, \tag{33.17}$$

and

$$\begin{aligned}(P'Q - PQ')^2 - 6PQ(P'Q + PQ') - 9e^{6\xi}Q^4 \\ + 18e^{3\xi}(\beta PQ^3 - P^3Q) - (C + 12g')P^2Q^2 = 0.\end{aligned} \tag{33.18}$$

Substituting the polynomials

$$P(\xi) = \sum_{j=0}^{n+1} p_j e^{\xi(n+1-j)}, \quad Q(\xi) = \sum_{j=0}^{n} q_j e^{\xi(n-j)}$$

into (33.17) and (33.18), we observe that $g'(\xi) = \sum_{j=0}^{4} \lambda_j \exp(j\xi)$ and there exist two possibilities for the coefficients:

(a) $p_{2j+1} = 0$, $j = 0, \dots, k$; $q_{2j+1} = 0$, $j = 0, \dots, k-1$, $q_0 = 1$, $p_n q_n \neq 0$, $n = 2k$, $k \in \mathbb{N} \cup \{0\}$;

(b) $p_{2j+1} = q_{2j+1} = 0$, $j = 0, \dots, k-1$, $p_{n+1}q_{n-1} \neq 0$, $q_0 = 1$, $n = 2k-1$, $k \in \mathbb{N}$.

Then, substituting into (33.18), we obtain for the coefficients of g' that $\lambda_1 = \lambda_3 = 0$. Comparing the coefficients in (33.18) of terms of degrees $2n+4$ and $2n+2$ in e^{ξ}, we get

$$\begin{cases} 9p_0 + 4\lambda_4 = 0, \\ 9p_2 + 9p_0q_2 + 2\lambda_2 + 8q_2\lambda_4 = 0. \end{cases} \tag{33.19}$$

Comparing further the coefficients of terms of degrees $4n+6, 4n+4, 4n+2, \dots, 2$ in (33.18), we get the equations

$$\begin{cases} 4p_0^2\lambda_4 + 3 + 6p_0^3 = 0, \\ 6q_2 + 3p_0^3q_2 + p_0(4p_2\lambda_4 - 3\beta) + p_0^2(9p_2 + 2\lambda_2 + 4q_2\lambda_4) = 0, \\ 12\lambda_0 + 5 + C + 12n - 3\beta^2 = 0, \\ 12\lambda_0 + 5 + C = 0. \end{cases} \tag{33.20}$$

Solving the equations (33.19), (33.20), we obtain

$$p_0^3 = 1, \ \ \lambda_4 = -9p_0/4, \ \ \lambda_2 = 3\beta p_0^2/2, \ \ \lambda_0 = -(5+C)/12, \ \ n = \beta^2/4. \tag{33.21}$$

The assertion of the theorem now follows from the last equality in (33.21). □

Theorem 33.5. *The equation* (33.5) *with* $\beta = 2k$, $k \in \mathbb{Z}$, *has exactly three rational solutions. They are generated from* (33.8) *by* (33.6), *depending on the value of* λ *in* (33.8).

Proof. By Theorem 33.1 and Theorem 33.3, it is sufficient to show that the equation (33.5) with $\beta = 0$ has no other rational solutions than those in (33.8). But by §30 (4), the coefficients of the series expansion of $w(\tau)$ with the leading term $\lambda\tau$ are uniquely determined, and there are no solutions other than them. □

§34 Connection between solutions of (P_3) and (P_5)

In this section, we shall consider the case $\gamma\delta \neq 0$ only, with the idea of constructing a transformation which carries (P_3) into (P_5), with parameters in (P_5) being expressed in terms of the parameters in (P_3). As (P_3) is invariant over the transformation $T_1(\sigma_2, \sigma_2)$ in (29.18), we may fix $\gamma = 1$ and $\delta = -1$, by taking $\sigma_1 = 1/(\sigma_2\sqrt{\gamma})$ and $\sigma_2 = \sqrt[4]{-1/\gamma\delta}$. Therefore, it suffices to consider (P_3) in the following form without loss of generality:

$$w'' = \frac{(w')^2}{w} - \frac{w'}{z} + \frac{1}{z}(\alpha w^2 + \beta) + w^3 - \frac{1}{w}. \tag{34.1}$$

To derive a transformation that relates solutions of (34.1) to solutions of an equation of type (P_5), we recall the pair (29.17) of differential equations equivalent to (P_3) for $\gamma \neq 0$. Using the notations $w_1 = v$, $\gamma = 1$, $\delta = -1$, $\sqrt{\gamma} = \varepsilon$, $\varepsilon^2 = 1$, we obtain

$$\begin{cases} zw' = (\alpha\varepsilon - 1)w + \varepsilon z w^2 + zv, \\ zwv' = \beta w - z + (\alpha\varepsilon - 2)wv + zv^2. \end{cases} \tag{34.2}$$

Of course, v may be expressed rationally in terms of w' and w. We next find a differential equation for v. Provided

$$zv' - \beta - (\alpha\varepsilon - 2)v \neq 0, \tag{34.3}$$

we obtain

$$w = z(v^2 - 1)/(zv' - \beta - (\alpha\varepsilon - 2)v) \tag{34.4}$$

from the second equation of (34.2). Substituting now (34.4) into the first equation of (34.2), we get

$$v'' = \frac{v}{v^2 - 1}(v')^2 - \frac{v'}{z} + \Theta(z, v), \tag{34.5}$$

where

$$\Theta(z, v) = \varepsilon(1 - v^2) - \frac{1}{z^2}\beta(\alpha\varepsilon - 2) - \frac{2\beta(\alpha\varepsilon - 2)}{z^2(v^2 - 1)} - \frac{\beta^2 + (\alpha\varepsilon - 2)^2}{z^2}\frac{v}{v^2 - 1}.$$

By the transformation

$$v = -\frac{\widetilde{u}(z) + 1}{\widetilde{u}(z) - 1}, \tag{34.6}$$

we obtain for $u(\tau) := \widetilde{u}(z) = \widetilde{u}(\sqrt{2\tau})$ from (34.5),

$$\begin{aligned} u'' = {} & \frac{3u - 1}{2u(u-1)}(u')^2 - \frac{u'}{\tau} \\ & + \frac{(u-1)^2}{32\tau^2}\left[(\beta - \alpha\varepsilon + 2)^2 u - \frac{(\beta + \alpha\varepsilon - 2)^2}{u}\right] - \frac{\varepsilon u}{\tau}. \end{aligned} \tag{34.7}$$

In (34.7), u' now stands for the differentiation with respect to τ. Moreover, it is an immediate observation that (34.7) is a special case of the fifth Painlevé equation (P_5). Hence, we have obtained the following theorem, see Gromak [2]:

Theorem 34.1. *Let $w = w(z)$ be a solution of the third Painlevé equation such that*

$$R := w' - \varepsilon w^2 - (\alpha\varepsilon - 1)\frac{w}{z} + 1 \neq 0.$$

Then the function

$$u(\tau) = 1 - \frac{2}{R(\sqrt{2\tau})} \tag{34.8}$$

is a solution of the fifth Painlevé equation with parameters

$$a = \frac{1}{32}(\beta - \alpha\varepsilon + 2)^2, \quad b = -\frac{1}{32}(\beta + \alpha\varepsilon - 2)^2, \quad c = -\varepsilon, \quad d = 0. \tag{34.9}$$

From the system (34.2) it follows that all solutions of the equation

$$w' = \varepsilon w^2 + (\alpha\varepsilon - 1)\frac{w}{z} - 1 \tag{34.10}$$

are solutions of (P_3) with

$$\beta - \alpha\varepsilon + 2 = 0, \tag{34.11}$$

and the equation

$$w' = \varepsilon w^2 + (\alpha\varepsilon - 1)\frac{w}{z} + 1 \tag{34.12}$$

generates solutions of (P_3) with

$$\beta + \alpha\varepsilon - 2 = 0. \tag{34.13}$$

Note, that (P_5) has the following property: if $u = u(t, a, b, c, d)$ is a solution of (P_5), then $u_1(t) = 1/u(t, -b, -a, -c, d)$ is a solution of (P_5) as well.

The following theorem is valid on the basis of Theorem 34.1 and the above-mentioned statement.

Theorem 34.2. *Let $w = w(z, \alpha, \beta)$ be a solution of (P_3) such that*

$$R_1(z) := w' - \varepsilon w^2 - (\alpha\varepsilon - 1)w/z - 1 \neq 0.$$

Then the function

$$u_1(t, a, b, c, d) = 1 + \frac{2}{R_1(\sqrt{2t})} \tag{34.14}$$

is the solution of (P_5) when

$$a = \frac{1}{32}(\beta + \alpha\varepsilon - 2)^2, \;\; b = -\frac{1}{32}(\beta - \alpha\varepsilon + 2)^2, \;\; c = \varepsilon, \;\; d = 0, \;\; \varepsilon^2 = 1. \tag{34.15}$$

Note, that one-parameter families of solutions of the third Painlevé equation, defined by (34.10) and (34.12) under the conditions (34.11), (34.13) correspond to (P_5)-solutions, namely, $u_1 = 0$, when $a = 0$, and $u = 0$, when $b = 0$, respectively.

Now we can find a solution w of (P_3) in terms of a solution u of (P_5):

Theorem 34.3. *Let $u(\tau)$ be a solution of (P_5) with the parameter values $a, b, c^2 = 1$ and $d = 0$, and define*

$$\Phi(\tau) := \tau u'(\tau) - \sqrt{2a}u(\tau)^2 + (\sqrt{2a} + \sqrt{-2b})u(\tau) - \sqrt{-2b} \neq 0.$$

Then the function

$$w(z) = \frac{\sqrt{2\tau}u}{\Phi(\tau)}, \quad z^2 = 2\tau \tag{34.16}$$

is a solution of (P_3) with the parameter values

$$\alpha = 2c(\sqrt{2a} - \sqrt{-2b} - 1), \quad \beta = 2\sqrt{2a} + 2\sqrt{-2b}, \quad \gamma = 1, \quad \delta = -1. \tag{34.17}$$

Remark 1. If $u = 0$ under $b = 0$, then according to (34.1) and (34.8) we have a one-parameter family of solutions of (34.1) given by the general solution of (34.12). If $u_1 = 0$ under $a = 0$, then we have solutions of (34.10).

Remark 2. Conditions $c = \pm 1$, $d = 0$ of the above-mentioned theorem are equivalent to $c \neq 0$, $d = 0$, since we can replace the independent variable $t \mapsto \pm t/c$ in (P_5), for any given $c \neq 0$.

Thus, these theorems establish a correspondence between solutions of (P_3) with parameters $\gamma\delta \neq 0$ and solutions of (P_5), for which $c \neq 0$, $d = 0$.

We continue to look at the solutions of (34.1) for which neither (34.10) nor (34.12) hold, and a solution $u \not\equiv 0$ of (P_5). As it follows immediately from Theorems 34.1–34.3, one solution of (P_3) generates two solutions of (P_5) on account of the choice of $\varepsilon = \sqrt{\gamma} = \pm 1$, while one solution of (P_5) generates four solutions of (P_3) because of the so-called parameter branching of $\sqrt{2a}$, $\sqrt{-2b}$. As a result of the parameter branching we can establish a correspondence between solutions of (34.1) with different values of the parameters, see Gromak [2]:

Theorem 34.4. *Let $w(z,\alpha,\beta)$ be a solution of (P_3) for which $\gamma\delta \neq 0$ such that*

$$R(z)(R(z)-2) \neq 0.$$

Then the function

$$\widetilde{w}(z,\widetilde{\alpha},\widetilde{\beta}) = \frac{2zR(z)(R(z)-2)}{2zR'(z)+(\varepsilon_3 A-\varepsilon_2 B)R(z)-2\varepsilon_3 A} \tag{34.18}$$

is a solution of (34.1) *as well with*

$$\widetilde{\alpha} = \varepsilon_1(\varepsilon_2 B-\varepsilon_3 A+4)/2, \quad \widetilde{\beta} = \varepsilon_2 B/2+\varepsilon_3 A/2, \tag{34.19}$$

where

$$B = \beta+\alpha\varepsilon_1-2, \quad A = \beta-\alpha\varepsilon_1+2, \quad \varepsilon_j^2 = 1, \quad j = 1,2,3.$$

The proof of this theorem immediately follows from Theorems 34.1–34.3. Indeed, since $R(z)(R(z)-2) \neq 0$, we may construct solutions of (34.1) by using solutions of (P_5). In the general case, the new solutions are different from the initial one, depending on choice of values. We get $\widetilde{w} = w$ if $\varepsilon_2 = \varepsilon_3 = 1$. The choice of ε_j may be limited using the transformations (29.18). Relations (34.18) and (34.19) may be obtained by the substitution of (34.18) and (34.9) into (34.16) and (34.17), respectively, where the choice of $\varepsilon_2, \varepsilon_3$ determines the choice of the branches of $\sqrt{(\beta+\alpha\varepsilon_1-2)^2}$ and $\sqrt{(\beta-\alpha\varepsilon_1+2)^2}$.

It is necessary to point out that the formulas (34.18) and (34.19) can be simplified. In order to show this, let us denote the transformation of Theorem 34.4 by $T_{\varepsilon_1,\varepsilon_2,\varepsilon_3}$: $w(z,\alpha,\beta) \mapsto \widetilde{w}(z,\widetilde{\alpha},\widetilde{\beta})$. It is not difficult to show that $T_{1,1,1} = T_{-1,1,1} = I$, where I is the identity transformation as before. In other cases of the choice of ε_j we get the following transformations:

$$T_{1,1,-1} : w(z,\alpha,\beta) \mapsto \widetilde{w}(z,2+\beta,-2+\alpha) = w + \frac{(\beta-\alpha+2)w^2}{zw'-w^2z-(1+\beta)w+z};$$

$$T_{1,-1,1} : w(z,\alpha,\beta) \mapsto \widetilde{w}(z,2-\beta,2-\alpha) = w + \frac{(\alpha+\beta-2)w^2}{w+z+w^2z-w\beta-zw'};$$

$$T_{1,-1,-1} : w(z,\alpha,\beta) \mapsto \widetilde{w}(z,4-\alpha,-\beta) = T_{1,-1,1} \circ T_{1,1,-1};$$

$$T_{-1,1,-1} : w(z,\alpha,\beta) \mapsto \widetilde{w}(z,-2-\beta,-2-\alpha) = w + \frac{(\alpha+2+\beta)w^2}{w^2z-w-w\beta+z+zw'};$$

$$T_{-1,-1,1} : w(z,\alpha,\beta) \mapsto \widetilde{w}(z,-2+\beta,2+\alpha) = w - \frac{(\beta-\alpha-2)w^2}{w^2z-w+w\beta-z+zw'};$$

$$T_{-1,-1,-1} : w(z,\alpha,\beta) \mapsto \widetilde{w}(z,-4-\alpha,-\beta) = T_{-1,-1,1} \circ T_{-1,1,-1}.$$

Moreover, after repeating $T_{\varepsilon_1,\varepsilon_2,\varepsilon_3}$, we obtain $T^2_{\varepsilon_1,\varepsilon_2,\varepsilon_3} = I$. Similarly as for (P_2) and (P_4), we may apply the Bäcklund transformations to solutions of the Riccati equation

(29.11) to obtain one-parameter families of solutions expressed in terms of Bessel functions, to be considered in the next section. In the next theorem, we assume that $w(z, \alpha, \beta)$ is outside of these families.

Theorem 34.5. *By successive applications of the transformations $T_{\varepsilon_1,\varepsilon_2,\varepsilon_3}$, T_2 and $T_1(\sigma_1, \sigma_2)$ with $\sigma_1^2 = \sigma_2^2 = \pm 1$ from (29.18) to a solution $w(z, \alpha, \beta)$ of (34.1), we obtain a solution $\widetilde{w}(z, \widetilde{\alpha}, \widetilde{\beta})$ with the pair $(\widetilde{\alpha}, \widetilde{\beta})$ of parameter expressed as either*

$$\widetilde{\alpha} = \varepsilon\alpha + 2n_1, \quad \widetilde{\beta} = \nu\beta + 2n_2 \tag{34.20}$$

or

$$\widetilde{\alpha} = \varepsilon\beta + 2n_1, \quad \widetilde{\beta} = \nu\alpha + 2n_2, \tag{34.21}$$

where (n_1, n_2) is a pair of integers satisfying $n_1 + n_2 \in 2\mathbb{Z}$, and $\varepsilon^2 = \nu^2 = 1$. Moreover, for an arbitrary pair of such integers (n_1, n_2) and for (ε, ν), there exists a composed transformation as above producing the substitution given by (34.20), *resp.* (34.21).

Proof. Let us consider the following transformations:

$$\begin{aligned}
&H_1 : w(z, \alpha, \beta) \mapsto \widetilde{w}(z, \alpha + 2, \beta + 2), && H_1 = T_2 \circ T_{-1,1,-1},\\
&H_2 : w(z, \alpha, \beta) \mapsto \widetilde{w}(z, \alpha - 2, \beta + 2), && H_2 = T_1(1, -1) \circ T_2 \circ T_{1,1,-1},\\
&H_3 : w(z, \alpha, \beta) \mapsto \widetilde{w}(z, \alpha + 2, \beta - 2), && H_3 = T_1(1, -1) \circ T_2 \circ T_{-1,-1,1},\\
&H_4 : w(z, \alpha, \beta) \mapsto \widetilde{w}(z, \alpha - 2, \beta - 2), && H_4 = T_2 \circ T_{1,-1,1}.
\end{aligned}$$

and denote $H_{n_1,n_2} := \prod_{j=1}^{4} H_j^{k_j}$, $k_j \in \mathbb{N} \cup \{0\}$. For arbitrary integers n_1, n_2 satisfying $n_1 + n_2 \in 2\mathbb{Z}$, we can choose $k_j \in \mathbb{N} \cup \{0\}$, $j = 1, \ldots, 4$, in such a way that $n_1 = k_1 - k_2 + k_3 - k_4$, $n_2 = k_1 + k_2 - k_3 - k_4$ and $n_1 + n_2 = 2(k_1 - k_4)$ hold. Then we have

$$H_{n_1,n_2} : w(z, \alpha, \beta) \mapsto \widetilde{w}(z, \alpha + 2n_1, \beta + 2n_2), \tag{34.22}$$

where $n_1 + n_2 = 2n_3$, $n_j \in \mathbb{Z}$, $j = 1, 2, 3$.

The second part of the statement now follows from (34.22) if we apply the transformations $T_1(\sigma_1, \sigma_2)$ with $(\sigma_1, \sigma_2) = (1, -1)$ or $(\sigma_1, \sigma_2) = (i, i)$ and T_2 to the solution $\widetilde{w}(z, \alpha + 2n_1, \beta + 2n_2)$ in (34.22). The first part, the closedness, is obvious. □

The transformation H_{n_1,n_2} and (29.18) allow us to deduce the fundamental domain in the space of the parameters of (34.1). In fact, the fundamental domain is given by

$$G := \{(\alpha, \beta) \mid \operatorname{Re}\beta \geq 0,\ \operatorname{Re}(\alpha - \beta) \geq 0,\ \operatorname{Re}(\alpha + \beta) \leq 2\}. \tag{34.23}$$

To construct auto-Bäcklund transformations of the equation (34.1), it follows from (29.18), that simple auto-Bäcklund transformations may be given by

$$\begin{aligned}
&S_1 : w(z, \alpha, \alpha) \mapsto -w^{-1}(z, \alpha, \alpha),\\
&S_2 : w(z, \alpha, -\alpha) \mapsto w^{-1}(z, \alpha, -\alpha).
\end{aligned} \tag{34.24}$$

Moreover, it is obvious that the transformations (34.24) generate solutions different from the initial ones unless $w = \lambda$, $\lambda^4 = 1$, and $\alpha\lambda^2 + \beta = 0$.

Theorem 34.6. *If $w(z, \alpha, \alpha\varepsilon + 4n)$, $n \in \mathbb{Z}$, $\varepsilon^2 = 1$, is a non-rational solution of the equation* (34.1), *then*

$$\widetilde{w}(z, \alpha, \alpha + 4n) = H_{0,2n} \circ S_1 \circ H_{0,-2n} w(z, \alpha, \alpha + 4n),$$
$$\widetilde{w}(z, \alpha, -\alpha + 4n) = H_{0,2n} \circ S_2 \circ H_{0,-2n} w(z, \alpha, -\alpha + 4n)$$

are solutions of the equation (34.1) *as well, different from the original one.*

Proof. The fact that $\widetilde{w}$ is a solution of equation (34.1) follows from its construction by means of the transformations S_1, S_2, H_{n_1,n_2}. To prove the distinction of the solutions $\widetilde{w}$ and w, observe that $S_j \circ H_{0,-2n} w(z, \alpha, \alpha\varepsilon + 4n) \neq \lambda$, $\lambda^4 = 1$, $\alpha\lambda^2 + \beta = 0$ as w is non-rational. This proves the observation above. □

Thus, the transformations S_j and $H_{0,2n} \circ S_j \circ H_{0,-2n}$, $j = 1, 2$ are auto-Bäcklund transformations, provided the parameters are related by $\beta + \alpha\varepsilon = 4n$, $n \in \mathbb{Z}$, $\varepsilon^2 = 1$.

Similar to the equations (P_2), (P_4), it is not difficult to construct nonlinear superposition formulas linking different solutions obtained by means of the Bäcklund transformation. Next we give examples of such formulas.

Let $w_{\alpha,\beta} = w(z, \alpha, \beta)$, $w_{-\beta-2,-\alpha-2} = T_{-1,1,-1} w_{\alpha,\beta}$, $w_{\beta-2,\alpha+2} = T_{-1,-1,1} w_{\alpha,\beta}$, $w_{-\alpha-4,-\beta} = T_{-1,-1,-1} w_{\alpha,\beta}$. Then the following relation is valid:

$$w^{-1}_{-\beta-2,-\alpha-2} + w^{-1}_{\beta-2,\alpha+2} = w^{-1}_{\alpha,\beta} + w^{-1}_{-\alpha-4,-\beta}.$$

In case of a repeated application of the Bäcklund transformation we get the nonlinear superposition formula

$$w_2 - w_1 = \frac{2 + \alpha + \beta}{2z} + \frac{-\alpha^2 + (2 + \beta)^2}{2z\,(-2 + 2\,w\,z + \alpha - \beta - 2\,z\,w_1)}$$

linking solutions $w = w(z, \alpha, \beta)$, $w_1 = w_1(z, 2 + \beta, -2 + \alpha) = T_{1,1,-1} w$, $w_2 = w_2(z, -\alpha, -4 - \beta) = T_{-1,1,-1} w_1$.

§35 Rational and classical transcendental solutions of (P_3) when $\gamma\delta \neq 0$

In this section too, we restrict ourselves to considering (P_3) under the condition $\gamma\delta \neq 0$. Therefore, as in §34, we may restrict to the special case (34.1) only. Let now $w(z) = P(z)/Q(z)$ be a rational solution of (34.1) with $P(z)$, $Q(z)$ being irreducible polynomials of degrees n and m, respectively. By (30.16), we readily observe that $n = m$. If $n = 0$, the equation (34.1) has a rational solution, in fact a constant,

$$w = \lambda, \quad \alpha\lambda^2 + \beta = 0, \quad \lambda^4 = 1. \tag{35.1}$$

The solution (35.1) is the unique rational solution, provided $\alpha\lambda^2 + \beta = 0$ and $\lambda^4 = 1$. This follows from the expansion (30.12) of the solution in a neighborhood of the point $z = \infty$ and the system (30.13). If in (30.13), $a_1 = 0$, then we find, respectively, that $a_j = 0$ for all $j \geq 2$. The statement in (35.1) to determine the coefficients in (30.13) may be obtained with the help of another method. Thus, if we take the initial solution (35.1) in (34.18), then we have four types of rational solutions in accordance with the choice of λ.

After the first step of the Bäcklund transformation (34.18) and (34.19) we get a new solution of (34.1):

$$w_1 = \frac{z+a}{\lambda z + a\lambda - \varepsilon}, \quad \alpha_1 = \lambda^2(2a\lambda - 3\varepsilon), \quad \beta_1 = -2a\lambda - \varepsilon, \quad \lambda^4 = \varepsilon^2 = 1. \tag{35.2}$$

The first such rational solutions are listed below.

Table 35.1. The first rational solutions of (P_3) with $\gamma = -\delta = 1$.

α	β	$w(z, \alpha, \beta), \lambda^4 = 1$
a	$-a\lambda^2$	λ
$2 - a\lambda^2$	$a - 2$	$\dfrac{\lambda^2(1-a) + z(1-\lambda^2)\lambda}{a\lambda^3 - \lambda + z(1-\lambda^2)}$
$2 + a\lambda^2$	$2 - a$	$\dfrac{\lambda^2(a-1) + z(1+\lambda^2)\lambda}{a\lambda^3 + \lambda + z(1+\lambda^2)}$
$a - 4$	$-a - 4$	$\big(16z^4 + (16+32a)z^3 + 24a(a+1)z^2 + 4(-3-2a+3a^2+2a^3)z + (a-1)^2(3+4a+a^2)\big)/\big(16z^4 + (-16+32a)z^3 + 24a(a-1)z^2 + 4(3-2a-3a^2+2a^3)z + (a+1)^2(3-4a+a^2)\big)$
$a - 4$	$a + 4$	$\big(16iz^4 + (16+32a)z^3 - 24ia(a+1)z^2 + 4(3+2a-3a^2-2a^3)z + i(a-1)^2(3+4a+a^2)\big)/\big(16z^4 + i(16-32a)z^3 - 24a(a-1)z^2 + 4i(3-2a-3a^2+2a^3)z + (a+1)^2(3-4a+a^2)\big)$

Note that $\alpha_1 + \beta_1 = -4\varepsilon$ in the case when $\lambda^2 = 1$ and $\alpha_1 - \beta_1 = 4\varepsilon$ in the case when $\lambda^2 = -1$. This property of the parameters of the rational solutions is also valid in the general case.

Theorem 35.1. *The equation* (34.1) *has rational solutions if and only if*

$$\alpha + \beta\varepsilon = 4n, \quad \varepsilon^2 = 1, \quad n \in \mathbb{Z}. \tag{35.3}$$

Proof. To prove the necessity, let $w(z, \alpha, \beta)$ be a rational solution of (34.1). By simple computation, this solution has the following expansion in a neighborhood of a pole $z_0 \neq 0$:

$$w(z, \alpha, \beta) = \frac{\mu}{z - z_0} - \frac{\alpha + \mu}{2z_0} + h(z - z_0) + \sum_{j=2}^{\infty} a_j (z - z_0)^j,$$

where h is arbitrary and $\mu^2 = 1$. In a neighborhood of $z = \infty$ the solution is of the form (30.12), satisfying (30.13) as well. It is easy to check that $z = \infty$ is a holomorphic point with residue $(\alpha + \beta\varepsilon)/4$, $\varepsilon^2 = 1$. Suppose now that $z = 0$ is not a pole of $w(z, \alpha, \beta)$. Then, by the residue theory, we obtain the condition (35.3). If the solution has a pole at $z = 0$, then applying the transformation T_2 from (29.18), we get a rational solution $1/w(z, -\beta, -\alpha)$ for which $z = 0$ is a simple zero. Consequently, α and β satisfy the condition (35.3).

The sufficiency of this theorem can be proved by induction. If $n = 0$ or $n = 1$ in (35.3), then the equation (34.1) has the solutions (35.1) and (35.2), respectively. Suppose that (34.1) has a rational solution when $\alpha + \beta\varepsilon = 4(n-1)$, $\varepsilon^2 = 1$, $n \in \mathbb{Z}$ and $R(z) \neq 0$. Then we apply (34.18) with $\varepsilon_j = -1$ and T_2 from (29.18). By (34.19), we get a rational solution with parameters $\alpha_1 + \beta_1\varepsilon = 4n$. If $R(z) = 0$ for some ε_1, then we take ε_1 to be of the opposite sign in (34.18). This procedure implies that the condition $R(z)(R(z) - 2) \neq 0$ holds. □

Corollary 35.2. *The simultaneous fulfillment of the conditions* (34.11), (35.3) *and* (34.13), (35.3) *is the necessary and sufficient condition for the existence of rational solutions of the equations* (34.10) *and* (34.12).

We next proceed to determine the number of poles with the residue equal to either 1 or -1 for any rational solution of the equation (34.1).

Let $w(z) = P(z)/Q(z)$ be a rational solution of (34.1) where $P(z)$ and $Q(z)$ are irreducible polynomials of degrees m. Then equation (32.1) has a solution $w(\zeta) = P(\zeta)/Q(\zeta)$, where $z = \exp(\zeta)$. Consequently, the system (32.4) to determine the entire functions $u(\zeta)$, $v(\zeta)$ has the solution

$$u(\zeta) = Q(\zeta)e^{g(\zeta)}, \quad v(\zeta) = P(\zeta)e^{g(\zeta)}, \tag{35.4}$$

where $g(\zeta)$ is an entire function. Substituting (35.4) into (32.4) and (32.5), we obtain

$$PP'' - (P')^2 + g''P^2 = \beta e^{\zeta} PQ - e^{2\zeta} Q^2, \tag{35.5}$$

$$\begin{aligned}(P'Q - PQ')^2 - 2(P'Q + PQ')PQ - e^{2\zeta}(P^4 + Q^4)\\ + 2e^{\zeta}(\beta PQ^3 - \alpha P^3 Q) = (C + 4g')P^2Q^2.\end{aligned} \tag{35.6}$$

Substituting now the expressions

$$P(\zeta) = \sum_{j=0}^{m} p_j e^{(m-j)\zeta}, \quad Q(\zeta) = \sum_{j=0}^{m} q_j e^{(m-j)\zeta}, \quad p_0 \neq 0, \quad q_0 = 1,$$

into (35.5), we get

$$g'(\zeta) = \lambda_2 e^{2\zeta} + \lambda_1 e^{\zeta} + \lambda_0.$$

Comparing the coefficients in (35.5), we obtain

$$p_0^4 = 1, \quad \lambda_2 = -p_0^2/2, \quad \lambda_1 = (\beta - \alpha p_0^2)/2p_0, \quad C + 4\lambda_0 = 0,$$

also verifying the validity of the condition

$$m = (\alpha + \beta p_0^2)^2/16.$$

From (35.3) it follows that $\alpha + \beta p_0^2 = 4n$, $n \in \mathbb{Z}$. Thus, any rational solution $w(z) = P(z)/Q(z)$ has $m = n^2$ poles and the same number of zeros. From (30.13) we have that the sum of the residues in the finite complex plane equals to $-n$ for any rational solution. Then, using the residue theory again, we see that any rational solution has $n(n-1)/2$ poles with the residue equal to 1 and $n(n+1)/2$ poles with the residue equal to -1.

One-parameter families of special solutions of (34.1) may be constructed in the same way as to it has been done for the equations (P_2) and (P_4). According to (29.11) and (29.13), such a family may be expressed in terms of Bessel functions. Indeed, if we take solutions of the Riccati equation

$$w' = -w^2 - \frac{\alpha+1}{z}w + 1, \quad \beta = \alpha + 2 \tag{35.7}$$

as the seed solutions for the transformation $T_{1,1,-1}$ above, then after the first step we obtain the equation

$$(u')^3 + Q_1(u')^2 + Q_2u' + Q_3 = 0, \tag{35.8}$$

where we have denoted $\widetilde{w} = u$ and where

$$\begin{aligned}
Q_1 &= -u^2 - \frac{1+\alpha}{z}u + 1,\\
Q_2 &= -u^4 - \frac{2(5+\alpha)}{z}u^3 - \frac{1}{z^2}(5 - 2z^2 + 2\alpha + \alpha^2)u^2 + \frac{2(\alpha-3)}{z}u - 1,\\
Q_3 &= u^6 + \frac{1}{z}(11 + 3\alpha)u^5 + \frac{1}{z^2}(23 - 3z^2 + 14\alpha + 3\alpha^2)u^4\\
&\quad + \frac{1}{z^3}(1+\alpha)(2\alpha + \alpha^2 - 3 - 6z^2)u^3 + \frac{1}{z^2}(3z^2 - 7 + 2\alpha - 3\alpha^2)u^2\\
&\quad + \frac{1}{z}(3\alpha - 5)u - 1.
\end{aligned}$$

All solutions of (35.8) are solutions of (34.1) at the same time under the condition $\widetilde{\alpha} = \alpha + 4$, $\widetilde{\beta} = \alpha - 2$. Note, that α is arbitrary and $\widetilde{\alpha} - \widetilde{\beta} = 6$. Thus, by (34.18), the solutions of (35.8) are expressed in terms of the solutions of (35.7) by means of the formula

$$u = w - \frac{2w^2}{zw^2 + (\alpha+2)w - z}.$$

By Theorem 34.1, we get conditions imposed upon the parameters of the equation (34.1) under which (34.1) admits one-parameter families of solutions expressed in terms of the Bessel functions:

Theorem 35.3. *The equation* (34.1) *with*

$$\alpha + \beta\varepsilon = 4n + 2, \quad \varepsilon^2 = 1, \quad n \in \mathbb{Z}$$

has one-parameter families of solutions expressed in terms of Bessel functions.

Proof. The proof follows from Theorem 34.5 by applying the transformations T_2, $T_1(\sigma_1, \sigma_2)$ with $\sigma_1^2 = \sigma_2^2 = \pm 1$, H_{n_1,n_2} to the solutions of the equation (35.7). □

To conclude this section, we collect real pairs of α, β from Theorem 35.1 and Theorem 35.3 in the following figure. In this figure, the gray area stands for a fundamental domain G, red parameter lines represent rational solutions from Theorem 35.1 and blue parameter lines one-parameter families of solutions from Theorem 35.3.

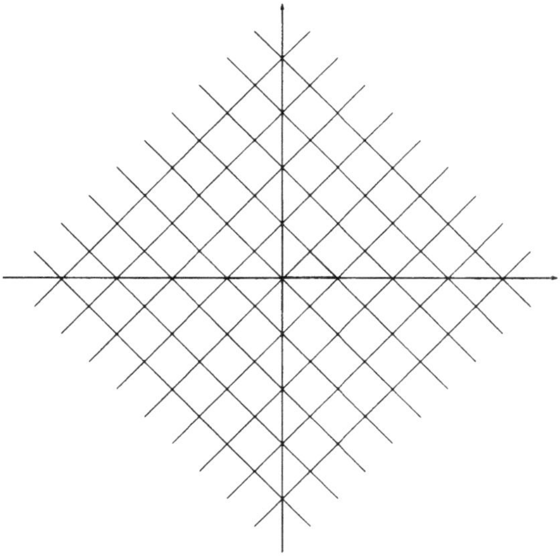

Figure 35.1.

Chapter 8

The fifth Painlevé equation (P_5)

The fifth Painlevé equation (P_5) is in many respects closely related with the third Painlevé equation (P_3). By the existence of two singular points for the equation, the transformation $z = \exp \zeta$ results in a modified form of (P_5), the solutions of which are all meromorphic in the complex plane, exactly as is the case for (P_3), as described in Chapter 1. Moreover, simple transformations may change some solutions of (P_3) to solutions of (P_5), and vice versa. This aspect, already considered in the previous chapter, will be pursued further in this chapter. Otherwise, the contents of this chapter follows the previous pattern, considering Bäcklund transformations, rational solutions and one-parameter families of solutions in detail.

§36 Preliminary observations

In this chapter, we consider the fifth Painlevé equation

$$w'' = \left(\frac{1}{2w} + \frac{1}{w-1}\right)(w')^2 - \frac{w'}{z} + \frac{(w-1)^2}{z^2}\left(\alpha w + \frac{\beta}{w}\right) + \frac{\gamma w}{z} + \frac{\delta w(w+1)}{w-1}, \tag{P_5}$$

where $\alpha, \beta, \gamma, \delta$ are arbitrary constant parameters, keeping the contents of the previous chapter as a model. Under some specific conditions upon the equation parameters, (P_5) may be solved in terms of elementary functions, or some of its solutions may be expressed in terms of some classical transcendental functions or certain non-classical functions like third Painlevé transcendents. As to a simple example of elementary solutions of (P_5) in a special case, we substitute $z = \exp \zeta$ in (P_5). Reverting to write z again instead of ζ after the transformation, we obtain, as already pointed out in §5, the modified fifth Painlevé equation

$$w'' = \left(\frac{1}{2w} + \frac{1}{w-1}\right)(w')^2 + (w-1)^2\left(\alpha w + \frac{\beta}{w}\right) + \gamma e^z w + \frac{\delta e^{2z} w(w+1)}{w-1}. \tag{$\widetilde{P}_5$}$$

Recalling §5, the auxiliary function

$$V := \frac{(w')^2}{w(w-1)^2} - 2\alpha w + \frac{2\beta}{w} + \frac{2\gamma e^z}{w-1} + \frac{2\delta e^{2z} w}{(w-1)^2} \tag{36.1}$$

immediately yields by differentiation,

$$V' = \frac{2\gamma e^z}{w-1} + \frac{4\delta e^{2z} w}{(w-1)^2}. \tag{36.2}$$

If now $\gamma = \delta = 0$, then V' vanishes identically, and so V has to be a constant, say $V = C \in \mathbb{C}$. Then we obtain, from (36.1), that

$$(w')^2 = (w-1)^2(2\alpha w^2 + Cw - 2\beta). \tag{36.3}$$

As mentioned in §5, and shown by Hinkkanen and Laine [3] in detail, (36.3) may be transformed into a Riccati differential equation with constant coefficients. Hence, all solutions of (36.3) may be expressed in terms of elementary transcendental functions. In fact, by a suitable mathematical software, one may verify that a representation for non-constant solutions of (36.3) may be written as

$$w(z) = \frac{K - 8\beta e^{\mu B(C_1-z)} + e^{2\mu B(C_1-z)} + 2Ce^{\mu B(C_1-z)}}{K - 8\alpha e^{\mu B(C_1-z)} + e^{2\mu B(C_1-z)} - 2Ce^{\mu B(C_1-z)}},$$

where C_1 is the constant of integration, $K := C^2 + 16\alpha\beta$, $B := \sqrt{2\alpha + C - 2\beta}$ and $\mu^2 = 1$. For a different representation of these solutions, see Kießling [1], p. 51–52.

By §5, all solutions of $(\widetilde{P}_5)$ are meromorphic in the complex plane. Therefore, all solutions of (P_5) are meromorphic functions in $\mathbb{C} \setminus \mathcal{L}$, where $\mathcal{L}$ is a Jordan curve from the origin to infinity. For subsequent considerations in this chapter, we next work out some necessary details of the local Laurent expansions for a solution $w(z)$ of (P_5), respectively of $(\widetilde{P}_5)$ at $z_0 \neq 0$.

Suppose first that $\alpha \neq 0$ and that $z_0 \neq 0$ is a pole of $w(z)$. Obviously, the pole now has to be a simple one. Substituting, in a neighborhood of z_0,

$$w(z) = \sum_{j=-1}^{\infty} a_j (z - z_0)^j \tag{36.4}$$

into (P_5), we obtain that $a_{-1}^2 = z_0^2/(2\alpha)$. Moreover, $a_0 =: h$ is an arbitrary complex constant, and all coefficients a_j, $j \geq 1$, may be uniquely expressed in terms of z_0, a_{-1}, h and the parameters $\alpha, \beta, \gamma, \delta$ of (P_5).

If next $\alpha = 0$ and $z_0 \neq 0$ is a pole of $w(z)$, then the pole has to be double and for the Laurent expansion at z_0, we get

$$w(z) = \frac{z_0 h}{(z-z_0)^2} + \frac{h}{z - z_0} + \sum_{j=0}^{\infty} a_j (z - z_0)^j, \tag{36.5}$$

where h is arbitrary and all coefficients a_j, $j \geq 0$, again follow uniquely in terms of $z_0, h, \beta, \gamma, \delta$.

Similarly, we may work out the Taylor expansion

$$w(z) = \sum_{j=k}^{\infty} a_j (z - z_0)^j \tag{36.6}$$

around a zero-point $z_0 \neq 0$ of $w(z)$ of multiplicity $k \in \mathbb{N}$. If now $\beta \neq 0$, then $k = 1$, and $z_0^2 a_1^2 = 2\beta$. If $\beta = 0$, then $k = 2$ and a_2 may be chosen arbitrarily.

Finally, we shall need the corresponding expansions

$$w(z) = 1 + \sum_{j=k}^{\infty} a_j (z - z_0)^j \tag{36.7}$$

around a one-point $z_0 \neq 0$ of $w(z)$ of multiplicity $k \in \mathbb{N}$. This time, if $\delta \neq 0$, then $k = 1$, and $a_1^2 + 2\delta = 0$, while if $\delta = 0$, then $k = 2$, and $2z_0 a_2 = -\gamma$.

Correspondingly, we may work out similar expansions for a solution $w(z)$ of $(\widetilde{P}_5)$, assuming that either $\gamma \neq 0$, or $\delta \neq 0$.

Let us consider the poles of $w(z)$ first. If $\alpha \neq 0$, any pole of $w(z)$, say at z_0, has to be simple. Substituting the expansion (36.4) in $(\widetilde{P}_5)$ results in $a_{-1}^2 = 1/(2\alpha)$. Moreover, a_0 may be chosen arbitrarily. Similarly, if $\alpha = 0$, any pole is double, the leading coefficient a_{-2} is arbitrary and $a_{-1} = 0$.

As to the zeros of $w(z)$, we substitute (36.6) into $(\widetilde{P}_5)$. If now $\beta \neq 0$, then $k = 1$ and $a_1^2 + 2\beta = 0$, while a_2 is arbitrary. Moreover, if $\beta = 0$, then $k = 2$ and the leading coefficient a_2 may be chosen arbitrarily. In addition, $a_3 = 0$.

Finally, looking at one-points of $w(z)$, we substitute (36.7) into $(\widetilde{P}_5)$. If $\delta \neq 0$, any one-point at z_0, say, is simple and $a_1^2 + 2\delta e^{2z_0} = 0$, while for $\delta = 0$, z_0 is a double point and $2a_2 + \gamma e^{z_0} = 0$.

To find out a representation for fifth Painlevé transcendents $w(z)$ as a quotient of two entire functions, we need to consider the modified equation $(\widetilde{P}_5)$. To this end, we have to find out an entire function $u(z)$ whose zeros are of multiplicity high enough to cancel the poles of $w(z)$. We may use the meromorphic function $G(z) := \frac{1}{2}V(z)$ for such a construction:

At a pole z_0 of $w(z)$, which is simple for $\alpha \neq 0$ and double for $\alpha = 0$, we infer from

$$G'(z) = \frac{1}{2}V'(z) = \frac{\gamma e^z}{w - 1} + \frac{2\delta e^{2z} w}{(w - 1)^2}$$

that G', hence G as well, is analytic in a neighborhood of z_0. The same situation applies at the zeros of $w(z)$. Looking then at an arbitrary one-point of $w(z)$, it is easy to deduce that

$$\frac{1}{2}V'(z) = -\frac{1}{(z - z_0)^2} + \cdots$$

whenever $\delta \neq 0$, and so

$$G(z) = \frac{1}{2}V(z) = \frac{1}{z - z_0} + \cdots .$$

On the other hand, if $\delta = 0$, then

$$\frac{1}{2}V'(z) = -\frac{2}{(z - z_0)^2} + \cdots ,$$

hence

$$G(z) = \frac{2}{(z - z_0)} + \cdots .$$

Towards the quotient representation, we now consider

$$H_\delta(z) := G(z) - \frac{w'}{w-1} = \frac{1}{2}V(z) - \frac{w'}{w-1}. \tag{36.8}$$

By the above construction, $H_\delta(z)$ is meromorphic, being analytic at all zeros and one-points of $w(z)$, and having poles exactly at the poles of $w(z)$. Moreover, the multiplicity of a pole of $w(z)$ is either $\mu = 1$ or $\mu = 2$, depending on whether $\delta \neq 0$ or $\delta = 0$, while the residue of $w'/(w-1)$ at these points equals to μ. Hence, there exists an entire function u such that $H_\delta(z) = u'(z)/u(z)$, and so $v = wu$ is entire as well.

We now construct a pair of differential equations satisfied by u and v. In fact, differentiating $H_\delta(z)$, using (36.8) and (36.2), and substituting $w = v/u$, we get

$$\begin{aligned} u'' = \big[&u^2(v')^2 - 2uvu'v' + v(2u-v)(u')^2 + 2(u-v)^2(\alpha v^2 + \beta u^2) \\ &- 2\gamma e^z u^2 v(u-v) + 2\delta e^{2z}u^2v^2\big]/\big(2uv(u-v)\big). \end{aligned} \tag{36.9}$$

Observing that $v'' = (wu)''$, and using (36.9), we obtain

$$\begin{aligned} v'' = \big[&-v^2(u')^2 + 2uvu'v' + u(u-2v)(v')^2 \\ &+ 2(u-v)^2(\alpha v^2 + \beta u^2) - 2\delta e^{2z}u^2v^2\big]/\big(2uv(u-v)\big). \end{aligned} \tag{36.10}$$

To illustrate the existence of one-parameter families of solutions for (P_5) by concrete examples, consider the Riccati differential equation

$$zw' = \sqrt{2\alpha}w^2 + (\sqrt{-2\delta}z - \sqrt{2\alpha} - \sqrt{-2\beta})w + \sqrt{-2\beta} \tag{36.11}$$

with some preassigned choices for the square roots in (36.11). Then one may verify by direct computation that all solutions of (36.11) satisfy (P_5) provided the parameters in (P_5) satisfy the condition

$$\gamma = \sqrt{-2\delta}(1 + \sqrt{-2\beta} - \sqrt{2\alpha}). \tag{36.12}$$

In fact, this now enables us to express the solutions in these one-parameter families in an explicit form. Using the notations $a := \sqrt{2\alpha}$, $b := \sqrt{-2\beta}$, $d := \sqrt{-2\delta}$, with the same square root branches as to above, and letting $F(A, B, z)$ stand for the confluent hypergeometric function, see below and Rainville [1], the general solution of (36.11) may be written as

$$\begin{aligned} w(z) = -\frac{1}{q(z)}\Big(&\frac{adC}{-1+a-b}z^{1-a}F(1-a, 2-a+b, dz) \\ &- aCz^{-a}F(-a, 1-a+b, dz) \\ &+ \frac{bd}{-1-a+b}z^{1-b}F(1-b, 2+a-b, dz) \\ &- bz^{-b}F(-b, 1+a-b, dz)\Big), \end{aligned} \tag{36.13}$$

provided that $a \neq 0, a-b-1 \notin \mathbb{Z}$, where $q(z) := aCz^{-a}F(-a, 1-a+b, dz) + az^{-b}F(-b, 1+a-b, dz)$, and C is an arbitrary complex parameter. Obviously, (36.13) possesses a non-algebraic branch point at $z=0$, provided α, respectively β, is chosen suitably. The expression (36.13) is obtained by the following procedure. Putting $w = -a^{-1}zy'/y$, (36.11) is taken into the equation

$$z(zy')' - (dz-a-b)(zy') + aby = 0.$$

By the further change of variables $y = z^{-b}Y$, $dz = t$, we obtain the Kummer equation

$$t\frac{d^2Y}{dt^2} + (a-b+1-t)\frac{dY}{dt} + bY = 0,$$

which admits linearly independent solutions

$$F(-b, a-b+1, t), \quad t^{b-a}F(-a, b-a+1, t)$$

with

$$F(A, B, t) = \sum_{n=0}^{\infty} \frac{(A)_n}{(B)_n n!} t^n, \quad (A)_n = \Gamma(A+n)/\Gamma(A).$$

Using these solutions yields (36.13). By the change of variables $Y = t^{-(a-b+1)/2}e^{t/2}Z$, the Kummer equation above is transformed into the Whittaker equation

$$\frac{dZ^2}{dt^2} + \left(-\frac{1}{4} + \frac{a+b+1}{2t} + \frac{1-(a-b)^2}{4t^2}\right)Z = 0.$$

As our final preliminary observation to (P_5), we again recall the connection of non-linear autonomous systems of differential equations to Painlevé equations, following Adler [1]. Assuming that $\delta \neq 0$, we may consider the system

$$\begin{cases} f_2' + f_1' = f_2^2 - f_1^2 + \alpha_1 \\ f_3' + f_2' = f_3^2 - f_2^2 + \alpha_2 \\ f_4' + f_3' = f_4^2 - f_3^2 + \alpha_3 \\ f_1' + f_4' = f_1^2 - f_4^2 + \alpha_4 \end{cases} \tag{36.14}$$

with constraints $f_1 + f_2 + f_3 + f_4 = \frac{1}{2}\bar{\alpha}z$, $\bar{\alpha} = \alpha_1 + \alpha_2 + \alpha_3 + \alpha_4$. By substituting $\zeta := z^2$, $g := f_1 + f_2$, $g_1 := f_2 + f_3$, $y(\zeta) = (2g(z) - \bar{\alpha}z)/(2g(z))$, $\phi(\zeta) = z(2g_1(z) - \frac{1}{2}\bar{\alpha}z)$ into (36.14), we obtain

$$\begin{cases} y'(\zeta) = \dfrac{\alpha_1 + \alpha_3}{\bar{\alpha}\zeta}(y(\zeta) - 1) - \dfrac{1}{2\zeta}\phi(\zeta)y(\zeta) + \dfrac{\alpha_1}{\bar{\alpha}\zeta}(y(\zeta)-1)^2 \\ \phi'(\zeta) = \left(\dfrac{\bar{\alpha}^2\zeta}{16} - \dfrac{\phi(\xi)^2}{4\zeta}\right)\dfrac{y(\zeta)+1}{y(\zeta)-1} + \dfrac{\alpha_1 + \alpha_3}{\bar{\alpha}z}\phi(\zeta) + \dfrac{\alpha_2 - \alpha_4}{2}. \end{cases}$$

Eliminating now ϕ from these equations we observe that y satisfies (P_5) with parameter values $\alpha := \alpha_1^2/(2\bar{\alpha}^2)$, $\beta := -\alpha_3^2/(2\bar{\alpha}^2)$, $\gamma := \frac{1}{4}(\alpha_4 - \alpha_2)$, $\delta := -\frac{1}{32}\bar{\alpha}^2$.

§37 The behavior of solutions near $z = 0$ and $z = \infty$

This section is devoted to considering the behavior of solutions of (P_5) in a neighborhood of the singular points $z = 0$ and $z = \infty$. We proceed to obtain necessary and sufficient conditions for the existence of solutions which are meromorphic at $z = 0$.

(1) We first work out necessary conditions for the existence of a solution analytic at $z = 0$. As in §30 above, we seek the formal expansion of such solutions as a Taylor series

$$w(z) = \sum_{j=0}^{\infty} a_j z^j. \tag{37.1}$$

Substituting (37.1) into the equation (P_5) and comparing the coefficients of the corresponding powers of z, we get the following system of equations:

$$\begin{aligned}
&(a_0 - 1)^3(\beta + \alpha a_0^2) = 0,\\
&(a_0 - 1)\big(3\beta(-1 + a_0) + a_0\big(-1 + \alpha(2 - 7a_0 + 5a_0^2)\big)\big)a_1 = -\gamma a_0^2(a_0 - 1),\\
&2(a_0 - 1)\big(-3\beta + (-4 + 2\alpha + 3\beta)a_0 - 7\alpha a_0^2 + 5\alpha a_0^3\big)a_2\\
&\quad = \sigma_2(a_0, a_1, \alpha, \beta, \gamma, \delta),\\
&\quad \dots,
\end{aligned} \tag{37.2}$$

where $\sigma_2(a_0, a_1, \alpha, \beta, \gamma, \delta) := -2\delta a_0^2(a_0 + 1) - a_1\big(2\gamma a_0(3a_0 - 2) + (a_0 - 1)(6\beta - 1 + 2\alpha(1 - 8a_0 + 10a_0^2))a_1\big)$ is a polynomial in terms of the parameters of the equation and of the coefficients a_0, a_1 in (37.1).

By the first equation in (37.2), two possibilities may take place: (1) $\beta + \alpha a_0^2 = 0$ or (2) $a_0 = 1$.

In the first case, suppose first that $a_0 \neq 0$. Since $\beta = -\alpha a_0^2$, we get the following system:

$$\begin{cases}
a_0(a_0 - 1)(1 - 2\alpha(a_0 - 1)^2)a_1 = \gamma a_0^2(a_0 - 1),\\
a_0(a_0 - 1)(n^2 - 2\alpha(a_0 - 1)^2)a_n\\
\quad = \sigma_n(a_0, \dots, a_{n-1}, \alpha, \beta, \gamma, \delta), \quad n \geq 2,
\end{cases} \tag{37.3}$$

where σ_n is a polynomial in the parameters of (P_5) and of the coefficients $a_0, \dots, a_{n-1}$ of the expansion (37.1).

By (37.3), two subcases appear again: either $a_0(a_0 - 1) \neq 0$, or $a_0(a_0 - 1) = 0$. In the first case, when $a_0(a_0 - 1) \neq 0$, the validity of $n^2 - 2\alpha(a_0 - 1)^2 \neq 0$ for all $n \in \mathbb{Z}$ implies that all coefficients a_k may be expressed in terms of the parameters of (P_5) and of the preceding coefficients $a_0, \dots, a_{k-1}$, implying further that the formal solution (37.1) exists. Moreover, if $\alpha \neq 0$ and there exists $n \in \mathbb{N}$ for which $n^2 - 2\alpha(a_0 - 1)^2 = 0$ holds, then the relation $\sigma_n(a_0, \dots, a_{n-1}, \alpha, \beta, \gamma, \delta) = 0$ is necessary for the existence of a series (37.1). In this case there exists a one-parameter family of formal solutions (37.1) which depends on an arbitrary constant a_n. On the other hand, if $\alpha = 0$, a necessary condition for the existence of a formal solution (37.1) is $\beta = 0$. Hence, the coefficient $a_0 \neq 0, 1$ in (37.1) may be chosen arbitrarily. On the other hand, if $a_0(a_0 - 1) = 0$, two further subcases $a_0 = 0$ and $a_0 = 1$ have to be considered. In the subcase $a_0 = 0$, it is necessary for the existence of (37.1) that $\beta = 0$. Then putting $w(z) = \sum_{n=k}^{\infty} a_n z^n$, $a_k \neq 0$, $k \geq 1$, we have

$$\begin{aligned} &(2\alpha - k^2)a_k^2 = 0, \\ &\big(2\alpha - (m^2 - km + k^2)\big)a_k a_m = \sigma_m(a_k, \dots, a_{m-1}, \alpha, \beta, \gamma, \delta), \end{aligned}$$

where $m > k$. This implies $\alpha = k^2/2$, and the coefficients are uniquely expressed in terms of a_k and the parameters α, β, γ, δ.

Let now $a_0 = 1$. Then to determine the coefficients a_n in (37.1) we get the following system of equations:

$$\begin{cases} \gamma a_1 + 2\delta = 0, \\ (\gamma^2 + 2(n-1)^2\delta)a_n = \sigma_n(a_0, a_1, \dots, a_{n-1}, \alpha, \beta, \gamma, \delta), \quad n = 2, \dots. \end{cases} \tag{37.4}$$

If now $\gamma^2 \neq -2(n-1)^2\delta$ for all $n \geq 1$, then the formal solution (37.1) exists as all coefficients are uniquely expressed in terms of the parameters of the equation (P_5). If $\gamma \neq 0$ and there exists some n for which $\gamma^2 = -2(n-1)^2\delta$, then the condition $\sigma_n = 0$ is necessary to obtain (37.1). In this case, there exists a one-parameter family of formal solutions (37.1), which depends on an arbitrary constant a_n. If $\gamma \neq 0$, $\delta = 0$, $a_0 = 1$ and $a_1 = 0$, then the solution (37.1) reduces to $w = 1$. Finally, if $\gamma = 0$, then by the first equation of (37.4), $\delta = 0$ as well. This special case was already considered in §36, and will be omitted here.

Before the proof of the convergence of the formal solutions constructed above, we note the following fact. Suppose that the equation

$$z(zu')' = F_2(z, u)(zu')^2 + F_1(z, u)(zu') + F_0(z, u) \tag{37.5}$$

admits a formal solution $u = U(z) = \sum_{j=1}^{\infty} c_j z^j$, where $F_j(z, u)$, $j = 0, 1, 2$ are analytic functions around $(z, u) = (0, 0)$ satisfying $F_0(0, 0) = 0$. Putting $v = zu'$, we obtain the system of equations

$$\begin{cases} zu' = v, \\ zv' = F_2(z, u)v^2 + F_1(z, u)v + F_0(z, u), \end{cases}$$

admitting a formal solution $(u, v) = (U(z), zU'(z))$. Then, by Theorem A.12, we see that the formal series $U(z)$ converges. In what follows, we show that all the cases treated above are reduced to (37.5).

Consider the formal solution $W(z) = \sum_{j=0}^{\infty} a_j z^j$ with $a_0 \neq 0, 1$. Then, by $u = w - a_0$, (P_5) takes the form

$$z(zu')' = \left(\frac{1}{2(u + a_0)} + \frac{1}{u + a_0 - 1}\right)(zu')^2 + G_0(z, u).$$

It is easy to check that $G_0(0, 0) = 0$, by using $\beta + \alpha a_0^2 = 0$, and that this has a formal solution $u = \sum_{j=1}^{\infty} a_j z^j$. Hence the convergence of $W(z)$ follows. In case $a_0 = 0$, note that $\beta = 0$. By the transformation $z = t^2$, $w = u^2$, the equation (P_5) takes the form

$$2t(tu')' = \frac{4u}{u^2 - 1}(tu')^2 + 4\alpha u(1 - u^2)^2 + 4\gamma t^2 u + 4\delta t^2 \frac{u(u^2 + 1)}{u^2 - 1},$$

which admits a formal solution $U(t) = W(t^2)^{1/2} = \sum_{j=1}^{\infty} \tilde{a}_j t^j$. From this fact the convergence of $W(z)$ follows. Finally, consider the case where $a_0 = 1$. Write (P_5) in the form

$$\begin{aligned} 2z(zw')'w(w - 1) = {} & (3w - 1)(zw')^2 + 2(w - 1)^3(\alpha w^2 + \beta) \\ & + 2\gamma z w^2(w - 1) + 2\delta z^2 w^2(w + 1). \end{aligned} \tag{37.6}$$

If $\gamma \neq 0, \delta = 0$, then $a_1 = 0$, and hence we may put $W(z) = 1 + a_l z^l + \cdots$, $l \geq 2$. Substituting this into (37.6), and comparing the terms of lowest order, we have $W(z) \equiv 1$. Let us consider the case where $\gamma\delta \neq 0$. We put $w = 1 + z(-2\delta/\gamma + u)$, and substitute into (37.6). Then we can check that the constant term vanishes, and we obtain the equation

$$z(zu')' = G_2(z, u)(zu')^2 + G_1(z, u)zu' + G_0(z, u),$$

where $G_j(z, u)$ $j = 0, 1, 2$ are rational functions of (z, u) analytic at $(0, 0)$ and satisfying $G_0(0, 0) = 0$. Since this equation admits the formal solution $u = \sum_{j=2}^{\infty} a_j z^{j-1}$, the convergence of $W(z)$ immediately follows. Denoting now $w_0 = w(0)$, $w_0' = w'(0)$, we have proved

Theorem 37.1. *The fulfillment of any of the following conditions is necessary and sufficient for the existence of a solution $w(z)$ of (P_5) analytic at $z = 0$:*

(a) $w_0 \neq 0, 1$, $\alpha w_0^2 + \beta = 0$, $\alpha \neq 0$, $(n^2 - 2\alpha(w_0 - 1)^2) \neq 0$ *for any* $n \in \mathbb{N}$, $w_0' = -\gamma w_0/(2\alpha(w_0 - 1)^2) - 1)$;

(b) $w_0 \neq 0, 1$, $\alpha w_0^2 + \beta = 0$, $\alpha \neq 0$, $\sigma_n(a_0, \dots, a_{n-1}, \alpha, \beta, \gamma, \delta) = 0$ *for some* $n \in \mathbb{N}$, *such that* $(n^2 - 2\alpha(w_0 - 1)^2) = 0$, $w_0' = -\gamma w_0/(2\alpha(w_0 - 1)^2) - 1)$, *if* $n \neq 1$ *and* w_0' *is arbitrary if* $n = 1$;

(c) $w_0 \neq 0, 1$, $\alpha = \beta = 0$, $w_0' = \gamma w_0$, $w_0 \neq 0, 1$;

(d) $w_0 = 1$, $\gamma w_0' + 2\delta = 0$, $\gamma \neq 0$, $\gamma^2 \neq -2(n - 1)^2\delta$ *for all* $n \in \mathbb{N}$;

(e) $w_0 = 1$, $\gamma w_0' + 2\delta = 0$, $\gamma \neq 0$, $\sigma_n(1, a_1, \dots, a_{n-1}, \alpha, \beta, \gamma, \delta) = 0$ *for some* $n \in \mathbb{N}$, *for which* $\gamma^2 = -2(n - 1)^2\delta$;

(f) $w_0 = \beta = 0$, $\alpha \neq k^2/2$, *in which case the solution vanishes identically;*

(g) $w_0 = w_0' = \beta = 0$, $\alpha = k^2/2$, $k \geq 1$ *and* $w''(0) = \cdots = w^{(k-1)}(0) = 0$, $w^{(k)}(0) \neq 0$ *for some* $k \in \mathbb{N}$;

(h) $w_0 = 1$, $w_0' = 0$, $\delta = 0$, $\gamma \neq 0$, $w = 1$.

We next continue to work out necessary and sufficient conditions for the existence of polar solutions of (P_5) in a neighborhood of $z = 0$. Hence, we have to consider

$$w(z) = \sum_{l=-n}^{\infty} a_l z^l, \quad a_{-n} \neq 0, \quad n \in \mathbb{N}, \tag{37.7}$$

a formal expansion for a solution of (P_5) in a neighborhood of the point $z = 0$. Observing that, by the transformation $w = 1/W$, the coefficients of (P_5) are changed as

$$(\alpha, \beta, \gamma, \delta) \mapsto (-\beta, -\alpha, -\gamma, \delta),$$

we obtain the following result from Theorem 37.1 (g):

Theorem 37.2. *The conditions* $\alpha = 0$, $n^2 + 2\beta = 0$, *where* $n \in \mathbb{N}$, *are necessary and sufficient for the existence of a family of polar solutions* (37.7) *of* (P_5) *in a neighborhood of* $z = 0$.

(2) A similar reasoning as above may be applied in a neighborhood of $z = \infty$ by defining $z = 1/t$, in (P_5). This results in

$$\begin{aligned} 2t^4 w(w-1)w'' &= t^4(3w-1)(w')^2 - 2t^3 w(w-1)w' \\ &\quad + 2t^2(\alpha w^2 + \beta)(w-1)^3 + 2\gamma t w^2(w-1) + 2\delta w^2(w+1). \end{aligned} \tag{37.8}$$

We seek for a formal solution of the equation (37.8) in the form of a series

$$w(t) = \sum_{j=0}^{\infty} a_j t^j. \tag{37.9}$$

By substituting the expansion (37.9) into (37.8) and comparing the coefficients of the powers of t, we get the following system to determine the coefficients:

$$\begin{aligned} &\delta a_0^2(1 + a_0) = 0, \\ &a_0\big(\gamma(-1 + a_0)a_0 + \delta(2 + 3a_0)a_1\big) = 0, \\ &(2\delta a_0 + 3\delta a_0^2)a_2 = -\big(\beta(-1 + a_0)^3 + \alpha(-1 + a_0)^3 a_0^2 - 2\gamma a_0 a_1 \\ &\qquad\qquad + 3\gamma a_0^2 a_1 + \delta a_1^2 + 3\delta a_0 a_1^2\big), \\ &\delta a_0(2 + 3a_0)a_n = P_n(a_0, a_1, \ldots, a_{n-1}, \alpha, \beta, \gamma, \delta). \end{aligned} \tag{37.10}$$

It follows from the first equation (37.10) that if $\delta = 0$, then either (a) $\gamma = 0$ and, consequently, we have the integrable case in §36, (b) $a_0 = 1$, $\gamma \neq 0$, $a_j = 0$ for all $j > 1$ or (c) $a_0 = 0$, $\gamma \neq 0$, $\beta = 0$ and $a_j = 0$ for all $j > 1$.

In the case $\delta \neq 0$, we have either $a_0 = 0$, or $a_0 = -1$. If $a_0 = 0$, then the third equation in (37.10) implies that $\beta = \delta a_1^2$. Hence, if $\beta \neq 0$, there exist two different formal solutions of the form (37.9) in a neighborhood of $t = 0$. The other coefficients a_j, $j \geq 2$, are uniquely determined. If $\beta = 0$, then $w = 0$ is the only solution of (37.8) analytic at $z = 0$.

Let $a_0 = -1$. Then $a_1 = 2\gamma/\delta$, $a_2 = -2\gamma^2/\delta^2 + 8(\alpha + \beta)/\delta$ and all coefficients a_j, $j \geq 3$, are obtained from the equation $\delta a_j = P_k(-1, a_1, \dots, a_{k-1}, \alpha, \beta, \gamma, \delta)$. Hence, when $a_0 = -1$, there exists exactly one formal expansion of the form (37.9) in a neighborhood of $t = 0$. Hence, we have obtained

Theorem 37.3. *The fulfillment of any of the following conditions is necessary and sufficient for the existence of a formal solution of* (P_5) *in power of* z^{-1} *around* $z = \infty$*:*

(a) $\delta = 0$, $\gamma \neq 0$, *in which case* $w(0) = 1$, *resulting in* $w = 1$*;*

(b) $\delta = 0$, $\gamma \neq 0$, $\beta = 0$, $w(0) = 0$, *resulting in* $w = 0$*;*

(c) $\delta \neq 0$, $\beta \neq 0$, $w(0) = 0$, $\delta(w'(0))^2 = \beta$*;*

(d) $\delta \neq 0$, $\beta = 0$, $w(0) = 0$, *resulting in* $w = 0$*;*

(e) $\delta \neq 0$, $w(0) = -1$, $w'(0) = 2\gamma/\delta$.

In each case above the coefficients of the series are determined uniquely up to $w(0)$ *or* $w'(0)$.

By using the transformation $w = 1/W$, we have the following

Theorem 37.4. *The conditions* $\alpha\delta \neq 0$ *and* $a_{-1}^2 = -\delta/\alpha$ *are necessary and sufficient for the existence of a formal polar solution of* (P_5) *in a neighborhood of* $z = \infty$. *The coefficients* a_j, $j \geq 0$, *are uniquely determined up to* a_{-1}.

In general, around the fixed singular point $z = \infty$ of irregular type, the formal series expansions of solutions are divergent, and represent asymptotically true solutions in appropriate sectors, see Takano [1], Yoshida [1] and Garnier [4]. Of course, in exceptional cases such as for rational solutions, they are convergent.

(3) To close this section, we now proceed to obtain necessary conditions under which solutions of (P_5) may have an algebraic branch point at $z = 0$. It is natural to expect that the equation (P_5) has solutions rational in $z^{1/s}$, where s is a natural number. In fact, if the points $z = 0$ and $z = \infty$ are algebraic branch points, the solution $w(z)$ is algebraic, as all other singularities in $\mathbb{C} \setminus \{0\}$ must be poles, see §5. Of course, if $s = 1$, then the algebraic solution is rational. Moreover, the order of the branching of $w(z)$ at the points $z = 0$ and $z = \infty$ is equal to two, whenever $\delta = 0$, since rational solutions of equation (P_3) generate algebraic solutions of the equation (P_5), see §34.

If $\delta \neq 0$, it appears that all algebraic solutions of (P_5) reduce to rational ones. This means that we have to prove that whenever $\delta \neq 0$ and $w(z)$ is an algebraic solution of (P_5), then its branch points at $z = 0$ and $z = \infty$ must be of multiplicity one.

To this end, we assume that $\delta = -1/2$, as we may do without restricting generality. Substituting $z = \tau^s$, (P_5) takes the form

$$\begin{aligned} &6s^2w^4\alpha - 2s^2w^5\alpha + 2s^2\beta + s^2w^3\big(-6\alpha - 2\beta + \tau^s(-2\gamma + \tau^s)\big) \\ &\quad + \tau^2(w')^2 + w^2\big(s^2(2\alpha + 6\beta + \tau^s(2\gamma + \tau^s)) + 2\tau(w' + \tau w'')\big) \\ &\quad - w\big(6s^2\beta + \tau(w'(2 + 3\tau w') + 2\tau w'')\big) = 0. \end{aligned} \tag{37.11}$$

Let now

$$w(\tau) = P(\tau)/Q(\tau) \tag{37.12}$$

be a rational solution of the equation (37.11), where $P(\tau)$ and $Q(\tau)$ are irreducible polynomials of the degrees p and q respectively. Then (37.12) generates an algebraic solution of (P_5). Substituting (37.12) into (37.11) and comparing the coefficients, we find that either (a) $p = q$ and s is arbitrary or (b) $q = p + s$. Observe that we may omit the case $q = p - s$ as (P_5) is invariant with respect to the transformation

$$T_2 : w(z, \alpha, \beta, \gamma, \delta) \mapsto 1/w(z, -\beta, -\alpha, -\gamma, \delta). \tag{37.13}$$

Let us now consider the equation (37.11) at infinity by putting $\tau = 1/t$. Then we obtain the equation

$$\begin{aligned} &-2s^2t^s(-1 + w)w^2\gamma + s^2w^2(1 + w) \\ &\quad - t^{2s}\big(-6s^2w^4\alpha + 2s^2w^5\alpha - 2s^2\beta + 2s^2w^3(3\alpha + \beta) \\ &\quad - t^2(w')^2 - 2w^2(s^2(\alpha + 3\beta) + t(w' + tw''))\big) \\ &\quad - t^{2s}\big(w(6s^2\beta + t(w'(2 + 3tw') + 2tw''))\big) = 0. \end{aligned} \tag{37.14}$$

Consider first the case $p = q$, with an arbitrary branching order s. Then the corresponding solution of (37.14) admits the following expression

$$w(t) = \sum_{k=0}^{\infty} b_k t^k, \quad b_0 \neq 0. \tag{37.15}$$

Substituting this into (37.14), we have $b_0^2(b_0+1)s^2 = 0$, and hence $b_0 = -1$. Then, the coefficients b_k, $k \geq 1$ are uniquely determined. On the other hand, recall the formal solution $\phi(z) = \sum_{j=0}^{\infty} a_j z^{-j}$ of (P_5) in (37.9) with $a_0 = -1$, whose coefficients are uniquely determined as well. Since $w(t) = \phi(t^{-s}) = \sum_{j=0}^{\infty} a_j t^{sj}$ also satisfies (37.14), this must coincide with (37.15), and hence $b_k = 0$ for every $k \notin s\mathbb{Z}$. Next consider the case $q = p + s$. Then, we write (37.15) in the form $w(t) = \sum_{k=s}^{\infty} b_k t^k$, $b_s \neq 0$. Substituting this into (37.14) yields $b_s^2 + 2\beta = 0$. Recalling the formal solution $\psi(z) = \sum_{j=1}^{\infty} a_j z^{-j}$, $a_1^2 + 2\beta = 0$ of (P_5), and considering the solution $w(t) = \psi(t^{-s}) = \sum_{j=1}^{\infty} a_j t^{sj}$ of (37.14), we deduce the same relation $b_k = 0$ for every $k \notin s\mathbb{Z}$. In both cases we conclude that (37.12) admits a series expansion of the form $\sum_{j=0}^{\infty} b_{sj}\tau^{-sj}$ around $\tau = \infty$. Hence, we have proved

Theorem 37.5. *If $\delta = 0$ in the equation (P_5), and if w is an algebraic solution of (P_5), then its branching order at $z = 0$ and $z = \infty$ has to be equal to two. If $\delta \neq 0$, then rational solutions are the only algebraic solutions of (P_5).*

§38 The special case $\delta = 0, \gamma \neq 0$

The fifth Painlevé equation (P_5) in the special case of $\delta = 0, \quad \gamma \neq 0$, written in the form

$$u'' = \frac{3u-1}{2u(u-1)}(u')^2 - \frac{u'}{z} + \frac{(u-1)^2}{z^2}(au + \frac{b}{u}) + \frac{c}{z}u, \tag{38.1}$$

where a, b, c are arbitrary complex parameters, may be reduced to (P_3) with parameters $\gamma\delta \neq 0$. In fact, this has been developed in §34, see Theorem 34.3 and a subsequent remark. Moreover, it is known from the results mentioned above that one solution of (P_3) generates two solutions of (P_5) due to the of $\pm\sqrt{\gamma}$, while one solution of (P_5) generates four solutions of (P_3), due to the parameter branching of $\pm\sqrt{2a}, \pm\sqrt{-2b}$. We take now $c^2 = 1$ in (38.1) without loss of generality. Taking into account the results in §34, connecting (P_3) and (P_5), it is easy to obtain the validity of the following statement, see Gromak [2]:

Theorem 38.1. *Let $u(z)$ be a solution of the equation* (38.1) *with some parameter values $a, b, c^2 = 1$. Then the function*

$$v(z) := 1 - 2\Phi(z)^2/G(z), \tag{38.2}$$

where

$$\Phi(z) := zu' - \sqrt{2a}u^2 + (\sqrt{2a} + \sqrt{-2b})u - \sqrt{-2b} \not\equiv 0, \tag{38.3}$$

and

$$\begin{aligned} G(z) := 2zu\Phi(z)' + \Phi(z)^2 - \big[2zu' - 2(\sqrt{2a} - \sqrt{-2b} - 2)u\big]\Phi(z) \\ + 2zcu^2 \not\equiv 0 \end{aligned} \tag{38.4}$$

is another solution of the equation (38.1) *with the parameter values*

$$a_1 := \frac{(\sqrt{2a}-1)^2}{2}, \quad b_1 := -\frac{(\sqrt{-2b}+1)^2}{2}, \quad c_1 := c. \tag{38.5}$$

Proof. The proof follows immediately from the corresponding theorems in §34, Theorem 34.2 and Theorem 34.3. The condition $R_1(z) \neq 0$ is satisfied provided $G(z) \neq 0$.□

Substituting the expression of $\Phi(z)$ from (38.3) into (38.4), we obtain that

$$\begin{aligned} G(z) &= 2z^2u(u')^2/(u-1) + 4(2a - 1 - \sqrt{2a}u)zuu' + 4au^4 \\ &+ 4(\sqrt{2a} - 3a)u^3 + 4(2a - \sqrt{2a} + \sqrt{-2b}(\sqrt{2a} - 1) + b + cz)u^2 \\ &- 4u\big[\sqrt{-2b}(\sqrt{2a} - 1) + b\big]. \end{aligned} \tag{38.6}$$

Substituting now $\Phi(z)$ and $G(z)$ into (38.1), it is not difficult to verify that (38.1) does not have a one-parameter family of solutions induced by the general solution of a Riccati equation. However, the solutions of

$$\begin{aligned} z^2(\omega')^2 - 2z(\omega^2-\omega)\omega' + \omega^4 - 2\omega^3 + (1+2b+2cz)\omega^2 \\ - 2(2b+cz)\omega + 2b = 0 \end{aligned} \tag{38.7}$$

are, at the same time, solutions of (38.1), when $a = 1/2$. Here, $\omega = (2z(s')^2 - 2\sqrt{-2b}s)/(2z(s')^2 - 2\sqrt{-2b}ss' + cs^2)$, $s = z^{\sqrt{-2b}}Z_{\sqrt{-2b}}(\sqrt{2cz})$, $c \neq 0$ and Z_ν is a cylinder function. Thus, in the case when $\sqrt{2a} \neq 1$, the function $G(z)$ is not vanishing identically which follows from (38.6) and (38.7).

The formulas (38.2)–(38.5) are convenient for constructing exact solutions of (38.1). We first observe that (38.1) may have solutions rational in $\sqrt{z}$. This fact follows from the corresponding theorems in §34 connecting solutions of (P_3) and (P_5) and from the statement that (P_3) has solutions rational in z.

Let us put $t^2 = 2z$ in (38.1) to simplify the calculations. Reverting to write z again in place of t, we get

$$u'' = \frac{3u-1}{2u(u-1)}(u')^2 - \frac{u'}{z} + \frac{4(u-1)^2}{z^2}\left(au + \frac{b}{u}\right) + 2cu, \quad c^2 = 1. \tag{38.8}$$

By direct calculation, we can verify that if $u = \varphi(z,a,b,c)$ is a solution of (38.8), then $\widetilde{\varphi}(z,a,b,c) = \varphi(-z,a,b,c)$, $\widetilde{\varphi}(z,a,b,-c) = \varphi(iz,a,b,c)$, $\widetilde{\varphi}(z,-b,-a,-c) = 1/\varphi(-z,a,b,c)$ are solutions of (38.8) as well. The process of constructing solutions of (38.1) rational in $\sqrt{t}$ is equivalent to constructing rational solutions of (38.8). □

Theorem 38.2. *Necessary and sufficient conditions for the existence of rational solutions of* (38.8) *are that either*

$$a = \frac{(2n-1)^2}{8}, \quad n \in \mathbb{N}$$

or

$$a \neq 0, \quad b = -\frac{(2n-1)^2}{8}, \quad n \in \mathbb{N}.$$

Proof. This follows immediately from similar statements for (P_3) and from the theorems in §34 linking solutions of (P_3) and (P_5). Recall first the condition on the existence of rational solutions in Theorem 35.1. By (34.17), if $c = 1$, respectively $c = -1$, then $\alpha + \beta = 4\sqrt{2a} - 2$ and $\alpha - \beta = -4\sqrt{-2b} - 2$, respectively $\alpha + \beta = 4\sqrt{-2b} + 2$ and $\alpha - \beta = -4\sqrt{2a} + 2$. Hence, under the supposition $c = 1$, respectively $c = -1$, the relation $\alpha + \varepsilon\beta \in 4\mathbb{Z}$ (35.3) holds for some ε, $\varepsilon^2 = 1$, if and only if either $2\sqrt{2a} \in 2\mathbb{Z}+1$ or $2\sqrt{-2b} \in 2\mathbb{Z}+1$ is valid. Furthermore, if

$a = (\beta - \alpha\varepsilon + 2)^2/32 = 0$, then by (34.11), the rational solution satisfies (34.10), and hence $R(z) \equiv 0$ in Theorem 34.1. By these facts, we arrive at the conclusion. □

The relations between the parameters are as follows:

$$2\sqrt{2a} = 2n + 1 - \alpha, \quad 2\sqrt{-2b} = 2n - 1, \quad c = -1, \tag{38.9}$$
$$2\sqrt{2a} = 2n + 1, \quad 2\sqrt{-2b} = 2n - \alpha - 1, \quad c = 1, \tag{38.10}$$
$$2\sqrt{2a} = 2n + 1, \quad 2\sqrt{-2b} = 2n + \alpha - 1, \quad c = -1, \tag{38.11}$$
$$2\sqrt{2a} = 2n + \alpha + 1, \quad 2\sqrt{-2b} = 2n - 1, \quad c = 1. \tag{38.12}$$

Table 38.1. The first rational solutions of the equation (38.8).

a	b	c	$w(z)$
$\frac{\alpha^2}{8}$	$-\frac{1}{8}$	-1	$1 + \frac{2z}{\alpha}$
$\frac{(\alpha-2)^2}{8}$	$-\frac{9}{8}$	-1	$\frac{8z^3 + 12(\alpha-2)z^2 + 6(\alpha-2)^2 z + \alpha(\alpha^2 - 6\alpha + 8)}{(\alpha-2)(4z^2 + 4\alpha z - 8z + \alpha^2 - 4\alpha)}$
$\frac{1}{8}$	$\frac{\alpha^2}{8}$	-1	$\frac{\alpha}{2z + \alpha}$
$\frac{9}{8}$	$\frac{(\alpha-2i)^2}{8}$	-1	$\frac{(\alpha - 2i)(\alpha^2 + 4\alpha(z-i) + 4z(z-2i))}{\alpha^3 + 6\alpha^2(z-i) + 8z(-3 - 3iz + z^2) + 4\alpha(-2 - 6iz + 3z^2)}$
$\frac{\alpha^2}{8}$	$-\frac{1}{8}$	1	$\frac{\alpha + 2iz}{\alpha}$
$\frac{(\alpha-2)^2}{8}$	$-\frac{9}{8}$	1	$\frac{-8iz^3 - 12(\alpha-2)z^2 + 6(\alpha-2)^2 iz + \alpha(\alpha^2 - 6\alpha + 8)}{(\alpha-2)(-4z^2 + 4\alpha iz - 8iz + \alpha^2 - 4\alpha)}$
$\frac{1}{8}$	$\frac{\alpha^2}{8}$	1	$\frac{\alpha}{2iz + \alpha}$

We complete this section by giving an extended example. Considering a special case

$$w'' = \frac{(w')^2}{w} - \frac{w'}{z} + \frac{1}{z}(\alpha w^2 + \beta) + w^3 - \frac{1}{w} \tag{38.13}$$

of (P_3), it is clear that $w(z) = 1$ is a solution provided when $\alpha + \beta = 0$. We may apply Theorem 34.1 that links solutions of (P_3) and (P_5) when $\alpha \neq 1, \varepsilon_1 = 1$. Consequently, we obtain two solutions of (38.8)

$$u_1^+ = \frac{2z+\alpha-1}{\alpha-1}, \quad a = \frac{(\alpha-1)^2}{8}, \quad b = -\frac{1}{8}, \quad c = -1, \quad \varepsilon_1 = 1; \qquad (38.14)$$

$$u_1^- = \frac{\alpha+1}{2z+1+\alpha}, \quad a = \frac{1}{8}, \quad b = -\frac{(\alpha+1)^2}{8}, \quad c = 1, \quad \varepsilon_1 = -1. \qquad (38.15)$$

The solutions u_1^+ and u_1^- now generate solutions w_1^+, w_1^- of (P_3). Taking only one solution $w_1(z) = (z+1)/(z+2)$ with $\alpha_1 = 5$, $\beta_1 = -1$, we obtain two solutions of (38.8)

$$u_1^+ = 1+z, \quad a = \frac{1}{2}, \quad b = -\frac{1}{8}, \quad c = -1, \varepsilon_1 = 1; \qquad (38.16)$$

$$u_1^- = \frac{2z^2+8z+6}{z^3+6z^2+12z+6}, \quad a = \frac{9}{8}, \quad b = -2, \quad c = 1, \quad \varepsilon_1 = -1. \qquad (38.17)$$

Taking $w = i$, when $\alpha - \beta = 0$, as the seed solution of (38.13), we obtain solutions of (38.8) with complex coefficients.

Using the established correspondence between solutions of (P_3) and (P_5), see §34, we shortly state the following properties of solutions of (P_5):

(1) If $c = \pm 1$, $d = 0$, the equation (38.1) has a one-parameter family of solutions expressed in terms of the elementary functions if either

$$a = \frac{m^2}{2} \neq 0, \quad b = -\frac{(2n+1)^2}{8}, \quad c = 1$$

or

$$a = \frac{(2n+1)^2}{8}, \quad b = -\frac{m^2}{2}, \quad c = -1,$$

where $m, n \in \mathbb{Z}$ and $m+n \in 1+2\mathbb{Z}$. This class of solutions is generated by the solution $w(z) = \tan(z+C)$ of (P_3) with parameter values $\alpha = \beta = \gamma = -\delta = 1$. The simplest solution of (38.8) from this class reads as

$$u(z) = 1 - \frac{2z\cos(z+C)^2}{2z+\sin\big(2(z+C)\big)}$$

with parameter values $a = 1/2$, $b = -1/8$, $c = 1$.

(2) The equation (P_3) with $\beta + \alpha\varepsilon = 4n+2$, $\varepsilon^2 = 1$, $n \in \mathbb{Z}$, has one-parameter families of solutions expressed in terms of the Bessel functions. Thus, with the help of the connection between solutions of (P_3) and (P_5), we come to the conclusion that (P_5) with $c = \pm 1$, $d = 0$ has one-parameter solutions expressed in terms of the Bessel functions if either

$$a = \frac{n^2}{2}, \quad n \in \mathbb{N},$$

or

$$b = -\frac{n^2}{2}, \quad n \in \mathbb{N}.$$

Of course, these parameter values contain the values obtained above when the Bessel functions degenerate into trigonometric functions.

Let us now denote the transformation in Theorem 38.1 as $T_{\varepsilon_1,\varepsilon_2} : u(z, a, b, c) \mapsto \widetilde{u}(z, \widetilde{a}, \widetilde{b}, \widetilde{c})$, where $\widetilde{a} = (\sqrt{2a} + \varepsilon_1)^2/2$, $\widetilde{b} = -(\sqrt{-2b} + \varepsilon_2)^2/2$, $\widetilde{c} = c$. Here $\varepsilon_1, \varepsilon_2$ with $\varepsilon_1^2 = \varepsilon_2^2 = 1$ determine the branch of $\sqrt{2a}$ and $\sqrt{-2b}$. As the solution u is admissible for successive applications of the transformation $T_{\varepsilon_1,\varepsilon_2}$, we have the following

Theorem 38.3. *By successive applications of the transformations $T_{\varepsilon_1,\varepsilon_2}$ to a solution $u(z, a, b)$, a solution $\widetilde{u}(z, \widetilde{a}, \widetilde{b})$ can be obtained with parameters*

$$\widetilde{a} = (\sqrt{2a} + n_1)^2/2, \quad \widetilde{b} = -(\sqrt{-2b} + n_2)^2/2, \quad \widetilde{c} = c, \tag{38.18}$$

where $n_1 + n_2 \in 2\mathbb{Z}$. Moreover, for every pair $(n_1, n_2) \in \mathbb{Z} \times \mathbb{Z}$ of such integers a transformation $T_{\varepsilon_1,\varepsilon_2}$ exists which carries (a, b) to $(\widetilde{a}, \widetilde{b})$ as in (38.18).

Proof. The assertion may be proved similarly as to Theorem 25.3 and Theorem 35.3. Indeed, the basic transformations in this case are $T_{1,1}, T_{-1,1}, T_{1,-1}, T_{-1,-1}$. For arbitrary $k_j \in \mathbb{N} \cup \{0\}$, composing of these transformations and the checking closedness, the assertion follows. □

We conclude here by a remark concerning the fundamental domain of (P_5). As in earlier sections, a domain G of the parameters is called a fundamental domain, it is sufficient to know the general solution of (P_5) for all $(a, b, c, d) \in G$ in order to construct the general solution of (P_5) for all parameter values a, b, c, d by using some known transformations. From the fundamental domain for (P_3) and applying the relations between (P_3) and (P_5), we infer that

$$G := \{(a, b, c) \mid 0 \le \sqrt{2a} \le 1,\ 0 \le \sqrt{-2b} \le 2,\ c = \pm 1\}.$$

§39 The Bäcklund transformations of (P_5)

As one may verify by direct computation, the equation (P_5) in the case of $\delta \neq 0$ is equivalent to the following pair of the differential equations:

$$\begin{cases} z\dfrac{du}{dz} = -a - (a + c)u - uv - u^2 v, \\ z\dfrac{dv}{dz} = \delta z^2 - \gamma z + (a + c)v + 2\delta z^2 u + \dfrac{v^2}{2} + uv^2, \end{cases} \tag{39.1}$$

where $a^2 = -2\beta$, $c^2 = 2\alpha$ and the function $u(z)$ is related to $w(z)$ through the transformation $u = w/(1-w)$. Differentiating the second equation of (39.1) and making use of the first equation, we obtain the following second order differential equation for v:

$$v'' = \frac{v}{v^2 + 2\delta z^2}(v')^2 - \frac{1}{z}\frac{v^2 - 2\delta z^2}{v^2 + 2\delta z^2}v' + \frac{1}{z}\theta(v, z), \tag{39.2}$$

where

$$\theta(v, z) = 2\delta z - \gamma - \frac{a(2\delta z^2 + v^2)}{z} - \frac{v(\gamma z - \delta z^2 - (a+c)v - v^2/2)^2}{z(v^2 + 2\delta z^2)}$$
$$+ \left(\frac{4\delta z}{2\delta z^2 + v^2} - \frac{(a+c+v)}{z}\right)\left(\gamma z - \delta z^2 - (a+c)v - \frac{v^2}{2}\right).$$

The equation (39.2) may further be reduced to an equation of type (P_5). In fact, as the function $v/(v^2 + 2\delta z^2)$ has two poles $v = kz$, $k^2 = -2\delta$, it is natural to apply the following transformation:

$$y(z) = \frac{v + kz}{v - kz}, \quad v \neq kz. \tag{39.3}$$

By this transformation, we obtain for $y(z)$ an equation of type (P_5) with parameter values

$$\begin{aligned} \alpha_1 &:= -\frac{1}{16\delta}(\gamma + k(1 - a - c))^2, \\ \beta_1 &:= \frac{1}{16\delta}(\gamma - k(1 - a - c))^2, \\ \gamma_1 &:= k(a - c), \\ \delta_1 &:= \delta. \end{aligned} \tag{39.4}$$

Therefore, taking into account the transformation (39.3), the validity of the following statement may be verified directly, see Gromak [3]:

Theorem 39.1. *Let $w = w(z, \alpha, \beta, \gamma, \delta)$, $\delta \neq 0$, be a solution of (P_5), where*

$$F_1(z) := zw' - cw^2 + (c - a + kz)w + a \not\equiv 0 \tag{39.5}$$

and $c^2 = 2\alpha$, $a^2 = -2\beta$, $k^2 = -2\delta$. Then the function

$$w_1(z, \alpha_1, \beta_1, \gamma_1, \delta_1) = 1 - 2kzw/F_1(z), \tag{39.6}$$

with parameter values (39.4), is a solution of (P_5) as well.

Proof. The following expression immediately follows from the first equation of (39.1):

$$v = [-zu' - a - (a+c)u](u+u^2)^{-1}, \tag{39.7}$$

where $u = -w/(w-1)$ and $v \not\equiv kz$. Substituting the value of v from (39.7) into (39.3), we see that y satisfies (39.6). Moreover, the condition $v \not\equiv kz$ is equivalent to (39.5) in this case. □

Similarly, w_1 may be expressed in terms of w. In fact, the inverse transformation to (39.6) is

$$w = 1 + 2kzw_1/F_2(z), \tag{39.8}$$

where

$$F_2(z) := zw_1' - c_1 w_1^2 + w_1(-a_1 + c_1 - k_1 z) + a_1 \not\equiv 0,$$

where now $c_1^2 = 2\alpha_1$, $a_1^2 = -2\beta_1$, $k_1^2 = -2\delta_1$, and

$$\begin{cases} \alpha = -\dfrac{1}{16\delta_1}\big(\gamma_1 - k_1(1 - a_1 - c_1)\big)^2, \\ \beta = \dfrac{1}{16\delta_1}\big(\gamma_1 + k_1(1 - a_1 - c_1)\big)^2, \\ \gamma = k_1(c_1 - a_1), \\ \delta_1 = \delta. \end{cases}$$

Hence, the formulas (39.6) and (39.8) establish a correspondence between solutions of (P_5) with different parameter values, provided $\delta \neq 0$. As we can see, the inverse transformation (39.8) has the same structure as (39.6) with suitable branches of a_1, c_1, k_1. This connection will be considered below.

It follows from Theorem 39.1 that a given solution of (P_5) generates, in the general case, eight different solutions of (P_5), due to the parameter branching. As for the hierarchy of solutions after repeated applications of the transformation in Theorem 31.1, see the following figure:

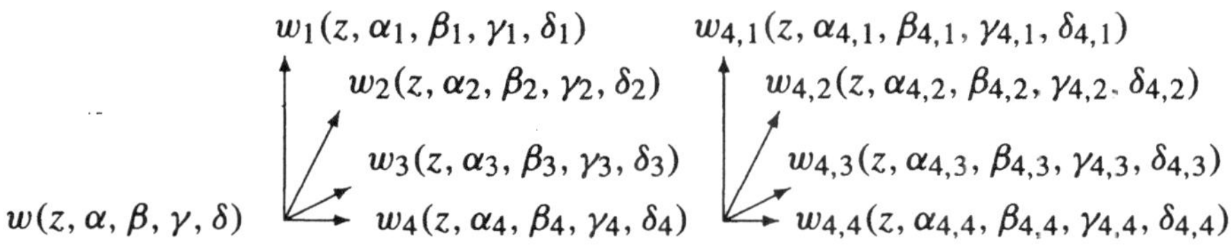

Figure 39.1.

The following restatement of Theorem 39.1 serves to studying the properties of Bäcklund transformations for (P_5) and to deduce nonlinear superposition formulas connecting different solutions of (P_5).

Theorem 39.2. *Let $w = w(z, \alpha, \beta, \gamma, \delta)$, be a solution of ($P_5$) with parameter values $\alpha, \beta, \gamma, \delta \neq 0$, such that*

$$F_1(z) := zw' - \varepsilon_1 c w^2 + (\varepsilon_1 c - \varepsilon_2 a + \varepsilon_3 k z)w + \varepsilon_2 a \not\equiv 0, \qquad (39.9)$$

where $c^2 = 2\alpha$, $a^2 = -2\beta$, $k^2 = -2\delta$. Then the transformation

$$T_{\varepsilon_1,\varepsilon_2,\varepsilon_3} : w(z, \alpha, \beta, \gamma, \delta) \to w_1(z, \alpha_1, \beta_1, \gamma_1, \delta_1) = 1 - 2\varepsilon_3 k z w / F_1(z) \qquad (39.10)$$

determines a solution $w_1(z, \alpha_1, \beta_1, \gamma_1, \delta_1)$ of the equation (P_5) with new parameter values

$$\begin{cases} \alpha_1 = -\dfrac{1}{16\delta}[\gamma + \varepsilon_3 k(1 - \varepsilon_2 a - \varepsilon_1 c)]^2, \\ \beta_1 = \dfrac{1}{16\delta}[\gamma - \varepsilon_3 k(1 - \varepsilon_2 a - \varepsilon_1 c)]^2, \\ \gamma_1 = \varepsilon_3 k(\varepsilon_2 a - \varepsilon_1 c), \\ \delta_1 = \delta, \end{cases} \qquad (39.11)$$

where $\varepsilon_i^2 = 1$, $i = 1, 2, 3$.

A solution $w(z, \alpha, \beta, \gamma, \delta)$ satisfying (39.9) generates eight different first level solutions of (P_5) $w_i(z, \alpha_i, \beta_i, \gamma_i, \delta_i)$, where $i = 1, \dots, 8$, according to the choice of the branches $c = \sqrt{2\alpha}$, $a = \sqrt{-2\beta}$, $k = \sqrt{-2\delta}$ determined by the choice of $\varepsilon_1, \varepsilon_2, \varepsilon_3$, respectively. Solutions generated by n times repeated applications of the transformation in Theorem 39.2 are called the solutions of the nth level. However, the value of ε_3 may be fixed. Therefore, we may consider the solutions of the nth depending on the choice of ε_1 and ε_2 only, fixing $\varepsilon_3 = 1$. The following lemmas essentially serve to prove this statement.

Lemma 39.3. *The change of the parameter ε_3 for the solutions of the first level is equivalent to the inversion $T_{\varepsilon_1,\varepsilon_2,1} w(z) = 1/T_{\varepsilon_1,\varepsilon_2,-1} w(z)$.*

Proof. Let $w(z)$ be a seed solution. Then from (39.9) for solutions of the first level we have

$$T_{\varepsilon_1,\varepsilon_2,\varepsilon_3} w(z) = \frac{zw' - \varepsilon_1 c w^2 + (\varepsilon_1 c - \varepsilon_2 a - \varepsilon_3 k z)w + \varepsilon_2 a}{zw' - \varepsilon_1 c w^2 + (\varepsilon_1 c - \varepsilon_2 a + \varepsilon_3 k z)w + \varepsilon_2 a}.$$

The assertion now immediately follows from this representation. □

Lemma 39.4. $T_{\sigma_1,\sigma_2,\sigma_3} \circ T_{\varepsilon_1,\varepsilon_2,\varepsilon_3} w = T_{\sigma_2,\sigma_1,-\sigma_3} \circ T_{\varepsilon_1,\varepsilon_2,-\varepsilon_3} w$, *where $\sigma_i^2 = \varepsilon_i^2 = 1$, $i = 1, 2, 3$.*

Proof. This follows immediately from Lemma 39.3 and the transformations (39.10), (39.11). In fact, if $\widetilde{w} := T_{\varepsilon_1,\varepsilon_2,\varepsilon_3} w$ and $\breve{w} := T_{\varepsilon_1,\varepsilon_2,-\varepsilon_3} w$, then $\breve{w} = 1/\widetilde{w}$. Thus, $T_{\sigma_2,\sigma_1,-\sigma_3}(\breve{w}) = T_{\sigma_1,\sigma_2,\sigma_3}(1/\widetilde{w})$. □

Observe that the relations in Lemma 39.4 demonstrate the commutativity of the construction of the solutions according to the choice of parameters values under the transformation (39.10), (39.11). Therefore, the second level solutions generated by a double application of the transformation (39.10), (39.11), with a fixed value of ε_3 coincide, provided the parameters $\alpha, \beta, \gamma, \delta$ remain fixed.

Lemma 39.5. *A double application of the transformation in Theorem 39.2 with the following choice of the values of the parameters $\varepsilon_1, \varepsilon_2, \varepsilon_3$ is equivalent to the inversion of the initial solution: $T_{\varepsilon_3,-\varepsilon_3,\varepsilon_3} \circ T_{\varepsilon_1,\varepsilon_2,\varepsilon_3} : w \mapsto w^{-1}$.*

Proof. The proof easily follows by applying of the transformation (39.10), (39.11).□

By using the preceding lemmas, we get the following statement:

Theorem 39.6. *Repeated applications of the transformation in Theorem 39.2 with the following choice of the parameter values $\varepsilon_1, \varepsilon_2, \varepsilon_3 = 1$ result in the following descriptions of the identity transformations:* (1) $I = T_{1,-1,-1} \circ T_{\varepsilon_1,\varepsilon_2,1} : w \mapsto w$ *and* (2) $I = T_{1,-1,1} \circ T_{\varepsilon_1,\varepsilon_2,1} \circ T_{1,-1,1} \circ T_{\varepsilon_1,\varepsilon_2,1} : w \mapsto w$.

Therefore, by virtue of Theorem 39.6, the value of the parameter ε_3 may be fixed and we may take $\varepsilon_3 = 1$ without loss of generality in what follows. Moreover, whenever $\delta \neq 0$ we may always take $\delta = -1/2$ without loss of generality by a gauge transformation.

We now proceed to studying nonlinear superposition formulas which link solutions of the equation (P_5) by repeated applications of the transformation (39.10), (39.11). Observe that the superposition formulas obtained below may be considered as an alternative version of discrete Painlevé equations, see §50. When $\varepsilon_3 = 1$ is fixed, the seed solution $w(z, \alpha, \beta, \gamma, -1/2)$ satisfying (39.9) generates four first level solutions $w_i(z, \alpha_i, \beta_i, \gamma_i, -1/2)$, $i = 1, 2, 3, 4$, of the equation (P_5), depending on the choice of the values of $\varepsilon_1, \varepsilon_2$:

$$T_{1,1,1}w = w_1 = 1 - 2zw(zw' - cw^2 + (c - a + z)w + a)^{-1}, \tag{39.12}$$

$$T_{1,-1,1}w = w_2 = 1 - 2zw(zw' - cw^2 + (c + a + z)w - a)^{-1}, \tag{39.13}$$

$$T_{-1,1,1}w = w_3 = 1 - 2zw(zw' + cw^2 + (-c - a + z)w + a)^{-1}, \tag{39.14}$$

$$T_{-1,-1,1}w = w_4 = 1 - 2zw(zw' + cw^2 + (-c + a + z)w - a)^{-1}. \tag{39.15}$$

Observe that for a general case, we may assume that $a \neq 0$, $c \neq 0$. Indeed, if $a = 0$, we have $w_1 = w_2$ and $w_3 = w_4$ and the superposition formula $1/(1 - w_1) - 1/(1 - w_3) = c(1 - w)/z$ is valid. If $c = 0$, we get $w_1 = w_3$, $w_2 = w_4$ and $1/(1 - w_1) - 1/(1 - w_2) = a(1 - w)/(zw)$. If $a = c = 0$, all first level solutions coincide provided $\varepsilon_3 = 1$ and a subsequent application of the transformation (39.10), (39.11) is possible if $w \neq C \exp(-\sqrt{-2\delta}z)$ for a constant C, assuming $\gamma = \sqrt{-2\delta}$.

Theorem 39.7. *A seed solution and any two first level solutions are algebraically dependent.*

Proof. Eliminating $w'(z)$ from the equations (39.12) through (39.15), we find the following explicit superposition formulas connecting the seed solution $w(z)$ and the first level solutions $w_i(z)$, where $i = 1, 2, 3, 4$.

$$\frac{1}{1-w_1} - \frac{1}{1-w_2} = \frac{a(1-w)}{zw}, \tag{39.16}$$

$$\frac{1}{1-w_1} - \frac{1}{1-w_3} = \frac{c(1-w)}{z}, \tag{39.17}$$

$$\frac{1}{1-w_1} - \frac{1}{1-w_4} = \frac{(1-w)(cw+a)}{zw}, \tag{39.18}$$

$$\frac{1}{1-w_2} - \frac{1}{1-w_3} = \frac{(1-w)(cw-a)}{zw}. \tag{39.19}$$

□

We remark that the solutions w_2 and w_4 are connected in the same way as the solutions w_1 and w_3 are in (39.17), and the solutions w_3 and w_4 as the solutions w_1 and w_2 in (39.16). Moreover, if we take $\varepsilon_3 = -1$, the relations from (39.16) to (39.19) remain are valid by replacing $w_i(z) \mapsto w_{i+4}(z)$ and $z \mapsto -z$. The formulas (39.16) to (39.19) immediately imply that

$$\frac{1}{1-w_1} - \frac{1}{1-w_2} = \frac{1}{1-w_3} - \frac{1}{1-w_4}. \tag{39.20}$$

Theorem 39.8. *Any three first level solutions are algebraically dependent.*

Proof. Eliminating $w(z)$ from the formulas (39.16) to (39.19), we obtain

$$w_4 = \frac{c(w_1-w_2)(w_2-1) + w_2(a(w_1-1)(w_2-1) + (w_1-w_2)z)}{a(w_1-1)(w_2-1) + (w_1-w_2)(c(w_2-1)+z)}, \tag{39.21}$$

$$w_3 = \frac{a(w_1-1)w_1(w_2-1) + (w_1-w_2)(c(w_1-1) + w_1 z)}{a(w_1-1)(w_2-1) + (w_1-w_2)(c(w_1-1)+z)}, \tag{39.22}$$

$$w_4 = \frac{a(w_1-w_3)(w_3-1) + w_3(c(w_1-1)(w_3-1) + (w_3-w_1)z)}{c(w_1-1)(w_3-1) + (w_1-w_3)(a(w_3-1)-z)}, \tag{39.23}$$

$$w_3 = \frac{a(w_2-w_4)(w_4-1) + w_4(c(1-w_2)(w_4-1) + (w_2-w_4)z)}{c(1-w_2)(w_4-1) + (w_2-w_4)(a(w_4-1)+z)}. \tag{39.24}$$

The relations (39.21) to (39.24) now link three arbitrary first level solutions, proving the statement. □

Let now $w_{i,j}(z)$, $i, j = 1, 2, 3, 4$, be the second level solutions obtained by repeated applications of the transformations (39.10) through (39.15) to the seed solution $w(z)$, i.e., $w_{i,j}(z, \alpha_{i,j}, \beta_{i,j}, \gamma_{i,j}, \delta_{i,j}) = T_{\sigma_1,\sigma_2,1} \circ T_{\varepsilon_1,\varepsilon_2,1} w(z, \alpha, \beta, \gamma, \delta)$, where

$\varepsilon_k^2 = \sigma_k^2 = 1$, $k = 1, 2$, see Figure 39.1 (p. 195). Taking into account (39.20) and Lemma 39.5, it is nothing but an exercise to obtain the nonlinear superposition formulas relating the seed solution with the second level solutions:

$$\frac{1}{1-w_{i,1}} + \frac{1}{1-w_{i,4}} + \frac{1}{w_{i,3}-1} = \frac{w}{w-1},$$

where $i = 1, 2, 3, 4$. To deduce nonlinear superposition formulas linking the seed solution with the first and second level solutions, let us denote $w_i := T_{\varepsilon_1,\varepsilon_2,1}w$ and $w_{i,j} := T_{\sigma_1,\sigma_2,1}w_i$. By a direct calculation, we observe that if $w_{i,1} = T_{1,1,1}w_i$, $w_{i,2} = T_{1,-1,1}w_i$, $w_{i,3} = T_{-1,1,1}w_i$, and $w_{i,4} = T_{-1,-1,1}w_i$, then the following relations are true:

$$\frac{w}{w-1} - \frac{1}{1-w_{i,1}} = -a_1\frac{(1-w_i)}{zw_i},$$
$$w_{i,2} = \frac{1}{w},$$
$$\frac{w}{w-1} - \frac{1}{1-w_{i,3}} = \frac{(1-w_i)(c_1w_i - a_1)}{zw_i},$$
$$\frac{w}{w-1} - \frac{1}{1-w_{i,4}} = \frac{c_1(1-w_i)}{z},$$

where the choice of the values of c_1, a_1 is fixed and $c_1 = \sqrt{2\alpha_1}$, $a_1 = \sqrt{-2\beta_1}$.

Note, that we may apply a sequence of Bäcklund transformations to a solution $w(z, \alpha, \beta, \gamma, -1/2)$. Similarly as for the equation (P_4), let us assume that this solution does not belong to the family of solutions generated from solutions of the Riccati equation (36.11). It is clear from (39.6) that this assumption enables us to apply successive Bäcklund transformations to a solution $w(z, \alpha, \beta, \gamma, -1/2)$.

Theorem 39.9. *Successive applications of the Bäcklund transformation $T_{\varepsilon_1,\varepsilon_2,1}$ to a seed solution $w(z, \alpha, \beta, \gamma, -1/2)$ of (P_5) leads to a solution $w_1(z, \alpha_1, \beta_1, \gamma_1, -1/2)$, where the new parameters take one of the following forms:*

$$\begin{aligned}
&(1)\ \alpha_1 = (\sqrt{2\alpha} + n_1)^2/2,\quad \beta_1 = -(\sqrt{-2\beta} + n_2)^2/2,\quad \gamma_1 = \gamma + n_3,\\
&(2)\ \alpha_1 = (\sqrt{-2\beta} + n_2)^2/2,\quad \beta_1 = -(\sqrt{2\alpha} + n_1)^2/2,\quad \gamma_1 = -\gamma - n_3,\\
&(3)\ \alpha_1 = (-\kappa\sqrt{2\alpha} - \nu\sqrt{-2\beta} + \gamma + 2n_1 + 1)^2/8,\\
&\quad\ \ \beta_1 = -(\kappa\sqrt{2\alpha} + \nu\sqrt{-2\beta} + \gamma + 2n_2 - 1)^2/8,\\
&\quad\ \ \gamma_1 = -\kappa\sqrt{2\alpha} + \nu\sqrt{-2\beta} + n_3,\\
&(4)\ \alpha_1 = (\kappa\sqrt{2\alpha} + \nu\sqrt{-2\beta} + \gamma + 2n_2 - 1)^2/8,\\
&\quad\ \ \beta_1 = -(-\kappa\sqrt{2\alpha} - \nu\sqrt{-2\beta} + \gamma + 2n_1 + 1)^2/8,\\
&\quad\ \ \gamma_1 = \kappa\sqrt{2\alpha} - \nu\sqrt{-2\beta} - n_3,
\end{aligned}\tag{39.25}$$

where $s = n_1 + n_2 + n_3 \in 2\mathbb{Z}$, $n_j \in \mathbb{Z}$, $j \in \{1, 2, 3\}$, $\kappa^2 = \nu^2 = 1$. *Moreover, for an arbitrary triple* (n_1, n_2, n_3) *of such integers, and for an arbitrary pair* (κ, ν) *of signs, there exist a composition of transformations* $T_{\varepsilon_1,\varepsilon_2,1}$ *which carries* $w(z, \alpha, \beta, \gamma, -1/2)$ *to* $w_1(z, \alpha_1, \beta_1, \gamma_1, -1/2)$ *with* $(\alpha_1, \beta_1, \gamma_1)$ *given by* (1), (2), (3), (4).

Proof. The proof is similar to the proof of the corresponding Theorem 25.3 for (P_4). In the present case, induction will be applied.

Let $\Lambda_{\varepsilon_1,\varepsilon_2}$, $\varepsilon_1^2 = \varepsilon_2^2 = 1$ denote the transformation of parameters $(\alpha, \beta, \gamma,) \mapsto (\alpha_1, \beta_1, \gamma_1)$, where $\alpha_1, \beta_1, \gamma_1$ are defined by (39.11), i.e.

$$\begin{aligned}\Lambda_{\varepsilon_1,\varepsilon_2} &: (p^2/2, -q^2/2, \gamma)\\ &\mapsto \big((-\varepsilon_1 p - \varepsilon_2 q + \gamma + 1)^2/8, -(\varepsilon_1 p + \varepsilon_2 q + \gamma - 1)^2/8, -\varepsilon_1 p + \varepsilon_2 q\big),\end{aligned}$$

where $p^2 = 2\alpha$, $q^2 = -2\beta$. As usual, the notation $\Lambda^k_{\varepsilon_1,\varepsilon_2}$ means that $\Lambda_{\varepsilon_1,\varepsilon_2}$ has been applied k times, $k \in \mathbb{N} \cup \{0\}$. In this case, the second step of the Bäcklund transformation results in

$$\begin{aligned}\Lambda_{\mu_1,\mu_2} \circ \Lambda_{\varepsilon_1,\varepsilon_2} &: (p^2/2, -q^2/2, \gamma)\\ &\mapsto \Big(\frac{1}{32}\big(\varepsilon_1(\mu_1 - \mu_2 - 2)p + \varepsilon_2(\mu_1 - \mu_2 + 2)q - (\mu_1 + \mu_2)\gamma\\ &\qquad + 2 - \mu_1 + \mu_2\big)^2, -\frac{1}{32}\big(\varepsilon_1(\mu_2 - \mu_1 - 2)p + \varepsilon_2(-\mu_1 + \mu_2 + 2)q\\ &\qquad + (\mu_1 + \mu_2)\gamma + \mu_1 - \mu_2 - 2\big)^2,\\ &\qquad\quad \frac{1}{2}\big((\mu_1 + \mu_2)(\varepsilon_1 p + \varepsilon_2 q) + (-\mu_1 + \mu_2)\gamma - \mu_1 - \mu_2\big)\Big).\end{aligned}$$

As in the case of (P_4), we first we prove, by induction, that specific combinations of the transformation give us the desired result. Let us introduce the following transformations:

$$\begin{aligned}S_\varepsilon &:= \Lambda^2_{1,1} \circ \Lambda_{1,\varepsilon} : (p^2/2, -q^2/2, \gamma) \mapsto (p^2/2, -(q - \varepsilon)^2/2, \gamma - 1),\\ R_\varepsilon &:= \Lambda^2_{-1-,1} \circ \Lambda_{\varepsilon,1} : (p^2/2, -q^2/2, \gamma) \mapsto ((p - \varepsilon)^2/2, -q^2/2, \gamma + 1),\\ S &:= \Lambda_{1,-1} \circ \Lambda_{\varepsilon_1,\varepsilon_2} : (p^2/2, -q^2/2, \gamma) \mapsto (q^2/2, -p^2/2, -\gamma).\end{aligned}$$

The choice of ε, $\varepsilon^2 = 1$ in the transformations S_ϵ and R_ε is independent. These transformations are similar as the transformations S_k in the proof of Theorem 25.3. Indeed, for arbitrary $k_j \in \mathbb{N} \cup \{0\}$, composing the transformations S_1, S_{-1}, R_1, R_{-1}, we obtain

$$S_{(k_1,k_2,k_3,k_4)} : (p^2/2, -q^2/2, \gamma) \mapsto ((p + n_1)^2/2, -(q + n_2)^2/2, \gamma + n_3),$$

where $n_1 = -k_3 + k_4$, $n_2 = -k_1 + k_2$, $n_3 = -k_1 - k_2 + k_3 + k_4$. Hence, for arbitrary n_j with $s = n_1 + n_2 + n_3 \in 2\mathbb{Z}$, we can choose non-negative integers k_j in such a way that the application of $S_{(k_1,k_2,k_3,k_4)}$ to $(p^2/2, -q^2/2, \gamma)$ yields the triplet corresponding to (1) with (n_1, n_2, n_3). Furthermore, to consider the case (3), we note the following:

$$\Lambda_{1,1} : (p^2/2, -q^2/2, \gamma) \mapsto \big((-p-q+\gamma+1)^2/8, -(p+q+\gamma-1)^2/8, -p+q\big).$$

We put $p = \kappa\sqrt{2\alpha}$, $q = \nu\sqrt{-2\beta}$, $\kappa^2 = \nu^2 = 1$. It is easy to see that a suitable composition of $\Lambda_{1,1}$ and $S_{(k_1,k_2,k_3,k_4)}$ yields the case (3) with an arbitrarily prescribed (n_1, n_2, n_3) and (κ, ν) satisfying $n_1 + n_2 + n_3 \in 2\mathbb{Z}$, $\kappa^2 = \nu^2 = 1$. Since, due to S, the cases (2) and (4) are reduced to (1) and (3), respectively, transformations corresponding to (2) and (4) are immediately obtained.

We next enter to verify the closedness. Now,

$$\begin{aligned}\Lambda_{\varepsilon_1,\varepsilon_2} : &\big((p+n_1)^2/2, -(q+n_2)^2/2, \gamma+n_3\big)\\ &\mapsto \big((-\varepsilon_1 p - \varepsilon_2 q + \gamma + N_1)^2/8,\ -(\varepsilon_1 p + \varepsilon_2 q + \gamma + N_2)^2/8,\\ &\qquad (-\varepsilon_1 p + \varepsilon_2 q + N_3)^2/8\big),\end{aligned}$$

where $N_1 = -\varepsilon_1 n_1 - \varepsilon_2 n_2 + n_3 + 1$, $N_2 = \varepsilon_1 n_1 + \varepsilon_2 n_2 + n_3 - 1$, $N_3 = -\varepsilon_1 n_1 + \varepsilon_2 n_2$, where $s = 2n_4$, $n_4 \in 2\mathbb{Z}$. Then $n_3 = 2n_4 - n_1 - n_2$. Hence we have for the constants N_1, N_2, N_3

$$\begin{aligned}N_1 - 1 &= \big(-(1+\varepsilon_1)n_1 - (1+\varepsilon_2)n_2 + 2n_4\big) \in 2\mathbb{Z},\\ N_2 + 1 &= \big((\varepsilon_1 - 1)n_1 + (\varepsilon_2 - 1)n_2 + 2n_4\big) \in 2\mathbb{Z}.\end{aligned}$$

Therefore $N_1 = 2m_1 + 1$, $N_2 = 2m_2 - 1$, $N_3 = m_3$, and $m_1 + m_2 + m_3 = 2(n_4 - n_1) \in 2\mathbb{Z}$. This means that the constants m_1, m_2, m_3 have the same properties as the numbers n_1, n_2, n_3 in (1) of (39.25). Therefore, $\Lambda_{\varepsilon_1,\varepsilon_2}$: (1) $\mapsto$ (3) with the condition $m_1 + m_2 + m_3 \in 2\mathbb{Z}$.

To prove now the closedness of the parameters (3) of (39.25), we apply

$$\begin{aligned}\Lambda_{\varepsilon_1,\varepsilon_2} : &\Big(\frac{1}{8}(-\kappa\sqrt{2\alpha} - \nu\sqrt{-2\beta} + \gamma + 2n_1 + 1)^2,\\ &\quad -\frac{1}{8}(\kappa\sqrt{2\alpha} + \nu\sqrt{-2\beta} + \gamma + 2n_2 - 1)^2,\ -\kappa\sqrt{2\alpha} + \nu\sqrt{-2\beta} + n_3\Big)\\ &\mapsto \Big(\frac{1}{8}\Big(-\frac{1}{2}(\varepsilon_1 - \varepsilon_2 - 2)\kappa\sqrt{2\alpha} - \frac{1}{2}(\varepsilon_1 - \varepsilon_2 + 2)\nu\sqrt{-2\beta}\\ &\quad + \frac{1}{2}(\varepsilon_1 + \varepsilon_2)\gamma + N_1\Big)^2, -\frac{1}{8}\Big(\frac{1}{2}(-\varepsilon_1 + \varepsilon_2 - 2)\kappa\sqrt{2\alpha}\\ &\quad + \frac{1}{2}(-\varepsilon_1 + \varepsilon_2 + 2)\nu\sqrt{-2\beta} + \frac{1}{2}(\varepsilon_1 + \varepsilon_2)\gamma + N_2\Big)^2,\\ &\qquad \frac{1}{2}(\varepsilon_1 + \varepsilon_2)(\kappa\sqrt{2\alpha} + \nu\sqrt{-2\beta}) + \frac{1}{2}(-\varepsilon_1 + \varepsilon_2)\gamma + N_3\Big),\end{aligned}$$

with $N_1 = \varepsilon_1 n_1 + \varepsilon_2 n_2 - n_3 + (\varepsilon_1 - \varepsilon_2)/2 - 1$, $N_2 = \varepsilon_1 n_1 + \varepsilon_2 n_2 + n_3 + (\varepsilon_1 - \varepsilon_2)/2 - 1$ and $N_3 = -\varepsilon_1 n_1 + \varepsilon_2 n_2 - (\varepsilon_1 + \varepsilon_2)/2$, $n_1 + n_2 + n_3 \in 2\mathbb{Z}$.

(a) If $\varepsilon_1\varepsilon_2 = -1$, then we have $\Lambda_{\varepsilon_1,\varepsilon_2}$: (3) $\mapsto$ (1) or (2). For example, if $\varepsilon_1 = -1$, $\varepsilon_2 = 1$, then $m_1 = N_1\kappa/2 = (\kappa/2)(-n_1 + n_2 - n_3 - 2)$, $m_2 = N_2\nu/2 = (\nu/2)(-n_1 + n_2 + n_3 - 2)$ and $m_3 = N_3 = n_1 + n_2$ are integers such that $m_1 + m_2 + m_3 = (-\kappa - \nu + 2)n_1/2 + (\kappa + \nu + 2)n_2/2 + (-\kappa + \nu)n_3/2 - \kappa - \nu \equiv -(\kappa + \nu)n_1 - (\kappa + 1)n_3 - (\kappa + \nu) \in 2\mathbb{Z}$.

(b) If $\varepsilon_1\varepsilon_2 = 1$, then we have $\Lambda_{\varepsilon_1,\varepsilon_2}$: (3) $\mapsto$ (3). For example, if $\varepsilon_1 = \varepsilon_2 = 1$, then $2l_1 + 1 = N_1 = n_1 + n_2 - n_3 - 1$, $2l_2 - 1 = N_2 = n_1 + n_2 + n_3 - 1$ and $l_3 = N_3 = -n_1 + n_2 - 1$, where $l_1 + l_2 + l_3 = 2n_2 - 2 \in 2\mathbb{Z}$, completing the proof. □

Theorem 39.9 permits us to construct auto-Bäcklund transformations that link different solutions of (P_5) with the same parameter values. There exist several trivial auto-Bäcklund transformations for (P_5), namely:

$$\begin{aligned}
S_0 &: w(z, \alpha, \beta, 0, \delta) \mapsto w(-z, \alpha, \beta, 0, \delta),\\
S_1 &: w(z, \alpha, -\alpha, 0, \delta) \mapsto 1/w(z, \alpha, -\alpha, 0, \delta),\\
S_2 &: w(z, \alpha, -\alpha, \gamma, \delta) \mapsto 1/w(-z, \alpha, -\alpha, \gamma, \delta),\\
S_3 &: w(z, \alpha, \beta, 0, 0) \mapsto w(\lambda z, \alpha, \beta, 0, 0).
\end{aligned}$$

The last transformation S_3 will be not considered in what follows, as it is concerned with the integrable case in §36. In general, the application of S_0, S_1, S_2 yields new solutions, as is clear from the following examples:

(1) $w = z - a, \alpha = 1/2, \gamma = a + 2, a^2 = -2\beta, a \neq 1, \delta = -1/2$;

(2) $w = kz + 1, \gamma \neq 0, \alpha = -\beta = 1/2, k^2 = -2\delta$.

We now apply compositions $T^{-k} \circ S_j \circ T^k$, $k \in \mathbb{Z}$, $j = 0, 1, 2$. This results in the following statement:

Theorem 39.10. *The Painlevé equation* (P_5) *with* $\delta = -1/2$ *admits auto-Bäcklund transformations of the form* $T^{-k} \circ S_j \circ T^k$ *with* $j = 0$ *and* $j = 2$, *respectively, when*

(1) $\gamma = n, n \in \mathbb{Z}$ *for* $j = 0$ *and*

(2) $-2\beta = (\sqrt{2\alpha} + n)^2, n \in \mathbb{N} \cup \{0\}$ *for* $j = 2$,

and auto-Bäcklund transformations of the form $T^{-k} \circ S_1 \circ T^k$, *when either*

(3) $\gamma = m, -2\beta = (\sqrt{2\alpha} + k)^2$, *where* $k \in \mathbb{N} \cup \{0\}$, $m \in \mathbb{Z}$ *and* $k + m \in 2\mathbb{Z}$ *or*

(4) $\gamma \notin \mathbb{Z}, 8\alpha = (2k - 1)^2, -8\beta = (2m - 1)^2, k, m \in \mathbb{N}$.

Proof. The proof of this theorem is easily deduced by combining the general structure of the parameters in the proof of Theorem 39.9 with the substitution of the parameters of S_0, S_1, S_2 into (39.25). □

Finally, we add a remark concerning the fundamental domain for (P_5) when $\delta = -1/2$. Define $X := (x, y, t)$, $x := \sqrt{2\alpha}$, $y := \sqrt{-2\beta}$, $t := \gamma$, $X_1 := (\sqrt{2\alpha_1}, \sqrt{-2\beta_1}, t_1)$, $L := (e_1/2, v_1/2, 0)$, $e_1^2 = v_1^2 = e^2 = v^2 = 1$,

$$A := \begin{pmatrix} -\frac{ee_1}{2} & -\frac{e_1 v}{2} & \frac{e_1}{2} \\ -\frac{ev_1}{2} & -\frac{vv_1}{2} & -\frac{v_1}{2} \\ -e & v & 0 \end{pmatrix}.$$

Then a Bäcklund transformation for the parameters (3) from (39.25) may be written in the form $X_1 = AX + L$, where $\det A = -ee_1vv_1 = \pm 1$. By Theorem 39.9 on repeated applications of Bäcklund transformations with parameters (3) from (39.25) we infer that the fundamental domain for (P_5) in the case $\delta = -1/2$ may be determined as

$$G := \{(x, y, t) \mid 0 \leqq x \leqq 1,\ 0 \leqq y \leqq 1,\ 0 \leqq t \leqq 1,\\ x + y \leqq 1,\ y + t \leqq 1,\ t + x \leqq 1\}.$$

§40 Rational and one-parameter families of solutions

In §36, we already observed that solutions of certain Riccati differential equations, see (36.11), generate one-parameter families of solutions of (P_5). Recalling Theorem 39.1, let us consider this differential equation

$$zw' - c_0 w^2 + (c_0 - a_0 + k_0 z)w + a_0 = 0, \tag{40.1}$$

where $c_0^2 = 2\alpha$, $a_0^2 = -2\beta$, $k_0^2 = -2\delta$, $\gamma = k_0(a_0 + c_0 - 1)$ more closely. Of course, we assume that the values of c_0, a_0, k_0 have been fixed. Taking $a = a_0 \neq 0$, $c = -c_0 \neq 0$, $k = -k_0$, $\delta = -1/2$ in the Bäcklund transformation (39.5), (39.6), we get the following parameter values and the following transformation:

$$\alpha_1 = \frac{1}{2}, \quad \beta_1 = -\frac{(c_0 + a_0)^2}{2}, \quad \gamma_1 = k_0(a_0 - c_0), \quad \delta_1 = -1/2, \tag{40.2}$$

$$w_1 = \frac{c_0 w^2 + (a_0 - c_0)w - a_0}{c_0 w^2 + (a_0 - c_0 - z)w - a_0}. \tag{40.3}$$

Hence, we obtain a new one-parameter family of solutions by substituting (39.8) into (40.1), resulting in

$$\begin{aligned}(w_1')^2 = {} & \frac{2(w_1^2 - w_1)}{z} w_1' - \frac{w_1^4}{z^2} + \frac{2w_1^3}{z^2} - \frac{1 + 2\beta_1 + 2\gamma_1 z - z^2}{z^2} w_1^2 \\ & + \frac{4\beta_1 + 2\gamma_1 z}{z^2} w_1 - \frac{2\beta_1}{z^2}.\end{aligned} \tag{40.4}$$

All solutions of (40.4) may be expressed in terms of the Whittaker functions and their derivatives, as one can easily see by (40.2) and (40.3). Applying the transformation T_2, see (37.14), to (40.4), we again find a new one-parameter family of (P_5)-solutions for the case

$$\alpha_2 = \frac{(c_0 + a_0)^2}{2}, \quad \beta_2 = -\frac{1}{2}, \quad \gamma_2 = -k(a_0 - c_0), \quad \delta_2 = -1/2. \tag{40.5}$$

The next step of application of the Bäcklund transformation results in

$$(w')^3 + P_1(z,w)(w')^2 + P_2(z,w)w' + P_3(z,w) = 0,$$

where

$$\begin{aligned}
P_1(z,w) &:= \frac{1}{z}\big(cw^2 + (kz - a - c)w + a\big),\\
P_2(z,w) &:= \frac{1}{z^2}\Big(-c^2w^4 + 2w^3\big(-2 + c^2 + a(2+c) - 2kz - c(2+kz)\big)\\
&\qquad + w^2(8 - 8a - a^2 + 8c - 4ac - c^2 + 2akz + 2ckz - k^2z^2)\\
&\qquad + 2w\big(a^2 - 2(1 + c - kz) + a(2 + c - kz)\big) - a^2\Big),\\
P_3(z,w) &:= \frac{1}{z^3}\Big(-c^3w^6 + cw^5\big(-4 - 4c + 3c^2 + a(4+3c) - 4kz - 3ckz\big)\\
&\qquad + w^4\big(-3c^3 - a^2(4+3c) + 6c^2(2+kz) - 4kz(3+kz)\\
&\qquad - 3c(-4 + k^2z^2) + a(4 - 9c^2 + 8kz + c(-8 + 6kz))\big)\\
&\quad + w^3\big(a^3 + c^3 + 24kz - k^3z^3 + 3a^2(4 + 3c - kz) - 3c^2(4+kz)\\
&\qquad + 3c(-4 + 4kz + k^2z^2) + 3a(-4 + 3c^2 - 4kz - 4ckz + k^2z^2)\big)\\
&\quad + w^2\big(-3a^3 - 3a^2(4 + 3c - 2kz) + 4(c + c^2 - 2ckz + kz(kz-3))\\
&\qquad + a(12 - 3c^2 - 3k^2z^2 + c(8 + 6kz))\big)\\
&\quad + aw\big(3a^2 + a(4 + 3c - 3kz) - 4(1 + c - kz)\big) - a^3\Big).
\end{aligned}$$

This equation determines solutions of (P_5) for $\gamma = k(a - c - 3)$, where $\alpha = c^2/2$, $\beta = -a^2/2$, $\delta = -k^2/2$.

More generally, we can prove

Theorem 40.1. *Equation (P_5) with $\delta = -1/2$ has one-parameter families of solutions expressed in terms of the Whittaker functions and their derivatives when either*

$$-2\beta = (\sqrt{2\alpha} + 2n - 1 + \eta\gamma)^2, \quad \eta^2 = 1, \quad n \in \mathbb{N}, \tag{40.6}$$

or

$$(2\alpha - n^2)(2\beta + n^2) = 0. \tag{40.7}$$

Proof. We prove this theorem by induction. If $n = 1$ in (40.6), then (P_5) has a one-parameter family of solutions determined by the Riccati equation (36.11) under the parameter condition (36.12). If $n = 1$ in (40.7), then we have either (40.4) when (40.3) holds or the equivalent family that can be obtained by means of the transformation T_2.

Let us now assume the validity of this theorem when $n = m$ in (40.6), using the notations $T : w(z) \mapsto w_1(z)$, $\Lambda : (\alpha, \beta, \gamma) \mapsto (\alpha_1, \beta_1, \gamma_1)$, where $w_1(z)$, α_1, β_1, γ_1 are determined as in Theorem 39.1. Note, that any application of Bäcklund transformations depends on the choice of the values of $\sqrt{2\alpha}$, $\sqrt{-2\beta}$ which may be indicated by ε, ν $(\varepsilon^2 = \nu^2 = 1)$. Applying the Bäcklund transformation to a one-parameter family of solutions with parameters (40.6) when $n = m$, we obtain (40.7) with $n = m$ and $\varepsilon_1 = \varepsilon\eta$, $\nu_1 = -\nu\eta$; $\varepsilon_1 = \varepsilon\eta$, $\nu_1 = \nu\eta$. Applying the Bäcklund transformation to a one-parameter family of solutions with parameters (40.7) with $n = m$, we get either (40.6) with $n = m + 1$ or with $n = m$ depending on the choice of parameters ε_1, ν_1. □

The relations (39.5), (39.6) may be used for the construction of the rational solution hierarchy for (P_5) in the case of $\delta \neq 0$. Indeed, it is known that (P_5) with $\delta \neq 0$ admits rational solutions of the form $P(z)/Q(z)$. These solutions are of the form

$$w(z) = \lambda z + \mu + \frac{P_{n-1}(z)}{Q_n(z)},$$

where λ, μ are constants and $P_{n-1}(z)$, $Q_n(z)$ are the polynomials of degrees $n - 1$ and n, respectively.

If $P_{n-1} = 0$, $Q_n \neq 0$, then (P_5) has the following rational solutions:

$$\begin{aligned}
&\text{(a) } w(z) = -1, \quad \gamma = 0, \quad \alpha + \beta = 0,\\
&\text{(b) } w(z) = -2\delta z/\gamma + 1, \quad \gamma \neq 0, \quad \beta = -1/2, \quad 4\alpha\delta + \gamma^2 = 0,\\
&\text{(c) } w(z) = kz + \sqrt{-2\beta}, \quad \alpha = 1/2, \quad \sqrt{-2\beta} \neq 1,\\
&\qquad \gamma = 2k - k\sqrt{-2\beta}, \quad k^2 = -2\delta.
\end{aligned} \tag{40.8}$$

Note, that the last two rational solutions satisfy the Riccati equation at the same time.

Taking now (40.8(a)) as a starting solution, we obtain the following hierarchy of rational solutions:

$$w_1(z) = \frac{kz + 4a}{-kz + 4a}$$

with $\alpha_1 = 1/8$, $\beta_1 = -1/8$, $\gamma_1 = 2ak$, $\delta_1 = \delta$, $k^2 = -2\delta$;

$$w_2(z) = \frac{-\delta z^2 + k(a_1 - c_1)z + 4a(1 - a_1 - c_1 - 2a)}{\delta z^2 + k(a_1 - c_1)z + 4\alpha(1 - a_1 - c_1 + 2a)},$$

with $\alpha_2 = (1 - a_1 - c_1 + 2a)^2/8$, $\beta_2 = -(-1 + a_1 + c_1 - 2a)^2/8$, $\gamma_2 = k(a_1 - c_1)$, $\delta_2 = \delta$, $a_1^2 = c_1^2 = 1/4$, $k_1 = k$.

Similar hierarchies of the solutions may be obtained for (40.8(b)) and for (40.8(c)).

Remark. Recall the expression (36.13). If $b = \sqrt{-2\beta} \in \mathbb{N}$, $b - a = \sqrt{-2\beta} - \sqrt{2\alpha} \notin \{1, 2, \ldots, \sqrt{-2\beta} + 1\}$, respectively $a = \sqrt{2\alpha} \in \mathbb{N}$, $a - b = \sqrt{2\alpha} - \sqrt{-2\beta} \notin \{1, 2, \ldots, \sqrt{2\alpha} + 1\}$, and if $C = 0$, respectively $C = \infty$, then a hierarchy of rational solutions immediately follows, expressed in terms of Laguerre polynomials. The first solutions of the hierarchies (40.8(a)) and (40.8(b)) may be found in Table 40.1 and Table 40.2.

More generally, let us consider the rational solutions generated by Bäcklund transformations from the starting solution (40.8(a)).

Theorem 40.2. *The equation* (P_5) *with* $\delta = -1/2$ *admits rational solutions generated by successive applications of Bäcklund transformations to the solution* (40.8(*a*)) *for all parameters which satisfy one of the following conditions:*

(1) $-2\beta = (\sqrt{2\alpha} + k)^2$, $\gamma = m$, $k \in \mathbb{N} \cup \{0\}$, $m \in \mathbb{Z}$ *and* $k + m \in 2\mathbb{Z}$,

(2) $8\alpha = (2k - 1)^2$, $-8\beta = (2m - 1)^2$, $\gamma \notin \mathbb{Z}$, $k, m \in \mathbb{N}$.

Table 40.1. The first rational solutions generated from $w = -1$ with $\alpha = -\beta, \gamma = 0, \delta = -1/2$.

α	β	γ	$w(z)$
a	$-a$	0	-1
$\frac{1}{8}$	$-\frac{1}{8}$	$2a$	$\frac{z + 4a}{-z + 4a}$
$\frac{(1+2a)^2}{8}$	$-\frac{(1-2a)^2}{8}$	1	$\frac{z^2 + 2z + 8a(1-2a)}{-z^2 + 2z + 8a(1+2a)}$
a	$-a$	-2	$\frac{z^2 - 4z + 4 - 32a}{-z^2 - 4z - 4 + 32a}$
$\frac{9}{8}$	$-\frac{1}{8}$	$2a$	$\frac{(4a - z)(-z^2 + 16a^2 - 4)}{-z^3 + 4a(3z^2 - 4) + 12z - 48a^2z + 64a^3}$
$\frac{9}{8}$	$-\frac{9}{8}$	$2a$	$\frac{(4a - z)\big(-z^4 - 8az^3 + 128a(a^2-1)z + 256a^2(a^2-1)\big)}{(4a + z)\big(-z^4 + 8az^3 - 128a(a^2-1)z + 256a^2(a^2-1)\big)}$
$\frac{1}{8}$	$-\frac{25}{8}$	-2	$-1 - \frac{16}{z-8} + \frac{24(12 - 6z + z^2)}{z^3 - 12z^2 + 36z - 48}$
0	$-\frac{9}{2}$	-1	$-1 + \frac{1}{z-8} + \frac{8}{(z-8)^2} - \frac{192}{z^3} + \frac{48}{z^2} + \frac{3}{z}$

Table 40.2. The first rational solutions generated from $w = az+1$ with $\alpha = 1/(2a^2), \beta = -1/2, \gamma = 1/a, \delta = -1/2$.

α	β	γ	$w(z)$
$\frac{1}{(2a^2)}$	$-\frac{1}{2}$	$\frac{1}{a}$	$az+1$
$\frac{1}{2}$	$-\frac{(a-1)^2}{(2a^2)}$	$-1-\frac{1}{a}$	$-z+1-\frac{1}{a}$
$\frac{(a-1)^2}{2a^2}$	-2	$\frac{1}{a}$	$\frac{-a^2z^2+(2az+1)(a-1)}{(a-1)(az+1)}$
2	$-\frac{(1-2a)^2}{2a^2}$	$-\frac{1+a}{a}$	$\frac{-a^2z^2+4a^2z-2az-2a^2+3a-1}{2a(az-a+1)}$
$\frac{1}{2a^2}$	$-\frac{1}{2}$	$-4+\frac{1}{a}$	$\frac{a\left(1+a(2z-3)+a^2(z^2-2z+2)\right)}{1+3a(z-1)+a^3(z-2)^2z+a^2(2-7z+3z^2)}$
$\frac{9}{2}$	$-\frac{(1-2a)^2}{2a^2}$	$-\frac{1}{a}$	$-\frac{(2a-1)\left(1+a\left(-1+(2+a(z-2))z\right)\right)}{1+a\left(3(z-1)+a\left(2+z\left(3(z-3)+a(6+(z-6)z)\right)\right)\right)}$
2	0	-5	$\frac{-z^2(6+4z+z^2)^2}{2(12+36z+36z^2+20z^3+7z^4+z^5)}$
8	-2	-1	$\frac{2(36+132z+132z^2+60z^3+13z^4+z^5)}{144+480z+564z^2+336z^3+104z^4+16z^5+z^6}$

Proof. By (40.6), the solution (40.8(a)) does not belong to the solutions generated by the Riccati equation (40.1). This means that we can apply Theorem 39.9. Substituting the parameters in (40.8(a)) into (39.25) and considering the two possible cases $\sqrt{-2\beta} = \pm\sqrt{2\alpha}$, we get the assertion. □

Making use of (39.25), it is possible to obtain conditions for the existence of rational solutions of (P_5) with $\delta \neq 0$, see Gromak and Lukashevich [1] and Kitaev, Law and McLeod [1]. Necessary conditions follow from the conditions of the existence of analytic and polar expansions at singularities $z = 0$, $z = \infty$. Sufficient conditions may be obtained by (39.4), (39.6) and the direct construction of the above-mentioned solution hierarchies using (40.8) as starting solutions. This results in the following theorem, see Kitaev, Law and McLeod [1]:

Theorem 40.3. *The equation (P_5) with $\delta = -1/2$ has a rational solution if and only if for some branch of λ_0, the parameters satisfy one of the following conditions with $k, m \in \mathbb{Z}$:*

(1) $2\alpha = (\lambda_0\gamma + k)^2, -2\beta = m^2, m > 0,$ *$k + m$ is odd, and $\alpha \neq 0$ when $|k| < m$;*

(2) $-2\beta = (\lambda_0\gamma + k)^2, 2\alpha = m^2, m > 0,$ *$k + m$ is odd, and $\beta \neq 0$ when $|k| < m$;*

(3) $-2\beta = (\alpha_1 + m)^2, \lambda_0\gamma = k, \alpha_1 = \sqrt{2\alpha}, m \geq 0,$ *and $k + m$ is even;*

(4) $8\alpha = k^2, -8\beta = m^2, \lambda_0\gamma \notin \mathbb{Z}, k, m > 0,$ *and k, m are both odd.*

§41 Connection between (P_3) and (P_5) revisited

We may Theorem 39.9 to extend the conditions on parameters, when (P_5) is reduced to (P_3), see Gromak [9]. Let u and w be solutions of (P_5) and (P_3) with parameters $\alpha, \beta, \gamma, \delta$ and a, b, c, d, respectively. Proving the following theorems is nothing but a direct computation:

Theorem 41.1. *Let u be a solution of (P_5) with parameters $a = b = 0$. Then the function w determined by the formula*

$$u = \left(\frac{w+1}{w-1}\right)^2,$$

satisfies (P_3) with the following parameter values:

$$\alpha = -\beta = -c/4, \quad \gamma = -\delta = -d/8.$$

Observe that the transformation in Theorem 41.1 also carries the Riccati equation of (P_5), see (36.11) into the Riccati equation of (P_3), see (29.11). This is again just a direct computation. We now proceed to extend values of the parameters such that (P_5) may be reduced to (P_3). We first find a connection between solutions of (P_5) when $\delta = 0$ and $\delta \neq 0$.

Theorem 41.2. *Let $w = w(t) = w(t, \alpha/4, -\alpha/4, 0, 4\gamma)$, $z = t^2$ be a solution of (P_5). Then the transformation*

$$R : w \mapsto \widetilde{w} = \frac{(w+1)^2}{4w} \tag{41.1}$$

determines a solution $\widetilde{w}(z, \alpha, 0, \gamma, 0)$ of (P_5).

Proof. This is an immediate consequence of the substitution of (41.1) into the equation (P_5). □

Remark. The transformation (41.1) links solutions of (P_5) with the values of the parameters $\delta = 0$ and $\delta \neq 0$.

We also derive another transformation connecting solutions of (P_5) with $\delta = 0$ and $\delta \neq 0$. For this purpose, we consider the following pair of the differential equations:

$$\begin{aligned} \frac{df}{dt} &= g(A + A^{-1}f^2) - (B+1)(f/t), \\ \frac{dg}{dt} &= f(A - A^{-1}g^2) + B(g/t), \end{aligned} \tag{41.2}$$

where B, $A \neq 0$ are arbitrary parameters.

Eliminating the function $g(t)$, we get the equation

$$(A^2 + f^2)(f' + tf'') = tf(f')^2 + (A^2 + f^2)^2 tf + (1+B)^2(fA^2/t), \tag{41.3}$$

which may be reduced to (P_5) by means of the substitution $\widetilde{w} = f^2/(A^2 + f^2)$, $z = t^2$, the parameters being $\widetilde{\alpha} = (1+B)^2$, $\widetilde{\beta} = 0$, $\widetilde{\gamma} = A^2/2$, $\widetilde{\delta} = 0$. Similarly, eliminating the function $f(t)$, we obtain

$$g'' = \frac{g}{g^2 - A^2}(g')^2 - \frac{g'}{t} + g(A^2 - g^2) + B^2A^2\frac{g}{(A^2 - g^2)t^2}. \tag{41.4}$$

Taking $g(t) = A(w(t)+1)/(w(t)-1)$, we achieve an equation (P_5) with parameters $\alpha = B^2/8$, $\beta = -B^2/8$, $\gamma = 0$, $\delta = 2A^2$, to determine the function $w(t)$.

Theorem 41.3. *Let $w(t)$ be a solution (P_5) with parameters $\alpha = B^2/8$, $\beta = -B^2/8$, $\gamma = 0$, $\delta = 2A^2$, $A \neq 0$, where B, $A \neq 0$ are arbitrary parameters. Then the transformation $G : w \mapsto \widetilde{w}$, where*

$$\widetilde{w}(z) = \frac{f(t)^2}{A^2 + f(t)^2}, \quad f(t) = \frac{w'(t)}{2w(t)} - \frac{B(w(t) - 1/w(t))}{4t}, \quad z = t^2$$

defines a solution $\widetilde{w}(z)$ of (P_5) with the values of the parameters $\widetilde{\alpha} = (1+B)^2/2$, $\widetilde{\beta} = 0$, $\widetilde{\gamma} = A^2/2$, $\widetilde{\delta} = 0$.

It is not difficult to show that the transformation G obtained in Theorem 41.3 is the composition of the transformations R and $T_{\varepsilon_1,\varepsilon_2,1}$:

Theorem 41.4. *The relation $G = R \circ T_{-1,-1,1}$ is valid.*

Proof. Let us consider the pair (41.2) and define $w_2 = (f - iA)/(f + iA)$, where i is the imaginary unit. Substituting in (41.3) and simplifying, we get for w_2 an equation (P_5) with parameters $\alpha_2 = (1+B)^2/8$, $\beta_2 = -(1+B)^2/8$, $\gamma_2 = 0$, $\delta_2 = 2A^2$. Moreover, we observe that $w_2 = T_{-1,-1,1}w$, $\widetilde{w} = Rw_2$. Hence, $\widetilde{w} = R \circ T_{-1,-1,1}w$, and $G = R \circ T_{-1,-1,1}$. □

Hence, the transformation R in Theorem 41.2 appears vitally important to obtain the relationship between (P_5) when $d = 0$ and (P_5) when $d \neq 0$. If $\delta \neq 0$ in (P_5), then Theorem 39.2 holds. Consequently, applying Theorem 39.9 to any solution of (P_5) with parameters

$$\alpha = B^2/8, \quad \beta = -B^2/8, \quad \gamma = 0, \quad \delta = 2A^2 \neq 0, \tag{41.5}$$

all parameters for which (P_5) may be reduced to (P_3) are found.

Theorem 41.5. *The Painlevé equation (P_5) with parameters taking one of the following forms:*

(1) $-2\beta = (\sqrt{2\alpha} + k)^2$, $\gamma = m$, $\delta = -1/2$, *where* $k \in \mathbb{N} \cup \{0\}$, $m \in \mathbb{Z}$ *and* $k + m \in 2\mathbb{Z}$,

(2) $8\alpha = (2k-1)^2$, $-8\beta = (2m-1)^2$, $\gamma \notin \mathbb{Z}$, $\delta = -1/2$ *where* $k, m \in \mathbb{N}$,

may be reduced to a Painlevé equation (P_3).

Proof. Let $w(z)$ be a solution of (P_5) with parameter values (41.5). Then by Theorem 41.2 it can be transformed to a solution $\widetilde{w}(t, B^2/2, 0, A^2/2, 0)$ of (P_5). By Theorem 34.3 and a subsequent remark, $\widetilde{w}$ may now be reduced to a solution of (P_3). The parameters (1) and (2) are now obtained similarly as in Theorem 40.2, by applying Theorem 39.9 and a scaling transformation to $w(z)$. □

Chapter 9
The sixth Painlevé equation (P_6)

The sixth Painlevé equation (P_6) is in some sense a master type for all Painlevé equations. Indeed, by a simple limiting process, all of the equations (P_1)–(P_5) may be obtained from (P_6). On the other hand, by the existence of three fixed singular points, no transformation of type $z = \phi(\xi)$, ϕ entire, can be applied to produce a modified form of (P_6) with meromorphic solutions only. However, the local behavior of sixth Painlevé transcendents shows several features similar as to what we are already familiar from the preceding chapters. To this end, we pay some attention to pairs of differential equations equivalent to (P_6) as well as to Riccati differential equations whose solutions may be used to generate one-parameter families of solutions of (P_6). We also offer some analysis for formal solutions of (P_6) around of their singular points. Similarly as to the previous chapters, Bäcklund transformations enable us to construct connection formulae between sixth Painlevé transcendents, treated only partially, as the intrinsic symmetries of (P_6) make it impossible to include a complete presentation here. The chapter will be closed by a rather extensive presentation of rational and algebraic solutions of (P_6).

§42 General properties of solutions

The sixth Painlevé equation

$$w'' = \frac{1}{2}\left(\frac{1}{w} + \frac{1}{w-1} + \frac{1}{w-z}\right)(w')^2 - \left(\frac{1}{z} + \frac{1}{z-1} + \frac{1}{w-z}\right)w' + \frac{w(w-1)(w-z)}{z^2(z-1)^2}\left(\alpha + \frac{\beta z}{w^2} + \frac{\gamma(z-1)}{(w-1)^2} + \frac{\delta z(z-1)}{(w-z)^2}\right) \tag{P_6}$$

with four arbitrary complex parameters α, β, γ, δ has three fixed singular points $z = 0, 1, \infty$. As indicated in §6, all singularities of $w(z)$, an arbitrary solution of (P_6), outside of $z = 0, 1, \infty$ are poles, see e.g. Okamoto and Takano [1] and, for another proof, Hinkkanen and Laine [4]. Therefore, the behavior of $w(z)$ at $z = 0, 1, \infty$ essentially determines the meromorphic nature of $w(z)$.

The sixth Painlevé equation (P_6) is the most general among the six classical Painlevé equations in the sense that (P_1)–(P_5) may be obtained by successive coalescences of the singular points of (P_6), see the classical treatise by Ince [1]. In fact, if we put $z \to 1 + \varepsilon z$, $\delta \to \delta\varepsilon^{-2}$, $\gamma \to \gamma\varepsilon^{-1} - \delta\varepsilon^{-2}$ in (P_6), then (P_6) $\to$ (P_5) as $\varepsilon \to 0$.

Similarly, $(P_5) \to (P_3)$ as $\varepsilon \to 0$, when $w \to 1 + \varepsilon w$, $\beta \to -\beta\varepsilon^{-2}$, $\alpha \to \alpha\varepsilon^{-1} + \beta\varepsilon^{-2}$, $\gamma \to \gamma\varepsilon$, $\delta \to \delta\varepsilon$;

$(P_5) \to (P_4)$ as $\varepsilon \to 0$, when $z \to 1 + \varepsilon\sqrt{2}z$, $\alpha \to \varepsilon^{-4}/2$, $\gamma \to -\varepsilon^{-4}$, $w \to \varepsilon\sqrt{2}w$, $\delta \to -(\varepsilon^{-4}/2 + \delta\varepsilon^{-2})$;

$(P_3) \to (P_2)$ as $\varepsilon \to 0$, when $z \to 1 + \varepsilon^2 z$, $w \to 1 + 2\varepsilon w$, $\gamma \to \varepsilon^{-6}/4$, $\delta \to -\varepsilon^{-6}/4$, $\alpha \to -\varepsilon^{-6}/2$, $\beta \to 2\beta\varepsilon^{-3} + \varepsilon^{-6}/2$;

$(P_4) \to (P_2)$ as $\varepsilon \to 0$, when $w \to 2^{2/3}\varepsilon w + \varepsilon^{-3}$, $z \to 2^{-2/3}\varepsilon z - \varepsilon^{-3}$, $\alpha \to -\varepsilon^{-6}/2 - \alpha$, $\beta \to -\varepsilon^{-12}/2$;

$(P_2) \to (P_1)$ as $\varepsilon \to 0$, when $w \to \varepsilon w + \varepsilon^{-5}$, $z \to \varepsilon^2 z - 6\varepsilon^{-10}$, $\alpha \to 4\varepsilon^{-15}$.

The analytic meaning of this procedure is as follows. After the substitution above, for example, (P_6) takes the form

$$\begin{aligned} w'' &= \left(\frac{1}{2w} + \frac{1}{w-1}\right)(w')^2 - \frac{w'}{z} + \frac{(w-1)^2}{z^2}\left(\alpha w + \frac{\beta}{w}\right) \\ &\quad + \frac{\gamma w}{z} + \frac{\delta w(w+1)}{w-1} + \varepsilon f(\varepsilon, z, w, w'). \end{aligned} \tag{$P_{6,\varepsilon}$}$$

Here the function $f(\varepsilon, z, w, w_1)$ possesses the following property: For every $(z_0, w_0, w_0') \in (\mathbb{C}\setminus\{0\}) \times (\mathbb{C}\setminus\{0, 1\}) \times \mathbb{C}$, $f(\varepsilon, z, w, w_1)$ is analytic around the point $(0, z_0, w_0, w_0')$. Let $w = \phi(\varepsilon, z)$ be the solution of $(P_{6,\varepsilon})$ satisfying $\phi(\varepsilon, z_0) = w_0$, $\phi'(\varepsilon, z_0) = w_0'$. By the Poincaré theory of regular perturbations, $\phi(\varepsilon, z)$ is analytic for $|\varepsilon| < \varepsilon_0$, $z \in D_0$, where D_0 is a domain containing z_0, hence $\phi(0, z)$ is a solution of (P_5). In every step of coalescence for Painlevé equations $(P_{j+1}) \to (P_j)$, $j = 1, \dots, 5$, as well, degeneration of solutions as above may be justified analytically. By the α-method of Painlevé combined with this fact, the non-existence of movable branch points for solutions of $(P_5), \dots, (P_1)$ follows from the same property for solutions of (P_6), see e.g. Painlevé [4, III], p. 115–121 and Okamoto–Takano [1]. It is quite interesting to study degeneration of solutions in a global domain or around fixed critical points; concerning degeneration of asymptotic solutions in the process $(P_2) \to (P_1)$, see Kapaev and Kitaev [1]. It is also known that the Painlevé equations are governed by monodromy preserving deformation of Fuchs' equations with regular or irregular singularities. Concerning these singularities, there is a degeneration scheme as well. Each step of this scheme gives rise to the coalescence procedure above for corresponding Painlevé equations, see Garnier [1], Okamoto [5] and Steinmetz [4].

Fuchs [1] presented (P_6) in the form

$$\begin{aligned} &z(1-z)\mathcal{L}_z \int_\infty^w \frac{dt}{\sqrt{t(t-1)(t-z)}} \\ &= \sqrt{w(w-1)(w-z)}\left(\alpha + \frac{\beta z}{w^2} + \frac{\gamma(z-1)}{(w-1)^2} + \left(\delta - \frac{1}{2}\right)\frac{z(z-1)}{(w-z)^2}\right). \end{aligned}$$

Here $\mathcal{L}_z$ is the Picard–Fuchs linear differential operator

$$\mathcal{L}_z = z(1-z)\frac{d^2}{dz^2} + (1-2z)\frac{d}{dz} - \frac{1}{4}.$$

Therefore, if we put

$$u = u(w, z) = \int_{\infty}^{w} \frac{dt}{\sqrt{t(t-1)(t-z)}},$$

then $w(u) = \wp(u; z)$. Here z is a parameter, and the periods of $\wp(u; z)$ are functions of z. Then (P_6) transforms into

$$\mathcal{L}_z u = \frac{1}{z(1-z)} \frac{\partial}{\partial u} \psi(u; z), \tag{42.1}$$

where $\psi(u; z) = \alpha \wp(u; z) - \beta \frac{z}{\wp(u;z)} + \gamma \frac{1-z}{\wp(u;z)-1} + \left(\frac{1}{2} - \delta\right) \frac{z(z-1)}{\wp(u;z)-z}$, see Painlevé [1] and Okamoto [6].

The equation (P_6) can also be rewritten in a canonical form, see Painlevé [1], Manin [1] and Takasaki [1], by means of the Weierstraß' function $\wp(q)$ with primitive periods 1 and τ. Following an idea of Manin, we apply a simultaneous change of the variables $(w, z) \mapsto (q, \tau)$, where $w := \frac{\wp(q)-e_1}{e_2-e_1}$, $z := \frac{e_3-e_1}{e_2-e_1}$ and $e_j = \wp(\omega_j)$, $j = 1, 2, 3$ are the values of $\wp(q)$ at three half-period points $\omega_1 = 1/2$, $\omega_2 = \tau/2$, $\omega_3 = (1+\tau)/2$. By this transformation, (P_6) takes the form

$$(2\pi i)^2 \frac{d^2 q}{d\tau^2} = \sum_{n=0}^{3} \alpha_n \wp'(q + \omega_n), \tag{42.2}$$

where $\alpha_0 = \alpha, \alpha_1 = -\beta, \alpha_2 = \gamma, \alpha_3 = -\delta + 1/2, \omega_0 = 0$ and $\wp'(q)$ is the derivative in q of the Weierstraß function

$$\wp(q) = \wp(q|1, \tau) = \frac{1}{q^2} + \sum_{(m,n)\neq(0,0)} \left(\frac{1}{(q+m+n\tau)^2} - \frac{1}{(m+n\tau)^2} \right).$$

In the special case $\alpha = \beta = \gamma = 0$, $\delta = 1/2$, see again Painlevé [1], Okamoto [6], the function $u(z)$ satisfies the Legendre differential equation

$$4z(z-1)\frac{d^2u}{dz^2} + 4(2z-1)\frac{du}{dz} + u = 0 \tag{42.3}$$

with two particular solutions of the form

$$\begin{aligned} K(z) &= \int_0^1 \frac{ds}{\sqrt{(1-s^2)(1-\mu^2 s^2)}}, \\ K'(z) &= \int_0^{1/\mu} \frac{ds}{\sqrt{(s^2-1)(1-\mu^2 s^2)}}, \end{aligned} \tag{42.4}$$

where $\mu^2 = z$. The functions $4K(z)$ and $2iK'(z)$ are, in fact, the periods of the function $\wp(u; z)$. Of course, the equation (42.3) is a particular case of the hypergeometric equation. Hence, the solutions (42.4) may be expressed by means of hypergeometric functions: $K(z) = \frac{\pi}{2}F(1/2, 1/2, 1, z)$, $K'(z) = \frac{\pi}{2}F(1/2, 1/2, 1, 1-z)$. As a consequence, the general solution of (P_6) with $\alpha = \beta = \gamma = 0$, $\delta = 1/2$ takes the form

$$w(z) = \wp\big(C_1F(1/2, 1/2, 1, z) + C_2F(1/2, 1/2, 1, 1-z); z\big). \tag{42.5}$$

As an introduction to Bäcklund transformations of (P_6), let $w(z) = w(z, \alpha, \beta, \gamma, \delta)$ be any solution of (P_6) with fixed parameters. Then the following transformations determine new solutions of (P_6):

$$T_j : w \mapsto w_j, \quad j = 1, 2, 3,$$

where $w_1(z, -\beta, -\alpha, \gamma, \delta) = 1/w(1/z)$, $w_2(z, -\beta, -\gamma, \alpha, \delta) = 1 - 1/w\big(1/(1-z)\big)$, $w_3(z, -\beta, -\alpha, -\delta + \frac{1}{2}, -\gamma + \frac{1}{2}) = z/w(z)$. Considering next different compositions of T_j, we obtain the following group of transformations:

$$w_4\left(z, \alpha, \beta, -\delta + \frac{1}{2}, -\gamma + \frac{1}{2}\right) = zw\left(\frac{1}{z}\right); \quad T_4 = T_1 \circ T_3;$$

$$w_5\left(z, \alpha, \delta - \frac{1}{2}, -\beta, -\gamma + \frac{1}{2}\right) = 1 - (1-z)w\left(\frac{1}{1-z}\right); \quad T_5 = T_2 \circ T_3;$$

$$w_6(z, \alpha, -\gamma, -\beta, \delta) = 1 - w(1-z); \quad T_6 = T_2 \circ T_1;$$

$$w_7\left(z, -\beta, \delta - \frac{1}{2}, \alpha, -\gamma + \frac{1}{2}\right) = 1 - (1-z)w\left(\frac{1}{1-z}\right); \quad T_7 = T_2 \circ T_4;$$

$$w_8\left(z, -\delta + \frac{1}{2}, -\gamma, -\beta, -\alpha + \frac{1}{2}\right) = \frac{z(w-1)}{w-z}; \quad T_8 = T_7 \circ T_7;$$

$$w_9\left(z, \gamma, \delta - \frac{1}{2}, \alpha, \beta + \frac{1}{2}\right) = \frac{w-z}{w-1}; \quad T_9 = T_3 \circ T_8;$$

$$w_{10}\left(z, \alpha, -\gamma, -\delta + \frac{1}{2}, \beta + \frac{1}{2}\right) = z - zw\left(\frac{z-1}{z}\right); \quad T_{10} = T_5 \circ T_5;$$

$$w_{11}\left(z, -\delta + \frac{1}{2}, -\alpha, \gamma, \beta + \frac{1}{2}\right) = z\left(1 - (1-z)w\left(\frac{1}{1-z}\right)\right)^{-1}; \quad T_{11} = T_5 \circ T_8;$$

$$w_{12}\left(z, \gamma, -\alpha, -\delta + \frac{1}{2}, \beta + \frac{1}{2}\right) = z\big(1 - w(1-z)\big)^{-1}; \quad T_{12} = T_3 \circ T_6;$$

$$w_{13}\left(z, \alpha, \delta - \frac{1}{2}, \gamma, \beta + \frac{1}{2}\right) = z + (1-z)w\left(\frac{z}{z-1}\right); \quad T_{13} = T_2 \circ T_{12};$$

$$w_{14}\left(z, -\delta + \frac{1}{2}, -\alpha, -\beta, -\gamma + \frac{1}{2}\right) = z\left(z + (1-z)w\left(\frac{z}{z-1}\right)\right)^{-1};$$

$$T_{14} = T_{13} \circ T_8;$$

$$w_{15}(z,\gamma,-\alpha,-\beta,\delta) = \left(1 - w\left(\frac{z-1}{z}\right)\right)^{-1}; \quad T_{15} = T_1 \circ T_6;$$

$$w_{16}(z,\gamma,\beta,\alpha,\delta) = w\left(\frac{z}{z-1}\right)\left(w\left(\frac{z}{z-1}\right)-1\right)^{-1}; \quad T_{16} = T_{15} \circ T_1;$$

$$w_{17}\left(z,-\delta+\frac{1}{2},\beta,\alpha,-\gamma+\frac{1}{2}\right) = zw\left(\frac{z-1}{z}\right)\left(zw\left(\frac{z-1}{z}\right)-z+1\right)^{-1};$$

$$T_{17} = T_{16} \circ T_4;$$

$$w_{18}\left(z,-\delta+\frac{1}{2},-\gamma,\alpha,\beta+\frac{1}{2}\right) = \left(z - zw\left(\frac{1}{z}\right)\right)\left(1 - zw\left(\frac{1}{z}\right)\right)^{-1}$$

$$T_{18} = T_{16} \circ T_{10};$$

$$w_{19}\left(z,\gamma,\delta-\frac{1}{2},-\beta,-\alpha+\frac{1}{2}\right) = \left(zw\left(\frac{1}{z}\right)-1\right)\left(w\left(\frac{1}{z}\right)-1\right)^{-1};$$

$$T_{19} = T_1 \circ T_8;$$

$$w_{20}\left(z,\gamma,\beta,-\delta+\frac{1}{2},-\alpha+\frac{1}{2}\right) = zw\left(\frac{1}{1-z}\right)\left(w\left(\frac{1}{1-z}\right)-1\right)^{-1};$$

$$T_{20} = T_2 \circ T_8;$$

$$w_{21}\left(z,-\delta+\frac{1}{2},\beta,\gamma,-\alpha+\frac{1}{2}\right) = zw(1-z)\big(w(1-z)-1+z\big)^{-1}; \quad T_{21} = T_6 \circ T_8;$$

$$w_{22}\left(z,-\beta,\delta-\frac{1}{2},\gamma,-\alpha+\frac{1}{2}\right) = \left(zw\left(\frac{z-1}{z}\right)-z+1\right)\Big/w\left(\frac{z-1}{z}\right);$$

$$T_{22} = T_{15} \circ T_8;$$

$$w_{23}\left(z,-\beta,-\gamma,-\delta+\frac{1}{2},-\alpha+\frac{1}{2}\right) = z\left(w\left(\frac{z}{z-1}\right)-1\right)\Big/w\left(\frac{z}{z-1}\right);$$

$$T_{23} = T_{16} \circ T_8;$$

completing the group by $T_{24} = I$, the identity transformation. Verification of the compositions described above is a direct computation. Observe that application of the transformations T_j implies the change of any two parameters by the scheme

$$\alpha \mapsto -\beta \mapsto \gamma \mapsto \frac{1}{2} - \delta \mapsto \alpha. \tag{42.6}$$

The transformations T_j constructed above will be applied below to extend the cases of solvability of (P_6) and to construct special classes of solutions.

§43 Pairs of differential equations equivalent to (P_6)

While investigating the local behavior of sixth Painlevé transcendents, at the fixed singular points $z = 0, 1, \infty$ in particular, it is often more convenient to consider pairs of first order differential equations equivalent to (P_6). In this section, we construct two such systems.

(1) We first proceed to construct such a pair of first order differential equations in the following Hamiltonian form:

$$\begin{aligned} w'(z) &= f(z,w) + \varphi(z,w)v \equiv \frac{\partial H_6(z,w,v)}{\partial v}, \\ v'(z) &= \psi(z,w) - f_w(z,w)v - \frac{1}{2}\varphi_w(z,w)v^2 \equiv -\frac{\partial H_6(z,w,v)}{\partial w}, \end{aligned} \tag{43.1}$$

where f, φ, ψ are polynomials in z and w.

Differentiating (43.1) and making use of (43.1) again, we obtain

$$w''(z) = \frac{1}{2}\varphi\varphi_w v^2 + (f\varphi_w + \varphi_z)v + ff_w + f_z + \varphi\psi. \tag{43.2}$$

Substituting w'' and w' from (43.2) and (43.1) into (P_6) and comparing coefficients of terms which are of the same degree in v, we obtain a system of differential equations to determine of functions f, φ and ψ. The system obtained may be written, after some computation, in the form

$$w(w-1)(w-z)\varphi_w = (3w^2 - 2zw - 2w + z)\varphi; \tag{43.3}$$

$$\begin{aligned} &z(z-1)w(w-1)(w-z)(\varphi_z + f\varphi_w) \\ &\quad = z(z-1)(3w^2 - 2zw - 2w + z)f\varphi - w(w-1)(2zw - w - z^2)\varphi; \end{aligned} \tag{43.4}$$

$$\begin{aligned} &2z^2(z-1)^2 w(w-1)(w-z)(f_z + ff_w + \varphi\psi) \\ &\quad = z^2(z-1)^2(3w^2 - 2zw - 2w + z)f^2 \\ &\qquad - 2z(z-1)w(w-1)(2zw - w - z^2)f \\ &\qquad + 2\alpha w^2(w-1)^2(w-z)^2 + 2\beta z(w-1)^2(w-z)^2 \\ &\qquad + 2\gamma(z-1)w^2(w-z)^2 + 2\delta z(z-1)w^2(w-1)^2. \end{aligned} \tag{43.5}$$

From (43.3) we obtain

$$\varphi(z,w) = a(z)w(w-1)(w-z), \tag{43.6}$$

where $a(z)$ is to be specified at once. Namely, substituting (43.6) into (43.3), we get

$$z(z-1)\big((w-z)a' - a\big) + (2zw - w - z^2)a = 0. \tag{43.7}$$

Integrating now (43.7), we conclude that $a(z) = C_1/[z(z-1)]$, where C_1 is an arbitrary constant. We may assume here that $C_1 \neq 0$.

Therefore, we have obtained, by (43.6),

$$\varphi(z, w) = C_1 w(w-1)(w-z)/\big[z(z-1)\big]. \tag{43.8}$$

The functions $f(z, w)$ and $\psi(z, w)$ may now be sought for in the following form:

$$f(z, w) = k(z)w^2 + l(z)w + m(z),\ \psi(z, w) \equiv \psi(z). \tag{43.9}$$

Substituting (43.8) and (43.9) into (43.5) and comparing the coefficients of w^5 and w^6, we infer that

$$2a\psi + k^2 = 2\alpha/[z^2(z-1)^2], \quad z(z-1)k' = -k(2z-1). \tag{43.10}$$

From (43.10) we obtain, by elementary computation, $k(z) = C_2/\big[z(z-1)\big]$, $\psi(z) = (2\alpha - C_2^2)/[2C_1 z(z-1)]$, where C_2 is an arbitrary constant. Equating constant terms in (43.5), we conclude that $m(z) = \eta_2/(z-1)$, where $\eta_2^2 = -2\beta$.

The equation which results by comparing coefficients of w, will be an identity. But comparing coefficients of w^2 and w^4, we obtain the following pair of equations:

$$\begin{aligned} &z^2(z-1)^2(2l' + l^2) + 2z(z-1)l\big[(\eta_2+1)z + \eta_2\big] \\ &\quad = 2\eta_2(z-1) - 2C_2\eta_2 z + C_2^2 z + 2\beta(z^2+z+1) \\ &\qquad + 2\gamma z(z-1) + 2\delta(z-1), \end{aligned} \tag{43.11}$$

$$\begin{aligned} &z^2(z-1)^2(2l' - l^2) + 2z(z-1)l\big[(2-C_2)z - 1 - C_2\big] \\ &\quad = 2\eta_2 C_2 z + C_2^2(z^2+z+1) - 2C_2 z(z-1) + 2\beta z \\ &\qquad + 2\gamma(z-1) + 2\delta z(z-1). \end{aligned} \tag{43.12}$$

Subtracting now (43.12) from (43.11), we easily obtain

$$\begin{aligned} &2z^2(z-1)^2 l^2 + 2z(z-1)l[(\eta_2 + C_2 - 1)z + \eta_2 + C_2 + 1] \\ &\quad = 2\eta_2(z-1) - 4C_2\eta_2 z - C_2^2(z^2+1) + 2C_2 z(z-1) \\ &\qquad + 2\beta(z^2+1) + 2\gamma(z-1)^2 - 2\delta(z-1)^2. \end{aligned} \tag{43.13}$$

But from (43.13), we conclude that

$$l(z) = \frac{Az + B}{2z(z-1)}, \tag{43.14}$$

where $A = p - \eta_2 - C_2 + 1$, $B = p + 1 + \eta_2 + C_2$ and $p^2 = 1 - C_2^2 + 2\eta_2 C_2 - 2\eta_2 + 2C_2 + 2\beta + 4\gamma - 4\delta$.

Substituting then (43.14) into (43.11), we get

$$\begin{aligned} &(\eta_2 A - 2\beta - 2\gamma + A^2/4)z^2 + (B - \eta_2 B + \eta_2 A - \frac{1}{2}AB - 2\eta_2 \\ &+ 2\eta_2 C_2 - C_2^2 - 2\beta + 2\gamma - 2\delta)z + \frac{B^2}{4} - B(1+\eta_2) \\ &+ 2(\eta_2 - \beta + \delta) = 0. \end{aligned} \tag{43.15}$$

The equation (43.15) thus obtained may become an identity, if

$$A^2 + 4\eta_2 A - 8\beta - 8\gamma = 0, \tag{43.16}$$

$$2(1 - \eta_2)B + 2\eta_2 A - AB - 4\eta_2 + 4\eta_2 C_2 - 2C_2^2 - 4\beta + 4\gamma - 4\delta = 0, \tag{43.17}$$

and

$$B^2 - 4(\eta_2 + 1)B + 8\eta_2 - 8\beta + 8\delta = 0. \tag{43.18}$$

We first observe that (43.17) is a consequence of (43.16) and (43.18). Indeed, adding (43.17) twice to (43.16) and (43.18), we get

$$(A - B)^2 + 8\eta_2(A - B) - 24\beta + 8\eta_2 C_2 - 4C_2^2 = 0,$$

from which either

$$A - B = -4\eta_2 \pm \sqrt{16\eta_2^2 + 24\beta - 8\eta_2 C_2 + 4C_2^2},$$

or

$$\eta_2 - C_2 = \pm\sqrt{C_2^2 - 2\beta - 2\eta_2 C_2}.$$

Consequently, it immediately follows from (43.16) and (43.18) that

$$A = 2(\eta_3 - \eta_2), \quad B = 2(\eta_4 + \eta_2 + 1), \quad \eta_3^2 = 2\gamma, \quad \eta_4^2 = 1 - 2\delta. \tag{43.19}$$

As $A = p + 1 - \eta_2 - C_2$ and $B = p + 1 + \eta_2 + C_2$, then, taking into account (43.19), we see that $p = \eta_4 + \eta_3$. The constant C_2 may be determined from the equation

$$C_2^2 - 2\eta_2 C_2 - 2(1 + \beta + \gamma - \delta - \eta_3 + \eta_4 - \eta_3\eta_4) = 0.$$

Therefore,

$$C_2 = 1 + \eta_2 - \eta_3 + \eta_4 \tag{43.20}$$

and, consequently, $l(z) = [(\eta_3 - \eta_2)z - 1 - \eta_2 - \eta_4]/[z(z-1)]$. This choice of $l(z)$ turns into an identity the equation that follows by comparing coefficients of w^3. Collecting now together what has been obtained above, implies that (P_6) may be written in the form of the following pair of differential equations:

$$\begin{aligned} z(z-1)w' &= \eta_2 z + \big[(\eta_3 - \eta_2)z - (1 + \eta_2 + \eta_4)\big]w + C_2 w^2 \\ &\quad + C_1 w(w-1)(w-z)v, \\ z(z-1)v' &= \frac{2\alpha - C_2^2}{2C_1} - \big[(\eta_3 - \eta_2)z - (1 + \eta_2 + \eta_4)\big]v - 2C_2 wv \\ &\quad - \frac{C_1}{2}(3w^2 - 2zw - 2w + z)v^2, \end{aligned} \tag{43.21}$$

where $\eta_2^2 = -2\beta$, $\eta_3^2 = 2\gamma$, $\eta_4^2 = 1 - 2\delta$, $C_1 \neq 0$, $C_2 = 1 + \eta_2 - \eta_3 + \eta_4$. In fact, it is a straightforward computation to verify that $w(z)$ satisfies (P_6), as well as to check the reversed conclusion for any sixth Painlevé transcendent.

(2) Since the pair (43.21) equivalent to (P_6) is a Hamiltonian system, with a Hamiltonian

$$\begin{aligned} H_6(z, w, v) = \frac{1}{z(z-1)} \Big\{ &\frac{C_1}{2} w(w-1)(w-z)v^2 \\ &+ \big(\eta_2 z + (\eta_3 z - \eta_2 z - 1 - \eta_2 - \eta_4)w + C_2 w^2\big)v - \frac{2\alpha - C_2^2}{2C_1} w \Big\} \end{aligned}$$

which is a polynomial in w and v, we add here a few remarks about the connection of Hamiltonian systems with the Painlevé equations, although this topics has been not considered otherwise in this book. It is clear that the Hamiltonian $H_6(z, w, v)$ above is not unique. In fact, the function

$$\begin{aligned} \widetilde{H}_6(z, w, v) = \frac{1}{z(z-1)} \Big\{ &\frac{C_1}{2} w(w-1)(w-z)v_1^2 + \big(\eta_2 z + (\eta_3 z - \eta_2 z - 1 \\ &- \eta_2 - \eta_4)w + C_2 w^2\big)v_1 - \frac{2\alpha - C_2^2}{2C_1} w \Big\} + F(z, w), \\ v_1 &= v + K(z, w), \\ F_w(z, w) &= K_z(z, w). \end{aligned}$$

with an arbitrary analytic function $K(z, w)$ is a Hamiltonian of (P_6) as well, i.e. it generates a pair of differential equations equivalent to (P_6), see (43.1). In particular, if $\int^w K_z(z, w)dw$ is a polynomial in w, then $\widetilde{H}_6(z, w, v)$ is a polynomial in w and v as well.

Representations of Painlevé equations in the Hamiltonian pairs were first considered by Malmquist [1] and later on by Gromak and Lukashevich [1] and by Okamoto [1]. This idea is closely related to the problem of monodromy preserving deformation for linear systems, see Jimbo and Miwa [1], [2], Jimbo, Miwa and Ueno [1], Its and Novokshenov [1] and Fokas and Its [1].

The same representation is valid for the other Painlevé equations (P_1)–(P_5) as well. Hence, the first five Painlevé equations may also be written in the form of equivalent Hamiltonian pairs, where corresponding Hamiltonians take the following forms:

$$\begin{aligned} H_1 &:= \frac{1}{2}v^2 - 2w^3 - zw, \\ H_2 &:= \frac{v^2}{2} - \left(w^2 + \frac{z}{2}\right)v - \left(\alpha + \frac{1}{2}\right)w, \\ H_3 &:= \frac{1}{z}\left\{w^2v^2 - [2r_1 zw^2 + (2r_2 + 1)w + 2r_3 z]v + 2r_1(r_2 + r_4)zw\right\} \end{aligned}$$

when $\alpha = -4r_1r_4$, $\beta = -4r_3(1+r_2)$, $\gamma = 4r_1^2$, $\delta = -4r_3^2$,

$$H_4 := 2wv^2 - (w^2 + 2zw + 2r_1)v + r_2w$$

when $\alpha = 2r_2 - r_1 + 1$, $\beta = -2r_1^2$,

$$H_5 := \frac{1}{z}\{w(w-1)^2v^2 - [r_1(w-1)^2 + r_2w(w-1) + r_3zw]v + \frac{1}{4}[(r_1+r_2)^2 - 2\alpha](w-1)\}$$

when $r_1^2 = -2\beta$, $\gamma = -r_3(1+r_2)$, $r_3^2 = -2\delta$.

We invite the reader to apply the idea of (43.1) with the above Hamiltonian expression, and to compare the result with the corresponding pairs of differential equations in the preceding chapters equivalent to (P_1)–(P_5).

(3) We close this section by constructing another pair of differential equations equivalent to (P_6). In fact, we proceed to construct an equivalent pair in the form:

$$\begin{aligned}(w')^2 &= A(z,w) + B(z,w)v,\\ v' &= P(z,w)v + Q(z,w).\end{aligned} \tag{43.22}$$

To shorten the computations, we denote in (P_6)

$$\begin{aligned}L(z,w) &\equiv \frac{1}{2}\left(\frac{1}{w} + \frac{1}{w-1} + \frac{1}{w-z}\right),\\ M(z,w) &\equiv -\frac{(2z-1)(w-z) + z(z-1)}{z(z-1)(w-z)},\\ N(z,w) &\equiv \frac{w(w-1)(w-z)}{z^2(z-1)^2}\left(\alpha + \beta\frac{z}{w^2} + \gamma\frac{z-1}{(w-1)^2} + \delta\frac{z(z-1)}{(w-z)^2}\right).\end{aligned}$$

Differentiating now the first equation of (43.22), we obtain that

$$2w'w'' = A_z + A_ww' + B_zv + B_ww'v + Bv'. \tag{43.23}$$

Replacing the expressions for w'' and $(w')^2$ from (43.23) and (43.22) into (P_6) and the first equation of (43.22), we obtain

$$\begin{aligned}&2M(A+Bv) + 2L(A+Bv)w' + 2Nw'\\ &= A_z + A_ww' + B_ww'v + B_zv + Bv'.\end{aligned} \tag{43.24}$$

Comparing the coefficients in (43.23) of $w'v$ and w', assuming $w' \neq 0$ as we may, we obtain the following pair of differential equations:

$$B_w = 2BL, \quad A_w = 2AL + 2N. \tag{43.25}$$

Solving (43.25), it is easy to obtain that

$$B(z,w) = a(z)w(w-1)(w-z),$$
$$A(z,w) = \frac{2w(w-1)(w-z)}{z^2(z-1)^2}\left(b(z) + \alpha w - \beta\frac{z}{w} - \gamma\frac{z-1}{w-1} - \delta\frac{z(z-1)}{w-z}\right),$$

where $a(z)$ and $b(z)$ are to be determined. The second equation of (43.22) now takes the form

$$v' = v\frac{2MB - B_z}{B} + \frac{2MA - A_z}{B}.$$

Taking $a(z) := 2/z^2(z-1)^2$ and $b(z) := \beta+\gamma-\alpha z-\delta(2z-1)$, a direct computation results in

$$\frac{B_z - 2MB}{B} = \frac{1}{w-z}, \qquad \frac{2MA - A_z}{B} = \frac{2\delta w(w-1)}{(w-z)^2}.$$

Then the pair (43.22) may be written in the form

$$\begin{aligned} \frac{z^2(z-1)^2(w')^2}{2w^2(w-1)^2(w-z)^2} &= \frac{\gamma}{(w-1)^2} - \frac{\beta}{w^2} - \frac{\delta}{(w-z)^2} \\ &\quad + \frac{\alpha+\beta-\gamma+\delta}{w(w-1)} + \frac{v}{w(w-1)(w-z)}, \\ v' &= -\frac{v}{w-z} + \frac{2\delta w(w-1)}{(w-z)^2}. \end{aligned} \tag{43.26}$$

As one may easily verify by direct computations, the first component of a solution $(w(z), v(z))$ of the pair (43.26) now satisfies (P_6), provided $w(z)$ is non-constant, see Garnier [2].

§44 A Riccati differential equation related to (P_6)

Similarly as in the case of all previous Painlevé equations, we can show that whenever the parameters α, β, γ, δ satisfy some specific conditions, then (P_6) admits a one-parameter family of special solutions, generated by the Riccati differential equation

$$w' = a(z)w^2 + b(z)w + c(z). \tag{44.1}$$

Substituting (44.1) into (P_6), we obtain by a standard coefficient comparison of powers of w the following system of equations:

$$z^2(z-1)^2a^2 = 2\alpha, \tag{44.2}$$

$$z^2(z-1)^2[a' - (1+z)a^2] + z(z-1)(2z-1)a + 2\alpha(z+1) = 0, \tag{44.3}$$

$$
\begin{aligned}
&z^2(z-1)^2[b^2+2ac-2b'-3za^2+2(z+1)a'+2(z+1)ab]\\
&\quad -2z(z-1)[(2z-1)b+(1-2z-z^2)a]+2\alpha(z^2+4z+1)\\
&\quad +2\beta z+2\gamma(z-1)+2\delta z(z-1)=0,
\end{aligned}
\tag{44.4}
$$

$$
\begin{aligned}
&z(z-1)^2[(z+1)b'+2bc-c'-za'-2abz]-(z-1)[(2z-1)c\\
&\quad +az^2+(1-2z-z^2)b]-2\alpha(z+1)-2\beta(z+1)\\
&\quad -2\gamma(z-1)-2\delta(z-1)=0,
\end{aligned}
\tag{44.5}
$$

$$
\begin{aligned}
&z(z-1)^2[3c^2-2bc(z+1)+2c'(z+1)-b^2z-2acz-2b'z]\\
&\quad -2(z-1)[(1-2z-z^2)c+bz^2]+2\alpha z+2\beta(z^2+4z+1)\\
&\quad +2\gamma z(z-1)+2\delta(z-1)=0,
\end{aligned}
\tag{44.6}
$$

$$(z-1)^2[zc'+c^2(z+1)]+cz(z-1)+2\beta(z+1)=0, \tag{44.7}$$

$$(z-1)^2c^2+2\beta=0. \tag{44.8}$$

Now, (44.2) and (44.8) immediately result in

$$a(z)=\frac{\eta_1}{z(z-1)},\qquad c(z)=\frac{\eta_2}{z-1},\qquad \eta_1^2=2\alpha,\qquad \eta_2^2=-2\beta. \tag{44.9}$$

It is now a direct computation to check that (44.9) turns (44.3) and (44.7) into identities. Substitution of (44.9) into (44.4) and (44.6) shows that these equations are Riccati differential equations with respect to $b(z)$, namely,

$$
\begin{aligned}
&-b'+b^2/2+\frac{1}{z(z-1)}(1+z(\eta_1-2)+\eta_1)b\\
&+\frac{1}{z^2(z-1)^2}\big(\alpha-\gamma+z(\alpha+\eta_1+\eta_1\eta_2+\beta+\gamma-\delta)\\
&\qquad\qquad +z^2(\alpha+\delta-\eta_1)\big)=0,
\end{aligned}
\tag{44.10}
$$

and

$$
\begin{aligned}
&-b'-b^2/2-\frac{1}{z(z-1)}\big(z+\eta_2(z+1)\big)b\\
&+\frac{1}{z^2(z-1)^2}\big(\beta-\eta_2+z^2(\beta+\gamma)-\delta\\
&\qquad\qquad +z(\alpha+\eta_2-\eta_1\eta_2+\beta-\gamma+\delta)\big)=0.
\end{aligned}
\tag{44.11}
$$

Moreover, (44.5) turns out to be a linear differential equation with respect to b:

$$
\begin{aligned}
&b'+\frac{1}{z(z^2-1)}\big(z^2-1+2z(1-\eta_1+\eta_2)\big)b\\
&-\frac{1}{z(z+1)(z-1)^2}\big(2\alpha(z+1)-\eta_1(z-1)-\eta_2+2\beta-2\gamma-2\delta\\
&\qquad\qquad +z(\eta_2+2\beta+2\gamma+2\delta)\big)=0.
\end{aligned}
\tag{44.12}
$$

Substituting now b' from (44.12) into (44.10) and (44.11) and eliminating b^2 from these equations, we conclude that

$$b(z) = \frac{\lambda z + \mu}{z(z-1)}, \tag{44.13}$$

where

$$\lambda = \frac{\eta_1 - (\alpha + \beta + \gamma + \delta)}{\eta_1 - \eta_2 - 1}, \quad \mu = \frac{\eta_2 - (\alpha + \beta - \gamma - \delta)}{\eta_1 - \eta_2 - 1}, \tag{44.14}$$

provided $\eta_1 - \eta_2 - 1 \neq 0$. Substituting then (44.13), (44.14) into (44.10), (44.11) and (44.12), we see that these equations reduce to identities, provided the following conditions for determining $b(z)$ in (44.1) hold:

$$\begin{aligned}(\alpha + \beta + \gamma + \delta)^2 + 2\eta_1(3\beta - \alpha + \gamma - \delta) + 2\eta_2(3\alpha - \beta - \gamma + \delta)\\ + 2\eta_1\eta_2(\beta - \alpha + \gamma - \delta - 1) + 2(\alpha - \beta - \gamma - 4\alpha\beta - 2\alpha\gamma - 2\beta\delta) = 0.\end{aligned} \tag{44.15}$$

Introducing the notations $\eta_3^2 = 2\gamma$, $\eta_4^2 = 1 - 2\delta$ and recalling $\eta_1^2 = 2\alpha$, $\eta_2^2 = 2\beta$ from (44.9), it is straightforward to see that (44.15) takes the form

$$\prod_{\mu=\pm 1, \nu=\pm 1} (\eta_1 - \eta_2 + \mu\eta_3 + \nu\eta_4 - 1) = 0.$$

By an elementary transformation, we may assume that

$$\eta_1 - \eta_2 + \eta_3 + \eta_4 = 1. \tag{44.16}$$

Therefore, (P_6) admits a one-parameter family of solutions generated by the solutions of the Riccati differential equation

$$w' = \frac{\eta_1}{z(z-1)} w^2 + \frac{\lambda z + \mu}{z(z-1)} w + \frac{\eta_2}{z-1}, \tag{44.17}$$

where λ and μ are taken as in (44.14) and provided that $\eta_1 - \eta_2 - 1 \neq 0$ and that α, β, γ, δ satisfy (44.16).

In order to get a similar result in the case of $\eta_1 - \eta_2 - 1 = 0$, it is clearly necessary that the numerators of λ, μ in (44.14) vanish. Therefore, the following conditions

$$\eta_1 = \alpha + \beta + \gamma + \delta; \quad \eta_2 = \alpha + \beta - \gamma - \delta \tag{44.18}$$

are to be fulfilled. But then, by (44.18), $\gamma + \delta = 1/2$. Solving (44.12) under this condition, we get

$$b(z) = \frac{K(z-1) - 2(\alpha + \beta)}{z(z-1)}, \tag{44.19}$$

where K is a constant. To determine K, (44.10) and (44.11) may be used to conclude that $K = \eta_3 - \alpha - \beta + 1/2$.

Remark. A one-parameter family of solutions of (P_6) generated by a Riccati differential equation may also be constructed by the pair of differential equations in (43.21) equivalent to (P_6). In fact, we may take $2\alpha = C_2^2$ and $v = 0$ in (43.21). The first equation in (43.21) then provides a Riccati differential equation to determine $w(z)$:

$$z(z-1)w' = \eta_2 z + \big[(\eta_3 - \eta_2)z - (1 + \eta_2 + \eta_4)\big]w + C_2 w^2. \tag{44.20}$$

In this case, the condition corresponding to (44.15) takes the form $2\alpha = (1 + \eta_2 - \eta_3 + \eta_4)^2$.

We now continue by considering the Riccati differential equation (44.17) in more detail, assuming that $\alpha \neq 0$. Defining an auxiliary function $v(z)$ from $w(z) = -\frac{z(z-1)}{\eta_1}\frac{v'}{v}$, it is immediate to see that v satisfies the differential equation

$$v'' + \frac{(2-\lambda)z - \mu - 1}{z(z-1)}v' + \frac{\eta_1\eta_2}{z(z-1)^2}v = 0. \tag{44.21}$$

Of course, (44.21) is a linear differential equation with three regular singular points $z = 0, 1, \infty$, admitting as a solution the Riemann P-function

$$v(z) = P\begin{Bmatrix} 0 & 1 & \infty & \\ 0 & -\eta_2 & 0 & z \\ -\mu & -\eta_1 & 1-\lambda & \end{Bmatrix}.$$

Here we have made use of $\lambda + \mu = -\eta_1 - \eta_2$, which follows from (44.14) and (44.9). By the transformation $t = (1-z)^{-1}$ of the independent variable in (44.21), we obtain the hypergeometric equation with parameters $-\eta_1, -\eta_2, \lambda$:

$$t(t-1)v'' + \big[(1 - \eta_1 - \eta_2)t - \lambda\big]v' + \eta_1\eta_2 v = 0. \tag{44.22}$$

Making use of this formulation, we may write the result of the above reasoning as the following statement, see Lukashevich and Yablonski [1]:

Theorem 44.1. *If $v(z)$ is any solution of hypergeometric equation* (44.22), *then the function*

$$w(z) = -\frac{1}{\eta_1}z(z-1)v_z'\big(1/(1-z)\big)/v\big(1/(1-z)\big)$$

is a solution of (P_6), provided $\alpha \neq 0$, $\eta_1 - \eta_2 - 1 \neq 0$, and the parameters α, β, γ, δ satisfy (44.16).

As to the case $\alpha = 0$, then the Riccati differential equation (44.17) reduces to a linear differential equation

$$w' = \frac{\lambda z + \mu}{z(z-1)}w + \frac{\eta_2}{z-1}.$$

The general solution of this equation may be expressed in the form

$$w(z) = z^{-\mu}(z-1)^{\lambda+\mu}\Big(C + \eta_2 \int^z t^{\mu}(t-1)^{-(\lambda+\mu+1)}dt\Big), \tag{44.23}$$

which may be further expressed in terms of beta functions. The solution representation (44.23) is an example of a case when the fixed singular points $z = 0, 1, \infty$ all may be transcendental singularities for a solution of (P_6).

To close this section, we remark that the equations (44.17) and so (P_6) as well may admit algebraic and rational solutions. For example, if we take $\beta = 0$ in (44.23), then $w(z) = Cz^{-\mu}$, $\lambda = \gamma + \delta$, $\mu = -(\gamma + \delta)$, $2\gamma = (\gamma + \delta)^2$. We also remark that (P_6) may have rational solutions depending on an arbitrary parameter different from the equation parameters $\alpha, \beta, \gamma, \delta$. For example, if we take $\lambda = -2$, $\mu = 1$ in (44.23) or $\alpha = 0$, $\beta = -1/2$, $\gamma + \delta = 1/2$ as well, then we obtain a one-parameter family of rational solutions $w(z) = (C + z^2)/2z(z-1)$, where C is an arbitrary constant. This situation is different to the case of the other Painlevé equations (P_1)–(P_5), which do permit finitely many rational solutions only for fixed values of the parameters. We shall consider rational solution of (P_6) in more detail in §48 below.

§45 A first order algebraic differential equation related to (P_6)

It is immediate to observe that we may put $\delta = 0, v = 0$ in the pair (43.26) of differential equations equivalent to (P_6). In this special case, the first equation of (43.26) reduces to

$$\frac{z^2(z-1)^2 w'^2}{2w^2(w-1)^2(w-z)^2} = \frac{\gamma}{(w-1)^2} - \frac{\beta}{w^2} + \frac{\alpha+\beta-\gamma}{w(w-1)}. \tag{45.1}$$

Of course, nonconstant solutions of (45.1) are solutions of (P_6) as well, as shown by Gromak and Prokasheva [1]. We shall consider the equation (45.1) more closely, rewriting it first in the form

$$(w')^2 = \frac{2(w-z)^2}{z^2(z-1)^2}\big(\alpha w^2 + (\beta+\gamma-\alpha)w - \beta\big). \tag{45.2}$$

We assume that $\alpha \neq 0$ and that the quadratic equation

$$\alpha s^2 + (\beta+\gamma-\alpha)s - \beta = 0 \tag{45.3}$$

has two distinct roots s_1 and s_2. Substituting

$$v^2 = (w-s_1)/(w-s_2) \tag{45.4}$$

in (45.2), we find that (45.2) takes the form

$$\frac{dv}{dz} = \frac{\eta_1}{2z(z-1)}\big((z-s_2)v^2 + s_1 - z\big), \quad \eta_1 = \sqrt{2\alpha}. \tag{45.5}$$

Transforming further by

$$v = -\frac{2}{\eta_1}\frac{z(z-1)}{z-s_2}\frac{u'}{u} \tag{45.6}$$

the equation (45.5) reduces to a second order linear differential equation

$$u'' + \frac{z^2 - 2s_2 z + s_2}{z(z-1)(z-s_2)}u' - \frac{\alpha}{2}\frac{(z-s_1)(z-s_2)}{z^2(z-1)^2}u = 0. \tag{45.7}$$

The linear equation (45.7) has four singular points $z = 0, 1, s_2, \infty$. Provided $s_2 \neq 0$ and $s_2 \neq 1$, the point $z = s_2$ is not a singular point for the starting equation (45.2), being created into (45.7) as a result of the transformation (45.6). Concerning the singular point at infinity, we may define $t = 1/z$, and then (45.7) takes the form

$$u'' + \left(\frac{2}{t} - \frac{t^2 s_2 - 2s_2 t + 1}{t(1-t)(1-s_2 t)}\right)u' - \frac{\alpha}{2}\frac{(1-s_1 t)(1-s_2 t)}{t^2(1-t)^2}u = 0.$$

The characteristic roots of the standard indicial equation related to the singular points $z = 0, 1, s_2, \infty$, may be directly computed resulting in $\rho_{1,2}^{(0)} = \rho_{1,2}^{(1)} = \pm\sqrt{(\alpha/2)s_1 s_2}$, $\rho_1^{(s_2)} = 0$, $\rho_2^{(s_2)} = 2$, and $\rho_{1,2}^{(\infty)} = \pm\eta_1/2$, respectively. It is not difficult to see that the root $\rho_2^{(s_2)} = 2$ corresponds to a solution holomorphic in a neighborhood of $z = s_2$, while the root $\rho_1^{(s_2)} = 0$ corresponds to a solution that does not have a logarithmic branch point at $z = s_2$, hence being holomorphic as well around $z = s_2$. This phenomenon has been observed by Ince [1]. If at least one of the roots of the quadratic (45.3) equals to $z = 0, 1$, then the linearized equation (45.7) reduces to a hypergeometric equation.

Assuming next that the roots of (45.3) are equal, then of course the parameters of (P_6) satisfy the condition

$$(\beta + \gamma - \alpha)^2 + 4\alpha\beta = 0. \tag{45.8}$$

In this case (45.2) splits into two Riccati differential equations, see (44.17).

We now proceed to show that the equation (45.1) can be reduced to a hypergeometric equation in general case as well, see Kitaev [2]. To this end, we substitute

$$w = \frac{2\beta + v^2}{2\beta + 2\gamma - 2\alpha - 2\eta_1 v} \tag{45.9}$$

assuming that

$$2\beta + 2\gamma - 2\alpha - 2\eta_1 v \neq 0. \tag{45.10}$$

In fact, if (45.10) is not true, then we must have $2\beta + v^2 = 0$ as well, and this implies (45.8), hence the situation reduces back to the case of equal roots of (45.3). It is now easy to verify that $v(z)$ satisfies a Riccati differential equation

$$2z(z-1)v' = \varepsilon\big(v^2 + 2\beta - z(2\beta + 2\gamma - 2\alpha - 2\eta_1 v)\big), \quad \varepsilon^2 = 1.$$

Applying the following linearizing change of variables

$$v = \eta_2 + (-\eta_2 - \eta_1 + \eta_3)z - 2\varepsilon z(z-1)\frac{u'}{u}, \quad \eta_2^2 = -2\beta,\ \eta_3^2 = 2\gamma, \tag{45.11}$$

it appears that the Riccati equation transforms into the hypergeometric equation

$$\begin{aligned} z(z-1)u'' &+ [-1 - \varepsilon\eta_2 + z(2 + \varepsilon\eta_2 - \varepsilon\eta_3)u' \\ &+ \frac{1}{2}(\varepsilon\eta_1 + \varepsilon\eta_2 - \varepsilon\eta_3 - \alpha - \beta + \gamma + \eta_3\eta_2)u = 0. \end{aligned} \tag{45.12}$$

We now state the following

Theorem 45.1. *Let $u(z)$ is any solution of the hypergeometric equation* (45.12) *for some fixed values of parameter α, β, γ such that* (45.10) *holds. Then the function*

$$w = \frac{2\beta u^2 + \big(((z-1)\eta_2 + z(\eta_1 - \eta_3))u + 2\varepsilon z(z-1)u'\big)^2}{2u\big((2z-1)\alpha + \beta + \gamma + \eta_1(\eta_2(z-1) - z\eta_3)u + 2\varepsilon\eta_1 z(z-1)u'\big)} \tag{45.13}$$

is a solution of (P_6) *with the above parameter values α, β, γ and with $\delta = 0$.*

Proof. The assertion immediately follows from the equivalence of (43.26) and (P_6) and the transformations (45.9) and (45.11). The expression in (45.13) is just a composition of (45.9) and (45.11). □

To complement the above considerations, we are looking at some special cases when (45.5) may be solved in a closed form. To this end, we try to find solutions of (45.5) in the form

$$v(z) = \frac{az + b}{z + c}. \tag{45.14}$$

Substituting into (45.5), we obtain the following system to determine the coefficients a, b, c, assuming $a^2 = 1$. If $a = 1$, then we obtain

$$\begin{aligned} 2(c-b) &= \eta_1(2b - s_2 - 2c + s_1), \\ -2(c-b) &= \eta_1(b^2 - 2s_2 b + 2s_1 c - c^2), \\ s_1 c^2 &= s_2 b^2. \end{aligned} \tag{45.15}$$

From the first and last equations of (45.15) we get

$$b = \frac{\eta_1(s_1 - s_2)}{2(1 + \eta_1)(\sqrt{s_2/s_1} - 1)}, \quad c = b\sqrt{s_2/s_1}. \tag{45.16}$$

Substituting now (45.16) into the second equation of (45.15), we obtain a relationship between parameters α, β, γ, provided a solution of the form (45.14) exists. After a straightforward simplification, this relationship takes the form

$$(\alpha - \beta - \gamma + 2 + 2\eta_1)^2 + 2\beta(2 + \eta_1)^2 = 0. \tag{45.17}$$

Thus, if $s_1 \neq s_2$ and the parameters α, β, γ satisfy (45.17), then the equation (45.5) has a solution (45.14), where $a = 1$ and b and c may be computed from (45.16). Let now $v(z)$ be such a solution. Substituting

$$v = \frac{z+b}{z+c} + y^{-1} \tag{45.18}$$

into (45.5), we get the following equation to determine y

$$z(z-1)\frac{dy}{dz} + \eta_1(z-s_2)\frac{z+b}{z+c}y + \frac{\eta_1(z-s_2)}{2} = 0. \tag{45.19}$$

The general solution of (45.19) may be represented in the form

$$y(z) = z^{k_1}(z-1)^{k_2}(z+c)^{k_3}\left(C - \frac{\eta_1}{2}\int (z-s_2)z^{-k_1-1}(z-1)^{-k_2-1}(z+c)^{-k_3}dz\right),$$

where $c \neq 0, 1$, $k_1 = -bs_2\eta_1/c$, $k_2 = -\eta_1(1+b)(1-s_2)/(c+1)$, $k_3 = \eta_1(b-c)(c+s_2)/(c(c+1))$.

The equation (45.5) also admits a particular solution of the form $w(z) = z(2z + \beta + \gamma - 2)/(\beta + \gamma)$, provided that $\alpha = 0$ and $\beta^2 + (\gamma - 2)^2 + 2\beta(\gamma + 2) = 0$. This may be verified by a reasoning which is quite similar to calculations above for the case (45.14). We leave this case as an exercise to the reader.

§46 Singular points of solutions of (P_6)

This section closely follows the previous pattern of presentation in §30 and §37 in the sense that we proceed to consider formal solutions of (P_6) in the form of a series

$$w(z) = \sum_{k=0}^{\infty} a_k z^k, \tag{46.1}$$

see e.g. Lukashevich [4]. For expressions of branched solutions, see Kimura [1], Takano [2], and Iwasaki, Kimura, Shimomura and Yoshida [1]. Furthermore, for a study on the behavior of solutions around fixed critical points, see Garnier [2].

Substituting the series (46.1) into (P_6) and comparing the coefficients, we obtain the following systems of equations:

$$\begin{aligned} \alpha(a_0-1)^2 &= \gamma, \quad a_0 \neq 0; \\ (a_0-1)(1-2\alpha a_0^2)a_1 &= \gamma a_0^2 + (a_0-1)^2(\beta-\delta); \\ a_0^2(a_0-1)(n^2-2\alpha a_0^2)a_n &= P_n(a_0, \ldots, a_{n-1}), \quad n \geq 2. \end{aligned} \tag{46.2}$$

Here P_n is a polynomial in all of its variables.

We first consider the case when the initial data is not singular, i.e. we assume that $w(0) \neq 0, 1$. We also assume that $\alpha\gamma \neq 0$. Then it readily follows from the first equation of (46.2) that

$$a_0 = 1 + \frac{\sqrt{2\gamma}}{\sqrt{2\alpha}}. \tag{46.3}$$

Substituting now (46.3) into (46.2), we obtain that

$$a_0^2(a_0 - 1)\big[n^2 - (\sqrt{2\alpha} + \sqrt{2\gamma})^2\big]a_n = P_n(a_0, \dots, a_{n-1}), \quad n \geq 1.$$

By an immediate inspection of the equations (46.2), we conclude that the following theorem holds:

Theorem 46.1. *The following properties are valid in a neighborhood of $z = 0$, provided that $\alpha\gamma \neq 0$ and $w(0) \neq 0, 1$:*

(a) *The equation (P_6) admits two formal solutions of the form (46.1) if the condition $n^2 \neq (\sqrt{2\alpha} + \sqrt{2\gamma})^2$ is fulfilled for all $n \in \mathbb{N}$.*

(b) *The equation (P_6) has two formal solutions depending on one arbitrary parameter a_n if the two conditions $n^2 = (\sqrt{2\alpha}+\sqrt{2\gamma})^2$ and $\sigma_n^*(\alpha, \beta, \gamma, \delta) = 0$ are satisfied for some $n \in \mathbb{N}$. Here σ_n^* stands for $P_n(a_0, \dots, a_{n-1})$, when $a_0, \dots, a_{n-1}$ have been replaced by their values expressed in terms of $\alpha, \beta, \gamma, \delta$.*

(c) *The equation (P_6) has no solutions analytic at $z = 0$, if the conditions $n^2 = (\sqrt{2\alpha} + \sqrt{2\gamma})^2$, $\sigma_n^*(\alpha, \beta, \gamma, \delta) \neq 0$ are satisfied for some $n \in \mathbb{N}$.*

As an example, if $\sqrt{2\alpha} + \sqrt{2\gamma} = 1$, then we get that $\sigma_1^* = \gamma(1/2 + \beta - \delta)/\alpha$ which follows by using the second equation of (46.2). The necessary condition for the existence of a formal solution (46.1) is then $\sigma_1^* = 0$; $\alpha\gamma \neq 0$.

Let us consider now the case $\alpha = 0$. A necessary and sufficient condition in this case is $\gamma = 0$. Hence, suppose that $\alpha = \gamma = 0$. Substituting now (46.1) into (P_6), we get the following system of equations:

$$\begin{aligned} a_1 &= (\beta - \delta)(a_0 - 1), \\ a_0^2(a_0 - 1)n^2 a_n &= P_n(a_0, \dots, a_{n-1}), \quad n \geq 2. \end{aligned} \tag{46.4}$$

From (46.4), it is immediate to conclude that a formal solution (46.1) exists and depends on one constant coefficient a_0.

Let us next suppose that $w(0) = 0$. Substituting the series (46.1) into (P_6) while $a_0 = 0$ and comparing coefficients, we get that

$$\begin{aligned} &2(a_1 - 1)^2\beta + a_1^2(1 - 2\delta) = 0, \\ &\big[a_1(a_1 - 1)(n - 1)^2 + 2\beta(a_1 - 1) + a_1(1 - 2\delta)\big]a_n \\ &\quad = P_n(a_1, \dots, a_{n-1}), \quad n \geq 2, \end{aligned} \tag{46.5}$$

where $P_n(a_1, \dots, a_{n-1})$ again is a polynomial in all its variables. From (46.5) we can obtain the following cases.

(a) If $a_1 = 0$, then $\beta = 0$. In this case, we have $\delta a_2 = 0$. If $\delta = 0$, then a_2 is arbitrary. If $a_1 = a_2 = \beta = 0$, then $(3 + 2\delta)a_3 = 0$. If again $3 + 2\delta = 0$, then a_3 is arbitrary. By the same consideration we have either $a_1 = a_2 = \dots = a_{k-1} = 0$, $\beta = 0$, $\delta = (1 - (k-1)^2)/2$ and a_k is arbitrary or $\beta = 0$ and $a_j = 0$, i.e., $w = 0$.

(b) If $a_1 = 1$, then $\delta = 1/2$. In this case, to determine a_2 we have $(1+2\beta)a_2 = 0$. If now $\beta = -1/2$, then a_2 is arbitrary. If $a_1 = a_2 = 0$, $\beta = -2$ and $\delta = 1/2$, then a_3 is arbitrary. Therefore, in this case, we have either $a_1 = a_2 = \dots = a_{k-1} = 0$, $\delta = 1/2$, $\beta = -(k-1)^2/2$, and a_k is arbitrary or $\delta = 1/2$ and $w = z$.

(c) If $\beta = 0$, $\delta = 1/2$, then the constant $a_1 \neq 0, 1$ is arbitrary and the coefficients a_j are determined uniquely by a_1 and the parameters α and γ. If $\beta = 0$, $\delta = 1/2$, $a_1 = 0$, or $a_1 = 1$, then $a_j = 0$.

(d) If $2\beta + 1 - 2\delta = 0$, then from (45.6) we have $\beta(1 - 2a_1) = 0$. If $\beta = 0$, then we have the case considered above. Let now $\delta = \beta + 1/2$ and $a_1 = 1/2$. If now $\beta \neq -(n-1)^2/8$, then a_n are uniquely determined by means of parameters α, β, γ. Otherwise, if $\beta = -(n-1)^2/8$, then the condition $\sigma_n^*(\alpha, \beta, \gamma, \delta) = 0$ is necessary for the existence of a formal solution (46.1). In this case the coefficients a_j, $j > n$, depend on one arbitrary constant a_n.

Thus, using (46.5), we may write:

Theorem 46.2. *The following conclusions hold in the neighborhood of $z = 0$, provided $w(0) = 0$, $w'(0) \neq 0, 1$ and $\beta(1 - 2\delta) \neq 0$, $2\beta + 1 - 2\delta \neq 0$.*

(1) *If $w'(0) = a_1$ and $\Delta := a_1(a_1 - 1)(n-1)^2 + 2\beta(a_1 - 1) + a_1(1 - 2\delta) \neq 0$ for all $n \in \mathbb{N}$, then there exist two formal solutions in the form of a series* (46.1) *in a neighborhood of $z = 0$.*

(2) *If there exist a_1 for which $\Delta = 0$, then the condition $\sigma_n^*(\alpha, \beta, \gamma, \delta) = 0$ is necessary for the existence of a formal solution* (46.1). *In this case the coefficients a_j, $j > n$, depend on one arbitrary constant a_n.*

Similarly, one may study the second case $a_0 = 1$ with singular initial data. We omit this consideration.

A similar analysis may be performed with respect to polar solutions as well. Substituting the series

$$w(z) = \sum_{j=-k}^{\infty} a_j z^j, \; k > 0 \tag{46.6}$$

into (P_6) and comparing the coefficients, it is easy to show that the following statement is true:

Theorem 46.3. *A formal solution* (46.6) *exists, if $\alpha = 0$, $\gamma \neq 0$ and $2\gamma = k^2$. The coefficient a_{-k} is arbitrary and the other coefficients may be uniquely expressed in terms of β, γ, δ and a_{-k}.*

Consequently, the point $z = 0$ is a pole of some solution of (P_6) whenever $\alpha = 0$, and the value of $\sqrt{2\gamma} \neq 0$ is an integer. The point $z = 0$ may also be an algebraic singularity. For example, $w = \sqrt{z}$ is a solution of (P6) provided that $\alpha + \beta = 0$ and $\gamma + \delta = 1/2$.

Using the transformations T_4 and T_6, we may also find necessary conditions for the existence of a analytic or polar solution in a neighborhood of the fixed singular points at $z = 1, \infty$. We again omit these considerations. In all cases, the proof for the convergence of the formal solutions constructed above, in a neighborhood of the singular point in question, may be performed by combining the Briot–Bouquet theory from Appendix A with the pair (43.21) equivalent with (P_6). This reasoning follows the pattern of previous reasoning in §30 and §37.

For completeness, we close this section by giving hints for the series expansions of solutions $w(z)$ of (P_6) around a point $z_0 \neq 0, 1, \infty$, assuming that $w(z)$ tends to $0, 1, z_0, \infty$ as $z \to z_0$. By §6, and the references cited therein, we know that $w(z)$ is locally meromorphic around z_0, hence we have nothing to do concerning the convergence of the corresponding Taylor, respectively, Laurent series around z_0.

(a) To start with, let first $z_0 \neq 0, 1$ be a pole of $w(z)$. Substituting the series

$$w(z) = \sum_{l=-k}^{\infty} a_l z^l, \quad k \in \mathbb{N}; \quad a_{-k} \neq 0, \tag{46.7}$$

into (P_6), the following statements may be verified directly:

(1) If $\alpha \neq 0$, then z_0 is must be a simple pole. The expansion (46.7) in a neighborhood of z_0 then takes the form:

$$w(z) = \frac{z_0(z_0 - 1)}{\sqrt{2\alpha}(z - z_0)} + \sum_{k=0}^{\infty} a_k (z - z_0)^k,$$

where a_0 is an arbitrary parameter.

(2) If $\alpha = 0$, then z_0 has to be a double pole and the polar expansion (46.7) is

$$w(z) = \frac{a_{-2}}{(z - z_0)^2} + \frac{(2z_0 - 1)a_{-2}}{4z_0(z_0 - 1)(z - z_0)} + \sum_{k=0}^{\infty} a_k (z - z_0)^k,$$

where the leading coefficient a_{-2} may be chosen arbitrarily.

(b) We next consider a point $z_0 \neq 0, 1$ such that $w(z_0) = 0$. Substituting now the Taylor series

$$w(z) = \sum_{l=k}^{\infty} a_l (z - z_0)^l, \quad k \in \mathbb{N}, \quad a_k \neq 0 \tag{46.8}$$

into (P_6) and comparing the coefficients, we immediately observe:

(1) If $\beta \neq 0$, then

$$w(z) = \frac{\sqrt{-2\beta}}{z_0 - 1}(z - z_0) + \sum_{k=2}^{\infty} a_k (z - z_0)^k,$$

where a_2 is an arbitrary parameter.

(2) If $\beta = 0$, then the zero at z_0 is a double zero and the leading coefficient a_2 in (46.8) may be chosen arbitrarily.

(c) Similarly as to the cases of zeros and poles, we may treat such points z_0 where $w(z_0) = 1$ by substituting the Taylor series

$$w(z) = 1 + \sum_{l=k}^{\infty} a_l (z - z_0)^l; \quad k \in \mathbb{N}; \quad a_k \neq 0, \tag{46.9}$$

into (P_6) and comparing the coefficients of the lowest possible degree of $z - z_0$. This again divides in two cases as follows:

(1) If $\gamma \neq 0$, then (46.9) has to be of the form

$$w(z) = 1 + \frac{\sqrt{2\gamma}}{z_0}(z - z_0) + \sum_{k=2}^{\infty} a_k (z - z_0)^k.$$

(2) If $\gamma = 0$, then $k = 2$ in (46.9) and the leading coefficients a_2 may again be chosen arbitrarily.

(d) Finally, if $w(z_0) = z_0$, then the Taylor series in a neighborhood of z_0 has to be

$$w(z) = z_0 + a_1 (z - z_0) + \sum_{k=2}^{\infty} a_k (z - z_0)^k, \tag{46.10}$$

where $a_1 = 1 - \sqrt{1 - 2\delta}$, and a_2 is an arbitrary parameter.

Remark. By using the transformations T_3, T_9 and T_8, respectively, we may derive the expansions in (b), (c) and (d) from those in (a).

§47 Connection formulae between solutions of (P_6)

This is section is devoted to considering Bäcklund transformations for the sixth Painlevé equation (P_6). In order to find the first such transformation, let us consider the pair (43.21) of differential equations equivalent to (P_6), fixing $C_1 = 1$ and defining a new function to be determined by the transformation

$$v = -\frac{2\eta_2}{w} - \frac{2\eta_4}{w - z} + \frac{\sigma}{w - u}, \tag{47.1}$$

where $\sigma = -1 + \eta_1 + \eta_2 + \eta_3 + \eta_4$. Then it is immediate to obtain the following pair of differential equations for w and u, see Okamoto [1] and Conte and Musette [1]:

$$z(z-1)w' = w(w-1)(w-z)\left(-\frac{\eta_2}{w} - \frac{\eta_3}{w-1} - \frac{\eta_4 - 1}{w-z} + \frac{\sigma}{w-u}\right), \tag{47.2}$$

$$z(z-1)u' = u(u-1)(u-z)\left(-\frac{\widetilde{\eta}_2}{u} - \frac{\widetilde{\eta}_3}{u-1} - \frac{\widetilde{\eta}_4 - 1}{u-z} + \frac{\sigma}{w-u}\right), \tag{47.3}$$

where

$$\sigma = -1 + \sum_{k=1}^{4} \eta_k, \quad \widetilde{\eta}_j = \eta_j - \frac{1}{2}\sum_{k=1}^{4} \eta_k + \frac{1}{2}. \tag{47.4}$$

It is now a simple computation to verify that (47.4) may also be rewritten in the form

$$\sigma = 1 - \sum_{k=1}^{4} \widetilde{\eta}_k = \sum_{k=1}^{4} (\eta_k^2 - \widetilde{\eta}_k^2), \quad \eta_j = \widetilde{\eta}_j - \frac{1}{2}\Big(\sum_{k=1}^{4} \widetilde{\eta}_k\Big) + \frac{1}{2}. \tag{47.5}$$

Similarly as in (43.21), it is easy to check that the pair (47.2), (47.3) is equivalent to (P_6) with respect to $w(z)$, provided the parameters in (P_6) are $2\alpha = \eta_1^2$, $2\beta = -\eta_2^2$, $2\gamma = \eta_3^2$, $1 - 2\delta = \eta_4^2$. Moreover, as one can immediately see, the pair (47.2), (47.3) is symmetric with respect to the transformation $w \mapsto u$, $\eta \mapsto \widetilde{\eta}$, which means that the pair (47.2), (47.3) is equivalent to (P_6) with respect to u with the parameters $2\widetilde{\alpha} = \widetilde{\eta}_1^2$, $2\widetilde{\beta} = -\widetilde{\eta}_2^2$, $2\widetilde{\gamma} = \widetilde{\eta}_3^2$, $1 - 2\widetilde{\delta} = \widetilde{\eta}_4^2$. Therefore, by (47.2) and (47.3), we have the following statement:

Theorem 47.1. *Let $w = w(z, \alpha, \beta, \gamma, \delta)$ be a solution of (P_6) such that $R(z, w) := z(z-1)w' + (\eta_2 + \eta_3 + \eta_4 - 1)w^2 - (z\eta_2 + z\eta_3 + \eta_2 + \eta_4 - 1)w + z\eta_2 \not\equiv 0$. Then the transformation*

$$T : w \to \widetilde{w} = w - \frac{\sigma w(w-1)(w-z)}{R(z,w)}, \tag{47.6}$$

where $\eta_1^2 = 2\alpha$, $\eta_2^2 = -2\beta$, $\eta_3^2 = 2\gamma$, $\eta_4^2 = 1 - 2\delta$, generates a solution $\widetilde{w}$ of (P_6) with parameters $2\widetilde{\alpha} = \widetilde{\eta}_1^2$, $2\widetilde{\beta} = -\widetilde{\eta}_2^2$, $2\widetilde{\gamma} = \widetilde{\eta}_3^2$, $1 - 2\widetilde{\delta} = \widetilde{\eta}_4^2$, where $\widetilde{\eta}_1, \widetilde{\eta}_2, \widetilde{\eta}_3, \widetilde{\eta}_4$ and σ are determined by (47.4).

Observe that the equation $R(z, w) = 0$ is a Riccati equation. Hence, by (44.16), $R(z, w)$ vanishes identically under condition $\sigma = 0$ only. Therefore, if we apply the Bäcklund transformation (47.6) to the solutions of the Riccati equation, we can always choose the signs of the parameters in such way that $R(z, w) \not\equiv 0$.

In repeated applications of the transformation T in (47.6), the choice of the signs of η_j is independent in each step. Let us denote the choice of the signs of η_j at the nth step by $\varepsilon_j^{(n)}$, where $(\varepsilon_j^{(n)})^2 = 1$, denoting the corresponding transformation (47.6) by $T_{\varepsilon_1,\varepsilon_2,\varepsilon_3,\varepsilon_4}$. By a direct verification, $T_{1,1,1,1} \circ T_{\varepsilon_1,\varepsilon_2,\varepsilon_3,\varepsilon_4} = I$.

We next obtain some nonlinear superposition formulas linking solutions of (P_6) obtained by applying the Bäcklund transformation (47.6). Let $w = w(z, \alpha, \beta, \gamma, \delta)$, $w_1 = T_{1,1,1,1}w$, $w_2 = T_{1,1,1,-1}w$, $\eta_1^2 = 2\alpha$, $\eta_2^2 = -2\beta$, $\eta_3^2 = 2\gamma$, $\eta_4^2 = 1 - 2\delta$, $n = -1 + \eta_1 + \eta_2 + \eta_3 + \eta_4$, $n_1 = -1 + \eta_1 + \eta_2 + \eta_3 - \eta_4$. Then we have an algebraic relation between solutions w, w_1 and w_2 at the initial and the first level of transformation:

$$w_2 = \frac{w(nw_1 - 2\eta_4 z) - n_1 z w_1}{n_1 w + 2\eta_4 w_1 - nz}.$$

To construct another way of linking solutions of (P_6) with different parameters, we may apply the following pair of differential equations with respect to (w, y), see Adler [1],

$$\begin{cases} w' = \dfrac{2s}{z-1} + 2p\dfrac{w}{z} - s\dfrac{z+1}{z(z-1)}w - \dfrac{2(w-z)(w-1)}{z(z-1)}u, \\ y' = -\dfrac{2s}{z-1} + 2q\dfrac{y}{z} + s\dfrac{z+1}{z(z-1)}y + \dfrac{2(y-z)(y-1)}{z(z-1)}u, \end{cases} \tag{47.7}$$

where

$$2s = \mu_3 + \mu_4 - 1, \ \ 4p = 1 + \mu_3 - \mu_4, \ \ 4q = 1 + \mu_4 - \mu_3, \ \ \mu_3^2 = 2\gamma, \ \ \mu_4^2 = 1 - 2\delta,$$

and

$$y = \frac{z}{w}\frac{\beta/2 + (u-s)^2}{u^2 - \alpha/2}.$$

The pair (47.7) of differential equations may be obtained from (43.21) by the transformation

$$v = -\frac{\Theta z + 2u(w-z)(w-1) - (\Theta + 2\eta_4 + \Theta z)w + C_2 w^2}{C_1(w-z)(w-1)w},$$

where $\Theta = 1 + \eta_2 - \eta_3 - \eta_4$, $\eta_j = \mu_j$, $j = 3, 4$. Eliminating y from (47.7), we get the equation (P_6). Observe now that the system (47.7) remains invariant under the substitution

$$w = \widetilde{y}, \quad y = \widetilde{w}, \quad u = -\widetilde{u}, \quad \mu_3 = 1 - \widetilde{\mu}_3, \quad \mu_4 = 1 - \widetilde{\mu}_4.$$

Therefore, y satisfies (P_6) with parameters

$$\widetilde{\alpha} = \alpha, \quad \widetilde{\beta} = \beta, \quad 2\widetilde{\gamma} = (\mu_3 - 1)^2, \quad 2\widetilde{\delta} = 1 - (\mu_4 - 1)^2. \tag{47.8}$$

Thus, the following statement is valid:

Theorem 47.2. *Let $w = w(z, \alpha, \beta, \gamma, \delta)$ be a solution of equation (P_6) such that $w(w-1)(w-z)(u^2 - \alpha/2) \not\equiv 0$, where u is determined by the first equation in (47.7). Then the transformation*

$$F_1 : w \to \widetilde{w} = \frac{z}{w}\frac{\beta/2 + (u-s)^2}{u^2 - \alpha/2} \tag{47.9}$$

also determines a solution of (P_6) with parameters (47.8).

We remark that the inverse transformation of F_1 is

$$F_1^{-1} : \widetilde{w} \to w = \frac{z}{\widetilde{w}} \frac{\beta/2 + (u-s)^2}{u^2 - \alpha/2}, \tag{47.10}$$

where u is determined by the second equation of (47.7) with $\widetilde{w} = y$.

A Bäcklund transformation of (P_6) was also constructed by Fokas and Yorsos [1] while investigating a system equivalent to (P_6) with respect to w:

$$\begin{cases} z(z-1)w' = -\mu_1 w^2 + \frac{1}{2}\lambda(z+1)w - \mu_2 z \\ + \left(\frac{1}{2} + \frac{\mu}{4} + f\right)(z-1)w, \\ z(z-1)(w-1)(w-z)f' = \big((z-1)J + k(z+1)f\big)\frac{w^2}{2} \\ - 2kzfw - \frac{z}{2}\big((z-1)J - k(z+1)f\big), \end{cases} \tag{47.11}$$

where

$$J = f^2 + \frac{f\mu}{2} + \nu, \qquad k = \mu_2 - \mu_1 - 1 \neq 0, \qquad \lambda = \mu_1 + \mu_2,$$

$$\mu = \frac{4}{k}\left(\frac{1}{2} - \gamma - \delta\right), \quad \nu = 2\delta - 1 + \left(\frac{\mu}{4} + \frac{k}{2}\right)^2, \quad \mu_1^2 = 2\alpha, \mu_2^2 = -2\beta.$$

The system (47.11) determines the transformations $H : w \mapsto f, G : f \mapsto w$ between the solutions of (P_6) and the solutions of the equation

$$z^2(z-1)^2\Omega^2 = \left((f')^2 + \frac{J^2 - k^2 f^2}{z(z-1)^2}\right)\psi^2, \tag{47.12}$$

where

$$\Omega = f'' + \frac{3z-1}{2z(z-1)}f' + \frac{1}{z(z-1)^2}(2fJ + \mu J/2 - k^2 f),$$

$$\psi = (z+1)f + \frac{\mu}{4}(z+1) + \frac{\lambda}{2}(z-1).$$

The equation (47.12) may be obtained from the system (47.11) by eliminating w. Moreover, the trivial symmetries in (47.12) generate nontrivial symmetries of the Painlevé equation (P_6). Considering the symmetry

$$S_1 : f(z, \lambda, \mu, \nu, k) \to f(z, \lambda, \mu, \nu, -k),$$

Fokas and Yorsos [1] obtained a Bäcklund transformation of the equation (P_6) according to the scheme $w \overset{H}{\mapsto} f \overset{S_1}{\mapsto} \widetilde{f} \overset{G}{\mapsto} \widetilde{w}$ of the form

$$F_2 : w \to \widetilde{w} = w + \frac{2kf\big((z+1)w - 2z\big)}{-2z(z-1)f' + (z-1)J - (z+1)kf}, \tag{47.13}$$

where f is determined by the first equation in (47.11). The relation between the parameters of solutions $w(z, \alpha, \beta, \gamma, \delta)$ and $\widetilde{w}(z, \widetilde{\alpha}, \widetilde{\beta}, \widetilde{\gamma}, \widetilde{\delta})$ is now determined by

$$2\widetilde{\alpha} = (\mu_2 - 1)^2, \quad 2\widetilde{\beta} = -(\mu_1 + 1)^2, \quad 2\widetilde{\gamma} = 1 - 2\delta, \quad 2\widetilde{\delta} = 1 - 2\gamma. \tag{47.14}$$

Considering other symmetries of the equation (47.12) one can obtain new forms of Bäcklund transformations. For example, with the help of the symmetry

$$S_2 : f(z, \lambda, \mu, \nu, k) \to -f(z, -\lambda, -\mu, \nu, -k)$$

Gromak and Tsegel'nik [2] obtained the Bäcklund transformation $F_3 = G \circ S_2 \circ H$, where

$$F_3 : w \to \widetilde{w} = w + \frac{\big(2(z+1) - 4w\big) f'}{2f' + J/z + \big(kf(z+1)\big)/\big(z(z-1)\big)} \tag{47.15}$$

with

$$2\widetilde{\alpha} = (\mu_1 + 1)^2, \quad 2\widetilde{\beta} = -(\mu_2 - 1)^2, \quad \widetilde{\gamma} = \gamma, \quad \widetilde{\delta} = \delta. \tag{47.16}$$

By a straightforward computation, one may easily verify that up to the choice of the signs of η_j, the transformations T, F_j, $j = 1, 2, 3$, and the symmetries T_j are connected by $F_1 = T_3 \circ T_{-1,-1,1,1} \circ T_{\varepsilon_1,\varepsilon_2,1,1}$, $F_2 = T_{1,1,-1,-1} \circ T_{-1,1,\varepsilon_3,\varepsilon_4}$ and $F_3 = T_3 \circ F_2$.

As a related matter, we remark that by applying the Landin transform for Weierstraß elliptic functions to the equation (42.2), Manin [1] found a new transformation for solutions of (42.2). In terms of the Manin variables τ and q, this transform reads as follows: If $\tau(q)$ is any solution of (42.2) with parameters $\alpha_1 = \alpha_3, \alpha_2 = \alpha_4$, then $\tau(2q)$ is a solution of (42.2) for $\widetilde{\alpha}_1 = 4\alpha_1, \widetilde{\alpha}_2 = 4\alpha_2, \widetilde{\alpha}_3 = \widetilde{\alpha}_4 = 0$.

Finally, we construct some auto-Bäcklund transformations of the sixth Painlevé equation (P_6). Indeed, some trivial auto-Bäcklund transformations can be obtained from the linear-fractional transformations in §42, namely from T_j, $j = 1, 4, 6, 13, 16, 21$:

$$\begin{aligned}
&S_1 : w(z, \lambda) \mapsto 1/w(1/z) = w_1(z, \lambda), \quad \lambda = (\alpha, -\alpha, \gamma, \delta),\\
&S_4 : w(z, \lambda) \mapsto zw(1/z) = w_4(z, \lambda), \quad \lambda = (\alpha, \beta, -\delta + 1/2, \delta),\\
&S_6 : w(z, \lambda) \mapsto 1 - w(1-z) = w_6(z, \lambda), \quad \lambda = (\alpha, \beta, -\beta, \delta),\\
&S_{13} : w(z, \lambda) \mapsto z + (1-z)w(z/(z-1)) = w_{13}(z, \lambda), \quad \lambda = (\alpha, \delta - 1/2, \gamma, \delta),\\
&S_{16} : w(z, \lambda) \mapsto w\big(z/(z-1)\big)\big(w(z/(z-1)) - 1\big)^{-1} = w_{16}(z, \lambda),\\
&\qquad \lambda = (\alpha, \beta, \alpha, \delta),\\
&S_{21} : w(z, \lambda) \mapsto zw(1-z)\big(w(1-z) + z - 1\big)^{-1} = w_{21}(z, \lambda),\\
&\qquad \lambda = (\alpha, \beta, \gamma, -\alpha + 1/2).
\end{aligned}$$

In the general case, the transformations S_j above lead to new solutions. For example,

$$w(z,\lambda)=2,\quad \lambda=(1/2,-2,1/2,0),\quad S_4w(z,\lambda)=2z,$$
$$w(z,\lambda)=h,\quad \lambda=\left(\frac{1}{2h^2},-\frac{1}{2},\frac{1}{2h^2}(h-1)^2,0\right),\quad S_{13}w(z,\lambda)=(1-h)z+h,$$
$$w(z,\lambda)=h,\quad \lambda=\left(1/2,-h^2/2,(h-1)^2/2,0\right),\quad S_{21}w(z,\lambda)=zh/(z+h-1),$$

where $h\neq 0,1$. It deserves to be mentioned, that for the other Painlevé equations (P_2)–(P_5), auto-Bäcklund transformations for other values of the parameters may be obtained from the superposition of transformations $(T^{-1})^n\circ Sj\circ T^n$, $n\in\mathbb{Z}$.

§48 Elementary solutions of (P_6)

In this section, we shall apply the Bäcklund transformations obtained in the previous §47 to construct a number of exact solutions of the equation (P_6). In fact, most of these exact solutions below are either rational or algebraic functions.

To start with, it is easy to obtain a parameter change formula for admissible solutions by the composed applications of Bäcklund transformations. As an example, the following theorem is easily found:

Theorem 48.1. *Let (n_1,n_2,n_3,n_4) be a quadruple of integers with $\sum_{j=1}^4 n_j\in 2\mathbb{Z}$, and let $(\eta_1^*,\eta_2^*,\eta_3^*,\eta_4^*)$ denote an arbitrary permutation of $(\eta_1,\eta_2,\eta_3,\eta_4)=(\sqrt{2\alpha},\sqrt{-2\beta},\sqrt{2\gamma},\sqrt{1-2\delta})$. Then, for arbitrary signs κ_i, $\kappa_i^2=1$, $1\le i\le 4$, there exist compositions of T in (47.6) and T_j in §42 which changes $w(z,\alpha,\beta,\gamma,\delta)$ into $\widetilde{w}(z,\widetilde{\alpha},\widetilde{\beta},\widetilde{\gamma},\widetilde{\delta})$ with the parameters, respectively, given by*

$$\begin{aligned}\widetilde{\alpha}&=(\eta_1^*+n_1)^2/2, & -\widetilde{\beta}&=(\eta_2^*+n_2)^2/2,\\ \widetilde{\gamma}&=(\eta_3^*+n_3)^2/2, & 1/2-\widetilde{\delta}&=(\eta_4^*+n_4)^2/2,\end{aligned}\tag{48.1}$$

and

$$\begin{aligned}\widetilde{\alpha}&=(\kappa_1\eta_1^*-\kappa_2\eta_2^*-\kappa_3\eta_3^*-\kappa_4\eta_4^*+2n_1+1)^2/8,\\ -\widetilde{\beta}&=(-\kappa_1\eta_1^*+\kappa_2\eta_2^*-\kappa_3\eta_3^*-\kappa_4\eta_4^*+2n_2+1)^2/8,\\ \widetilde{\gamma}&=(-\kappa_1\eta_1^*-\kappa_2\eta_2^*+\kappa_3\eta_3^*-\kappa_4\eta_4^*+2n_3+1)^2/8,\\ 1/2-\widetilde{\delta}&=(-\kappa_1\eta_1^*-\kappa_2\eta_2^*-\kappa_3\eta_3^*+\kappa_4\eta_4^*+2n_4+1)^2/8.\end{aligned}\tag{48.2}$$

Proof. A suitable composition of the transpositions T_1, T_{16}, T_{21} gives rise to an arbitrary permutation $(\eta_1^*,\eta_2^*,\eta_3^*,\eta_4^*)$. Hence we may suppose that $\eta_1^*=\eta_1$, $\eta_2^*=\eta_2$, $\eta_3^*=\eta_3$, $\eta_4^*=\eta_4$. For each (ε,ν) such that $\varepsilon^2=\nu^2=1$, the transformation F_3 in (47.15) with a suitable choice of signs of η_j yields

$$H_{\varepsilon,\nu}=F_{3,\varepsilon,\nu}:(\alpha,-\beta,\gamma,1/2-\delta)\mapsto\left((\eta_1+\varepsilon)^2/2,(\eta_2+\nu)^2/2,\gamma,1/2-\delta\right).$$

Furthermore,

$$I_{\varepsilon,\nu} = T_{16} \circ F_{3,\varepsilon,\nu} \circ T_{16} : (\alpha, -\beta, \gamma, 1/2 - \delta) \\ \mapsto \big(\alpha, (\eta_2 + \nu)^2/2, (\eta_3 + \varepsilon)^2/2, 1/2 - \delta\big),$$

and

$$J_{\varepsilon,\nu} = T_9 \circ F_{3,\varepsilon,\nu} \circ T_9 : (\alpha, -\beta, \gamma, 1/2 - \delta) \mapsto \big(\alpha, -\beta, (\eta_3 + \varepsilon)^2/2, (\eta_4 + \nu)^2/2\big).$$

Then, for nonnegative integers k_j, $1 \le j \le 8$, we have

$$H_{1,1}^{k_1} \circ H_{-1,-1}^{k_2} \circ I_{1,1}^{k_3} \circ I_{-1,-1}^{k_4} \circ J_{1,1}^{k_5} \circ J_{-1,-1}^{k_6} \circ J_{1,-1}^{k_7} \circ J_{-1,1}^{k_8} :$$
$$(\alpha, -\beta, \gamma, 1/2 - \delta) \mapsto \big((\eta_1 + K_1)^2/2, (\eta_2 + K_2)^2/2, (\eta_3 + K_3)^2/2, (\eta_4 + K_4)^2/2\big),$$

where

$$\begin{cases} K_1 = k_1 - k_2, \\ K_2 = k_1 - k_2 + k_3 - k_4, \\ K_3 = k_3 - k_4 + k_5 - k_6 + k_7 - k_8, \\ K_4 = k_5 - k_6 - k_7 + k_8, \end{cases}$$

and so

$$\begin{cases} k_1 - k_2 = K_1, \\ k_3 - k_4 = K_2 - K_1, \\ k_5 - k_6 = (K_1 - K_2 + K_3 + K_4)/2, \\ k_7 - k_8 = (K_1 - K_2 + K_3 - K_4)/2. \end{cases}$$

This implies that, for an arbitrary quadruple (n_1, n_2, n_3, n_4), $n_j \in \mathbb{Z}$ satisfying $\sum_{j=1}^4 n_j \in 2\mathbb{Z}$, we can choose $k_j \in \mathbb{N} \cup \{0\}$, $1 \le j \le 8$, in such a way that $K_l = n_l$, $1 \le l \le 4$. The formula (48.2) is obtained by combining (47.4) with (48.1). □

Remark. By the above theorem we can obtain the fundamental domain:

$$G := \big\{(\eta_1, \eta_2, \eta_3, \eta_4) \mid 0 \le \eta_j \le 1,\ 0 \le \eta_j + \eta_k \le 1,\ \ j, k = 1, 2, 3, 4,\ j \ne k\big\}.$$

To concretize the above reasoning, let us have a look at the solutions expressed in terms of elliptic and hypergeometric functions, see Okamoto [6], Gromak and Tsegel'nik [2].

We first consider elliptic solutions with the two sets of parameters $(\alpha, \beta, \gamma, \delta) = (0, 0, 0, 1/2)$ and $(\alpha, \beta, \gamma, \delta) = (1/8, -1/8, 1/8, 3/8)$ as the initial solutions for the transformation T. Note that the second parameter set may be obtained by applying the transformation T in (47.6) to the elliptic solutions with the first set of parameters. As to the representation of elliptic solutions for this set of parameters, see Kitaev and Korotkin [1]. We may state the results as

Theorem 48.2. *The equation (P_6) with the parameters either*

$$2\alpha = n_1^2, \quad -2\beta = n_2^2, \quad 2\gamma = n_3^2/2, \quad 1-2\delta = n_4^2,$$
$$n_j \in \mathbb{N} \cup \{0\}, \quad \sum_{j=1}^{4} n_j \in 2\mathbb{Z} \tag{48.3}$$

or

$$\alpha = \frac{1}{8}(2n_1-1)^2, \quad \beta = -\frac{1}{8}(2n_2-1)^2, \quad \gamma = \frac{1}{8}(2n_3-1)^2,$$
$$1-2\delta = \frac{1}{4}(2n_4-1)^2, \quad n_j \in \mathbb{N} \tag{48.4}$$

may be integrated in terms of elliptic functions.

Proof. We first remark that for $(\alpha, \beta, \gamma, \delta) = (0, 0, 0, 1/2)$, the condition $\sigma \neq 0$ is satisfied. This means that we can apply the transformation S_{k_1,k_2,k_3,k_4} associated with the parameters (48.1). As result we have (48.3). We now apply the transformation T from Theorem 48.1 to the solutions of (P_6) with parameters $(0, 0, 0, 1/2)$, expressed in terms of elliptic functions. Then $T_{\varepsilon_1,\varepsilon_2,\varepsilon_3,\varepsilon_4} : w(z, 0, 0, 0, 1/2) \mapsto w(z, 1/8, -1/8, 1/8, 3/8)$. Let us now fix $\eta_1 = -1/2$, $\eta_2 = -1/2$, $\eta_3 = -1/2$, $\eta_4 = -1/2$. Then $\sigma \neq 0$. The next step of transformation T gives $w(z, 9/8, -1/8, 1/8, 3/8)$. Applying now the transformations

$$H_1 := T^2_{1,-1,-1,-1} \circ T_{\varepsilon_1,\varepsilon_2,\varepsilon_3,\varepsilon_4} : w(\alpha, \beta, \gamma, \delta) \mapsto w(z, (\eta_1 - 2\varepsilon_1)^2/2, \beta, \gamma, \delta),$$
$$H_2 := T^2_{-1,1,-1,-1} \circ T_{\varepsilon_1,\varepsilon_2,\varepsilon_3,\varepsilon_4} : w(\alpha, \beta, \gamma, \delta) \mapsto w(z, \alpha, -(\eta_2 - 2\varepsilon_2)^2/2, \gamma, \delta),$$
$$H_3 := T^2_{-1,-1,1,-1} \circ T_{\varepsilon_1,,\varepsilon_2,\varepsilon_3,\varepsilon_4} : w(\alpha, \beta, \gamma, \delta) \mapsto w(z, \alpha, \beta, (\eta_3 - 2\varepsilon_3)^2/2, \delta),$$
$$H_4 := T^2_{-1,-1,-1,1} \circ T_{\varepsilon_1,,\varepsilon_2,,\varepsilon_3,\varepsilon_4} : w(\alpha, \beta, \gamma, \delta) \mapsto w(z, \alpha, \beta, \gamma, \frac{1}{2} - (\eta_4 - 2\varepsilon_4)^2/2).$$

to the solutions $w(z, 1/8, -1/8, 1/8, 3/8)$, $w(z, 9/8, -1/8, 1/8, 3/8)$, we get (48.4). □

We next take as the initial solutions the solutions of the Riccati equation (44.17) under the conditions (44.16), and the solutions of the equation (45.1) expressed in terms of the hypergeometric function. In this case, applying the transformations H_j above to these solutions, we obtain the following statement.

Theorem 48.3. *The equation (P_6) with the parameters*

$$2\alpha = (\eta_2 + \eta_3 + \eta_4 + 2n - 1)^2, \quad n \in \mathbb{N}, \tag{48.5}$$

or

$$\left(\alpha - \frac{n^2}{2}\right)\left(\beta + \frac{n^2}{2}\right)\left(\gamma - \frac{n^2}{2}\right)\left(\delta - \frac{1}{2} + \frac{n^2}{2}\right) = 0, \quad n \in \mathbb{N} \tag{48.6}$$

has one-parameter families of solutions expressed in terms of the hypergeometric functions.

We next proceed to consider examples of how to construct solution hierarchies of the equation (P_6) from some given initial solutions. The examples have been collected in the subsequent tables from Table 48.1 to Table 48.5. As the first example, to construct a rational solution hierarchy, we may take the solution

$$\begin{gathered} w(z) = h, \quad \alpha h^2 + \beta = 0, \quad \gamma = \alpha(h-1)^2, \\ \delta = 0, \quad h \neq 0, 1; \quad \eta_1 = \sqrt{2\alpha}, \end{gathered} \tag{48.7}$$

as the initial solution to obtain Table 48.1.

As the second example, if we take the solution

$$w(z) = 1 - a(1-z)^2, \tag{48.8}$$

with $\alpha = \gamma = \delta = 0$, $\beta = -2$, as the initial solution, we obtain the hierarchy of rational solutions of (P_6) in Table 48.2. For a complete analysis of rational solutions of (P_6), see the recent papers by Mazzocco [1] and Yuan and Li [1].

To construct an explicit example of algebraic solutions, we may take the solution

$$w = z^n, \quad \alpha = \beta = 0, \quad \gamma = n^2/2, \quad \delta = -n(n-2)/2 \tag{48.9}$$

as the initial solution. As the result, we now get Table 48.3.

Algebraic solutions of (P_6) in implicit form, under some special values of the parameters, have been found by Dubrovin and Mazzocco [1]. In this paper, all algebraic solutions of the equation (P_6) were constructed, provided $\alpha = (2s-1)^2/2$, $\beta = \gamma = 0$, $\delta = 1/2$, where s is an arbitrary complex parameter satisfying the condition $2s \notin \mathbb{Z}$. However, such algebraic solutions can be built for other parameter values as well. Taking as the initial solution

$$w(s) = (1+s)^2(2+s)/\big(4(1+2s^2)\big), \quad z(s) = (s+2)^2/(8s) \tag{48.10}$$

with $\alpha = 2$, $\beta = \gamma = 0$, $\delta = 1/2$, we obtain the solution family in Table 48.4.

We close this chapter by giving, in the final Table 48.5 below, some further examples of algebraic solutions for the sixth Painlevé equation.

Table 48.1. Rational solutions obtained from (48.7) by (47.6).

α	β	γ	δ	$w(z)$
$\frac{(-1+(-1+h)\eta_1)^2}{2}$	$-\frac{1}{2}$	$\frac{(-1+\eta_1)^2}{2}$	$\frac{1-h^2\eta_1{}^2}{2}$	$\frac{z}{1+(-h+z)\eta_1}$
$\frac{h^2\eta_1{}^2}{2}$	$\frac{-\eta_1{}^2}{2}$	2	$\frac{-((-1+h)\eta_1(2+(-1+h)\eta_1))}{2}$	$\frac{z(z+(h-z)\eta_1)}{h+h(h-z)\eta_1}$
$\frac{1}{2}$	$\frac{-(1+(-1+h)\eta_1)^2}{2}$	$\frac{(-1+h\eta_1)^2}{2}$	$\frac{1-\eta_1{}^2}{2}$	$z+(h-z)\eta_1$
$\frac{(1+h\eta_1)^2}{2}$	$\frac{-(-1+\eta_1)^2}{2}$	$\frac{1}{2}$	$\frac{1}{2}-\frac{(-1+h)^2\eta_1{}^2}{2}$	$1+\frac{-1+z}{1+(h-z)\eta_1}$
$\frac{(-1+h\eta_1)^2}{2}$	$\frac{-(-1+\eta_1)^2}{2}$	$\frac{1}{2}$	$\frac{2-2(-1+h)^2\eta_1{}^2}{4}$	$\frac{z(-1+\eta_1)}{-1+h\eta_1}$
$\frac{(-2+\eta_1)^2}{2}$	$\frac{-(h^2\eta_1{}^2)}{2}$	$\frac{(-1+h)^2\eta_1{}^2}{2}$	0	$z+\frac{(-1+z)z}{1-2z+(-h+z)\eta_1}$
2	$\frac{-((-1+h)^2\eta_1{}^2)}{2}$	$\frac{h^2\eta_1{}^2}{2}$	$\frac{-((-2+\eta_1)\eta_1)}{2}$	$-(\frac{(-1+h)z(z+(h-z)\eta_1)}{h-2hz+z^2-(h-z)^2\eta_1})$
$\frac{(1+(-1+h)\eta_1)^2}{2}$	$-\frac{1}{2}$	$\frac{(-1+\eta_1)^2}{2}$	$\frac{1-h^2\eta_1{}^2}{2}$	$\frac{z+(h-z)\eta_1}{1+(-1+h)\eta_1}$
$\frac{(-1+h)^2\eta_1{}^2}{2}$	-2	$\frac{\eta_1{}^2}{2}$	$\frac{-(h\eta_1(-2+h\eta_1))}{2}$	$-(\frac{h+(-2+z)z-(h-z)^2\eta_1}{(-1+h)(-1+(h-z)\eta_1)})$
$\frac{(-1+\eta_1)^2}{2}$	$\frac{-(-1+h\eta_1)^2}{2}$	$\frac{(1+(-1+h)\eta_1)^2}{2}$	$-\frac{3}{2}$	$\frac{(-1+h\eta_1)(z+(h-z)\eta_1)}{(-1+\eta_1)(-1+2z+(h-z)\eta_1)}$
$\frac{1}{2}$	$\frac{-(-1+(-1+h)\eta_1)^2}{2}$	$\frac{(1+h\eta_1)^2}{2}$	$\frac{-((-3+\eta_1)(-1+\eta_1))}{2}$	$\frac{(z+(h-z)\eta_1)(-h+2z-z^2+(h-z)^2\eta_1)}{h-2hz+z^2-(h-z)^2\eta_1}$

Table 48.2. Rational solutions obtained from (48.8) by (47.6)

α	β	γ	δ	$w(z)$
0	-2	0	0	$1-a(1-z)^2$
$\frac{1}{2}$	$-\frac{1}{2}$	$\frac{1}{2}$	$-\frac{3}{2}$	$-(\frac{(-1+a(-1+z)^2)z}{-1+a(-1+z)^2+2z})$
0	0	2	0	$\frac{az^2}{-1+a}$
$\frac{9}{2}$	$-\frac{1}{2}$	$\frac{1}{2}$	$\frac{1}{2}$	$\frac{(-1+a(-1+z)^2)z}{-1+a(-1+z)^2(1+2z)}$
2	0	0	0	$\frac{z^2}{-1+a(-1+z)^2+2z}$
$\frac{1}{2}$	$-\frac{9}{2}$	$\frac{1}{2}$	$\frac{1}{2}$	$\frac{-2+3z+a(2-3z+z^3)}{-1+a(-1+z)^2+2z}$
0	0	0	-4	$\frac{(-1+a(-1+z)^2)(1+a^2(-1+z)^2-a(2-2z+z^2))}{a(-1+a(-1+z)^2+2z)^2}$
2	-2	2	0	$\frac{(-1+a(-1+z)^2)z(-2+3z+a(2-3z+z^3))}{(-1+a(-1+z)^2+2z)(-1+a(-1+z)^2(1+2z))}$
$\frac{9}{2}$	$-\frac{9}{2}$	$\frac{1}{2}$	$-\frac{3}{2}$	$\frac{(-1+a-4az^3+3az^4)(-2+3z+a(2-3z+z^3))}{(-1+a(-1+z)^2(1+2z))(-3+4z+a(3-4z+z^4))}$
0	0	2	-4	$\frac{-1+a}{az^2}$
8	-2	0	0	$\frac{z(-2+3z+a(2-3z+z^3))}{-3+4z+a(3-4z+z^4)}$
$\frac{1}{2}$	$-\frac{1}{2}$	$\frac{1}{2}$	$-\frac{15}{2}$	$\frac{-2+3z+a(2-3z+z^3)}{-1+a(-1+z)^2(1+2z)}$

Table 48.3. Algebraic solutions obtained from (48.9) by (47.6).

α	β	γ	δ	$w(z)$
0	0	$\frac{n^2}{2}$	$-\frac{n(n-2)}{2}$	z^n
$\frac{1}{2}$	$-\frac{1}{2}$	$\frac{(-1+n)^2}{2}$	$\frac{-((-1+n)(1+n))}{2}$	$-(\frac{z-nz^n+(-1+n)z^{1+n}}{-1+n-nz+z^n})$
$\frac{(-2+n)^2}{2}$	$\frac{-n^2}{2}$	0	0	$\frac{z(z-nz^n+(-1+n)z^{1+n})}{z^2-(-1+n)z^n+(-2+n)z^{1+n}}$
$\frac{(1+n)^2}{2}$	$\frac{-(-1+n)^2}{2}$	$\frac{1}{2}$	$\frac{1}{2}$	$\frac{z-nz^n+(-1+n)z^{1+n}}{1-(1+n)z^n+nz^{1+n}}$
$\frac{n^2}{2}$	$\frac{-(-2+n)^2}{2}$	0	0	$\frac{(-2+n)z-(-1+n)z^2+z^n}{-1+n-nz+z^n}$
$\frac{(-1+n)^2}{2}$	$\frac{-(1+n)^2}{2}$	$\frac{1}{2}$	$\frac{1}{2}$	$\frac{n-nz+z(-1+z^n)}{-1+n-nz+z^n}$
$\frac{(-1+n)^2}{2}$	$\frac{-(-1+n)^2}{2}$	$\frac{1}{2}$	$-\frac{3}{2}$	$\frac{((-2+n)z-(-1+n)z^2+z^n)(z-nz^n+(-1+n)z^{1+n})}{(-1+n-nz+z^n)(z^2-(-1+n)z^n+(-2+n)z^{1+n})}$
0	0	$\frac{(-2+n)^2}{2}$	$\frac{-(n(2+n))}{2}$	$\frac{(z-nz^n+(-1+n)z^{1+n})^2}{z^n(-1+n-nz+z^n)^2}$
$\frac{n^2}{2}$	$\frac{-n^2}{2}$	2	0	$\frac{(z-nz^n+(-1+n)z^{1+n})(n-nz+z(-1+z^n))}{(-1+n-nz+z^n)(1-(1+n)z^n+nz^{1+n})}$

Table 48.4. Some implicit algebraic solutions of equation (P_6).

α	β	γ	δ	$z(s)=\frac{(2+s)^2}{8s}$, $w(s)$
2	0	0	$\frac{1}{2}$	$\frac{(1+s)^2(2+s)}{4(1+2s^2)}$
$\frac{9}{8}$	$-\frac{1}{8}$	$\frac{1}{8}$	$\frac{3}{8}$	$\frac{(1+s)(2+s)}{6s}$
$\frac{1}{8}$	$-\frac{9}{8}$	$\frac{9}{8}$	$-\frac{5}{8}$	$\frac{-2+s+5s^2+2s^3}{2(s+2s^3)}$
0	$-\frac{1}{2}$	$\frac{1}{2}$	$-\frac{3}{2}$	$\frac{-((-3+s)(1+s)(2+s))}{12}$
$\frac{9}{8}$	$-\frac{25}{8}$	$\frac{1}{8}$	$\frac{3}{8}$	$\frac{5(1+s)(2+s)(3+2s)}{6(4+9s+8s^2+2s^3)}$
$\frac{49}{8}$	$-\frac{1}{8}$	$\frac{1}{8}$	$\frac{3}{8}$	$\frac{(2+s)(3+2s)}{14s}$
$\frac{9}{2}$	$-\frac{1}{2}$	$\frac{1}{2}$	0	$\frac{(2+s)(-1+2s)(3+2s)}{-12+24s^2}$
$\frac{1}{2}$	0	$\frac{9}{2}$	$\frac{1}{2}$	$-(\frac{(1-2s)^2(2+s)}{s(1+2s^2)(-4+9s-8s^2+2s^3)})$
$\frac{25}{8}$	$-\frac{9}{8}$	$\frac{9}{8}$	$-\frac{5}{8}$	$\frac{(2+s)(-1+2s)(1+4s+4s^3+4s^4)}{10s(-1+4s^4)}$
$\frac{1}{2}$	0	0	-4	$\frac{(1-2s)^2(2+s)}{4+8s^2}$
$\frac{9}{8}$	$-\frac{1}{8}$	$\frac{25}{8}$	$-\frac{21}{8}$	$\frac{(6+7s+2s^2)(1+4s+4s^3+4s^4)}{6(-1+2s^2)(4+9s+8s^2+2s^3)}$

Table 48.5. Other examples of algebraic solutions of (P_6).

α	β	γ	δ	$w(z)$	$z(s)$
$\frac{25}{18}$	0	0	$\frac{1}{2}$	$\frac{(2-s)(1+s)(-3+s^2)^2}{(2+s)(9-10s^2+5s^4)}$	$\frac{(2-s)^2(1+s)}{(1-s)(2+s)^2}$
$\frac{2}{9}$	$-\frac{1}{2}$	0	$\frac{1}{2}$	$\frac{3(-2-s+s^2)}{(2+s)(-3+s^2)}$	$\frac{(2-s)^2(1+s)}{(1-s)(2+s)^2}$
$\frac{1}{18}$	$-\frac{8}{9}$	$\frac{8}{9}$	$-\frac{7}{18}$	$-(\frac{(-3+s^2)(-18+9s+10s^2-5s^3+50s^4-25s^5-10s^6+5s^7)}{(-1+s)(2+s)(3+s^2)(9-10s^2+5s^4)})$	$\frac{(2-s)^2(1+s)}{(1-s)(2+s)^2}$
$\frac{8}{9}$	$-\frac{1}{18}$	$\frac{1}{18}$	$\frac{4}{9}$	$-(\frac{(-2+s)(-3+s^2)}{(-1+s)(2+s)(3+s^2)})$	$\frac{(2-s)^2(1+s)}{(1-s)(2+s)^2}$
$\frac{8}{9}$	$-\frac{1}{18}$	$\frac{1}{18}$	$\frac{4}{9}$	$\frac{-6+3s+2s^2-s^3}{(-1+s)(2+s)(3+s^2)}$	$\frac{(2-s)^2(1+s)}{(1-s)(2+s)^2}$
$\frac{1}{2}$	0	0	$\frac{5}{18}$	$\frac{2-s}{2-s+s^2+s^3}$	$\frac{(2-s)^2(1+s)}{(1-s)(2+s)^2}$
$\frac{2}{9}$	$-\frac{1}{18}$	$\frac{1}{18}$	$\frac{4}{9}$	$\frac{-2+s}{-2+s+s^2}$	$\frac{(2-s)^2(1+s)}{(1-s)(2+s)^2}$
$\frac{1}{18}$	$-\frac{2}{9}$	$\frac{8}{9}$	$\frac{5}{18}$	$\frac{2+3s-2s^2}{(2+s)(1+s+s^2)}$	$\frac{(2-s)^2(1+s)}{(1-s)(2+s)^2}$
2	0	0	$\frac{1}{2}$	$\frac{3(3-s)(1+s)(-3+s^2)^2}{(3+s)^2(9-9s^2+3s^4+s^6)}$	$\frac{(3-s)^3(1+s)}{(1-s)(3+s)^3}$
$\frac{1}{2}$	$-\frac{1}{2}$	0	$\frac{1}{2}$	$-(\frac{(-3+s)^2(1+s)}{(3+s)(-3+s^2)})$	$\frac{(3-s)^3(1+s)}{(1-s)(3+s)^3}$

Table 48.5. Continued.

α	β	γ	δ	$w(z)$	$z(s)$
$\frac{9}{8}$	$-\frac{1}{8}$	$\frac{1}{8}$	$\frac{3}{8}$	$\frac{-(27-9s-3s^4+s^5)}{3(-1+s)(3+s)^2(1+s^2)}$	$\frac{(3-s)^3(1+s)}{(1-s)(3+s)^3}$
$\frac{1}{8}$	$-\frac{9}{8}$	$\frac{9}{8}$	$-\frac{5}{8}$	$\frac{(-3+s^2)(27-9s+18s^2-6s^3-108s^4+36s^5+18s^6-6s^7-3s^8+s^9)}{(-1+s)(3+s)^2(1+s^2)(9-9s^2+3s^4+s^6)}$	$\frac{(3-s)^3(1+s)}{(1-s)(3+s)^3}$
0	-2	$\frac{1}{2}$	0	$\frac{(-3+s)(3+s^2)(-9-6s^2+36s^4-6s^6+s^8)}{9(-1+s)^2(1+s)(3+s)^2(1+s^2)^2}$	$\frac{(3-s)^3(1+s)}{(1-s)(3+s)^3}$
$\frac{1}{8}$	$-\frac{25}{8}$	$\frac{1}{8}$	$\frac{3}{8}$	$\frac{5(1+s)(3+s^2)(-63+21s-117s^2+39s^3-9s^4+3s^5-3s^6+s^7)}{(3+s)^2(1+s^2)(-105-45s^2+21s^4+s^6)}$	$\frac{(3-s)^3(1+s)}{(1-s)(3+s)^3}$
$\frac{1}{2}$	-2	0	0	$\frac{(3+s^2)(-945-1575s-4860s^2-2220s^3+1242s^4+270s^5-60s^6-60s^7+15s^8+s^9)}{(3+s)^2(3+3s+9s^2+s^3)(-105-45s^2+21s^4+s^6)}$	$\frac{(3-s)^3(1+s)}{(1-s)(3+s)^3}$

Chapter 10

Applications of Painlevé equations

This chapter is devoted to considering Painlevé differential equations from the point of view of applications. During the last three decades both the wave theory in nonlinear media and the analytic theory of differential equations have simultaneously experienced a considerable development under mutual influence. This development has been, essentially, stimulated by the needs of physics. Indeed, interaction of large amplitude waves has frequently appeared in different fields of physics such as plasma physics, nonlinear optics, and ferromagnetism from the beginning of 1960's. Therefore, it is natural that the applications of Painlevé equations mostly fall in the fields of partial differential equations and of difference equations, both belonging to main mathematical tools of physics. This chapter aims to offer a short overview of the connections between Painlevé differential equations and their physical applications. In fact, interdisciplinary research in this type of field necessary needs that some ideas of applications should be familiar to mathematicians, while physicists should be aware of the usefulness of the mathematical machinery behind of Painlevé equations. However, we feel it necessary to emphasize that this chapter is an overview only, not trying at all to be a complete presentation.

§49 Partial differential equations related to Painlevé equations

By experience, despite of the large number of nonlinear problems appearing in physics, the key role among them seems to belong to a relatively small number of universal mathematical models. Moreover, many of them possess the property of integrability, making possible a more deep and efficient investigation. In 1967, Gardner, Green, Kruskal, and Miura [1] proposed the inverse scattering transform method (IST) to analyze the question of integrability. This method made possible to find the connection between the Korteweg–de Vries equation and the linear Schrödinger equation on a line, proposing an exact method of solving some nonlinear partial differential equations. Since then, this method has been known as the IST-method, see Debnath [1], 363–386.

Applying the IST-method permits to establish an exact integrability of a large number of both nonlinear partial differential equations and of ordinary differential equations. Of course, this integrability property has appeared to be extremely important in different fields of science. Moreover, by the IST-method, there appears to be a close connection between nonlinear partial differential equations admitting to use this method and ordinary differential equations admitting the Painlevé property. In particular, this includes the Painlevé differential equations. In fact, the well-known conjecture, due to Ablowitz, Ramani and Segur [1], on the connection between ordinary differential equations of Painlevé type and nonlinear partial differential equations of IST-type, the group analysis method for partial differential equations and the method of Painlevé analysis as modified by Weiss [1] and Weiss, Tabor and Carnevale [1] gave rise to several new physically significant models.

Perhaps we should remark here that the transcendental solutions of Painlevé equations, the Painlevé transcendents, seem to have in nonlinear theoretical physics very much the same role as the classical special functions possess in the corresponding linear theory.

This section has been organized according to the general type of the partial differential equations in question. Although this means that the corresponding Painlevé equations appear in a somewhat nonlogical manner, we hope that this is not too confusing, as this section remains relatively short anyway.

A. Korteweg–de Vries type equations

1. The *Burger equation*

$$u_t = u_{xx} + 2uu_x \tag{49.1}$$

is the simplest nonlinear model equation for diffusive waves in fluid dynamics, hence being of considerable interest in physics applications relying on nonlinear evolution equations. It may be regarded as one-dimensional reduction of the Navier–Stokes equations, see Calogero and Degasperis [1]. For an overview of basic properties of the Burgers equation, see Debnath [1], p. 284–301. Writing now (49.1) in the form

$$u_t - (u_x + u^2)_x = 0 \tag{49.2}$$

and considering (49.2) as a compatibility condition for a function Ψ, we obtain

$$u = \Psi_x, \quad u_x + u^2 = \Psi_t \tag{49.3}$$

and further

$$\Psi_{xx} + \Psi_x^2 = \Psi_t. \tag{49.4}$$

Applying now the Cole–Hopf transformation $\Psi = \log \Phi$, we first get from (49.3) $u = \Psi_x = \Phi_x/\Phi$ and by partial differentiation of this,

$$\Psi_{xx} = \frac{\Phi_{xx}}{\Phi} - \left(\frac{\Phi_x}{\Phi}\right)^2, \quad \Psi_t = \frac{\Phi_t}{\Phi}. \tag{49.5}$$

Substituting (49.5) into (49.4) results in the linear diffusion equation

$$\Phi_t = \Phi_{xx}. \tag{49.6}$$

If we are now interested for solutions of (49.1) of type $u(x+t)$, we may denote $z = x + t$, $v(z) = u(x+t)$, obtaining first from (49.1) $v' = v'' + 2vv'$. Integrating then results in a Riccati differential equation

$$v' = v - v^2 + K, \tag{49.7}$$

K being a constant of integration. We remark that we have included the Burgers equation here due to the close connection of the Riccati differential equation to Painlevé equations.

2. The *Korteweg–de Vries equation*

$$u_t - 6u\, u_x + u_{xxx} = 0, \tag{KdV}$$

see Debnath [1], p. 347-356, was introduced in hydrodynamics to describe wave evolution in the shallow water; the KdV-equation describes the balance between time evolution, nonlinearity (uu_x) and dispersion (u_{xxx}) for a solitary wave. As the KdV-equation also describes plasma waves in a dispersive medium, it is natural that the main applications of the (KdV) appear in hydrodynamics and in plasma physics. The connection to Painlevé equations comes easily out from the fact that (KdV) admits stationary (travelling wave) solutions $u(x,t) = U(x-ct)$, where $U(z)$ satisfies the equation

$$(U'' - 3U^2 - cU) = K_1, \tag{49.8}$$

where K_1 is an arbitrary constant of integration. Multiplying (49.8) by $2U'$ and integrating we obtain

$$(U')^2 - 2U^3 - cU^2 = 2K_1U + K_2. \tag{49.9}$$

This immediately shows that a stationary solution of (KdV) may be represented by the Weierstraß $\wp$-function in the form

$$u(x,t) = -\frac{c}{6} + 2\wp\left(x - ct - x_0,\ \frac{c^2}{12} - K_1,\ -\frac{1}{4}K_2 + \frac{c}{12}K_1 - \left(\frac{c}{6}\right)^3\right). \tag{49.10}$$

In particular, if $K_1 = K_2 = 0$, we obtain

$$u(x,t) = -\frac{c}{2}\operatorname{sech}^2(\sqrt{c}(x - ct - x_0)/2). \tag{49.11}$$

On the other hand, if we substitute

$$u(t,x) = -w(z) + \lambda t,\ z = x + 3\lambda t^2$$

into (KdV), we obtain

$$w'''(z) = -6w(z)w'(z) + \lambda.$$

An immediate integration results in the first Painlevé equation in the form

$$w'' = -3w^2 + \lambda z + K. \tag{49.12}$$

Writing the KdV-equation in the form

$$2u_t + 2uu_x + u_{xxx} = 0, \tag{49.13}$$

it is a straightforward computation to reduce (49.13) to the equation

$$f''' + 2ff' - 2\alpha\tau f' - 4\alpha f = 0 \tag{49.14}$$

by using the substitution

$$\begin{cases} u = \beta/(2\alpha) + (3\alpha t + \delta)^{-2/3} f(\tau), \\ \tau = (3\alpha t + \delta)^{-1/3}\big(x - \beta t/(2\alpha) + \gamma/\alpha - \beta\delta/(2\alpha^2)\big). \end{cases}$$

By a subsequent transformation

$$f = 3\alpha\tau - 6\alpha^{2/3} F, \quad z = -\alpha^{1/3}\tau$$

in (49.14), we get the equation

$$FF'' = (F')^2/2 + 4F^3 + 2zF^2 + K, \tag{49.15}$$

where K is an arbitrary parameter of integration. Substituting further

$$F = (w^2 - w' - z)/2$$

in (49.15), we obtain

$$w'' = 2w^3 - 2zw + \alpha \tag{49.16}$$

with respect to $w(z)$, where $\alpha = (-8K)^{1/2} - 1$. Of course, (49.16) is a slightly modified form of the second Painlevé equation. Note, that if $K = 0$ in (49.15), then we may apply the substitution $F = w^2$ to obtain (P_2) with respect to $w(z)$ in its standard form with $\alpha = 0$.

3. The *modified Korteweg–de Vries equation*

$$v_t - 6v^2 v_x + v_{xxx} = 0 \tag{mKdV}$$

follows from the KdV-equation by the Miura transformation $u = v_x + v^2$. Indeed, substituting this transformation in (KdV), we get

$$2vv_t + v_{xt} - 6(v^2 + v_x)(2vv_x + v_{xx}) + 6v_x v_{xx} + 2vv_{xxx} + v_{xxxx} = 0,$$

which may be written in the form

$$\left(2v + \frac{\partial}{\partial x}\right) Mv = \left(2v + \frac{\partial}{\partial x}\right)(v_t - 6v^2 v_x + v_{xxx}) = 0. \tag{49.17}$$

Clearly, if v satisfies (mKdV), then u is a solution of (KdV). The connection of (mKdV) with the Painlevé equations is due to Ablowitz and Segur [1]. In fact, assuming that a solution of (mKdV) may be written in the form

$$v(x, t) = (3t)^{-1/3} w(z),$$

where $z = x(3t)^{-1/3}$, then a simple computation results in

$$w''' - 6w^2w' - (zw)' = 0.$$

Clearly, the second Painlevé equation now follows by integration.

4. The *cylindrical Korteweg–de Vries equation*

$$2u_t + \frac{1}{t}u + 2uu_x + u_{xxx} = 0 \tag{cKdV}$$

reduces to the first Painlevé equation by the transformation

$$\begin{aligned} \tau &= xt^{-1/2} + (\delta + 2\beta t^{1/2})/(2\gamma t), \\ u(x,t) &= x/(2t) + \delta t^{-3/2}/(2\gamma) + \beta/(2\gamma t) + f(\tau)/t, \end{aligned}$$

see Tajiri and Kawamoto [1]. In fact, integration now leads to the equation

$$f'' + f^2 - \delta\tau/(2\gamma) = K, \tag{49.18}$$

where K is a parameter of integration.

There exists another transformation leading to the second Painlevé equation. Indeed, the transformation

$$\begin{aligned} z &= \alpha t + 2\gamma t^{3/2}/3, \\ \tau &= \Big(x + \delta/\alpha + 2\gamma\big(\delta/\alpha^2 - \beta/(\alpha\gamma)\big)t^{1/2}\Big)z^{-1/3}, \\ u(x,t) &= \big(\gamma x + (3\beta - 2\gamma\delta/\alpha)\big)t^{1/2}(3z)^{-1} + z^{-2/3}f(\tau) \end{aligned}$$

applied in (cKdV) results in

$$f''' + 2ff' - \frac{2}{3}\alpha\tau f' - \frac{\alpha}{3}f = 0,$$

the solutions of which may be expressed in terms of the solutions $w(\xi)$ of the equation (P_2) with $\alpha = 0$ by using the transformation

$$w^2 = -6^{-1/3}\alpha^{-2/3}f, \quad \xi = (\frac{\alpha}{6})^{1/3}\tau.$$

5. *Higher order Korteweg–de Vries equations.* The connection of higher order Korteweg–de Vries equations ($_m$KdV) with the second Painlevé equation and its higher analogues has been treated above, see §22. Therefore, we omit these considerations from this section.

B. The Boussinesq and Kadomtsev–Petviashvili type equations

1. The *Boussinesq equation*

$$u_{tt} = c^2 \left(u_{xx} + \frac{3}{2}\left(\frac{u^2}{h}\right)_{xx} + \frac{1}{3}h^2 u_{xxxx} \right) \tag{B}$$

has a similar hydrodynamical background as the Korteweg–de Vries equation above, as it may be used to describe the propagation of long water waves (with speed c) in shallow water (of depth h). Closely related to (B) is the Kadomtsev–Petviashvili equation

$$(u_t - 6uu_x + u_{xxx})_x + 3\mu^2 u_{yy} = 0, \quad \mu^2 = 1, \tag{KP}$$

originally used to describe slowly varying nonlinear waves in a dispersive medium. The Kadomtsev–Petviashvili equation may be considered as a two-dimensional generalization of the Korteweg–de Vries equation. By a simple normalization, (KP) may be written as

$$v_{tx} + (v^2 + v_{xx})_{xx} + v_{yy} = 0. \tag{49.19}$$

Using now the transformation

$$x \to x - \frac{1}{3}y - \frac{10}{9}t, \quad t \to y + \frac{2}{3}t, \quad v \to u,$$

we obtain

$$u_{tt} - u_{xx} + \varepsilon(u^2 + u_{xx})_{xx} = 0, \quad \varepsilon^2 = 1, \tag{49.20}$$

with $\varepsilon = 1$, while for $\varepsilon = -1$, (49.20) is nothing but the Boussinesq equation, see Nishitani and Tajiri [1] and Bassom, Clarkson and Hicks [2] for more details.

To find out connection of (49.20) with Painlevé equations, first observe that substituting $z = x - ct$ into (49.20) leads to

$$(c^2 - 1)u + \varepsilon(u^2 + u'') = K_1 z + K_2.$$

The solutions of this equation may be expressed by the Weierstraß function $\wp(z)$ or by the first Painlevé transcendents, depending on the constants K_1, K_2 of integration.

On the other hand, substituting

$$\varsigma = x - \left(\frac{\delta}{2\theta}\right)t^2, \quad f = \varepsilon\left(\frac{\delta}{\theta}t\right)^2 + 2u$$

into (49.20) and integrating once, we obtain

$$f''' + ff' - \varepsilon f' - \varepsilon\left(\frac{\delta}{\theta}\right)f - 2\left(\frac{\delta}{\theta}\right)^2 \varsigma = K_1. \tag{49.21}$$

But now, applying the transformation,

$$z = -\varepsilon \left(\frac{\delta}{2\theta}\right)^{1/3} \left[\varsigma + \frac{1}{2}\frac{\theta}{\delta}\left(1 + K_1 \frac{\theta}{\delta}\right)\right],$$

$$F = -\frac{1}{12}\left(2\frac{\theta}{\delta}\right)^{2/3} \left(f - 2\varepsilon\left(\frac{\delta}{\theta}\right)\varsigma - 2\varepsilon - K_1 \varepsilon \frac{\theta}{\delta}\right),$$

the equation (49.21) reduces to (49.15), and therefore, further, to the second Painlevé equation (P_2).

2. The *generalized Boussinesq equation*

$$u_{xxxx} + p u_t u_{xx} + q u_x u_{xt} + r u_x^2 u_{xx} + u_{tt} = 0, \tag{49.22}$$

where p, q and $r \neq 0$ are constants, arises in several physical applications, see Ludlow and Clarkson [1]. The connection to Painlevé equations now appears in special cases of constants in (49.22).

For example, substituting into (49.22)

$$u(x,t) = w(z) + \lambda_1 \log t, \quad z = xt^{-1/2}, \quad W(z) = w'(z),$$

where λ_1 is an arbitrary parameter, we get the equation

$$W''' - \frac{1}{2}(p+q)zWW' + rW^2W' + \left(\frac{1}{4}z^2 + p\lambda_1\right)W' - \frac{1}{2}qW^2 + \frac{3}{4}zW - \lambda_1 = 0.$$

An additional transformation

$$W(z) = -(3^{3/4}y(x) - z)/p, \quad x = -\frac{1}{2}3^{1/4}z, \tag{49.23}$$

with $q = 0$ and $r = -\frac{1}{2}p^2$ now results in the fourth Painlevé equation (P_4) for $y(x)$ with $\alpha = p\lambda_1/\sqrt{3}$ and β being a constant of integration.

Similarly, if $q = 0$ and $r = -\frac{1}{2}p^2 \neq 0$, then by the substitution

$$u(x,t) = w(z) + 2\lambda_1 xt + \lambda_2 \log t, \quad z = xt^{-1/2} + \lambda_1 p t^{3/2},$$
$$W(z) = w'(z),$$

into (49.22), where λ_1, λ_2 are arbitrary parameters, we get the equation

$$W''' - \frac{1}{2}p^2W^2W' - \frac{1}{2}pzWW' + \left(\frac{1}{4}z^2 + p\lambda_2\right)W' + \frac{3}{4}zW - \lambda_2 = 0.$$

Again applying the additional transformation (49.23), we observe that $y(x)$ satisfies the equation (P_4) with $\alpha = p\lambda_2/\sqrt{3}$ and with β being a constant of integration.

Moreover, if $r = \frac{1}{4}q(p+q) \neq 0$, then taking

$$u(x,t) = w(z) - 4\lambda_1 xt + \lambda_2 t - \frac{8}{3}\lambda_1^2 q t^3, \quad z = x + \lambda_1 q t^2,$$
$$W(z) := w'(z),$$

where $\lambda_1 \neq 0$ and λ_2 are arbitrary constants, we obtain

$$W''' + \frac{1}{4}q(p+q)W^2 W' + (-4\lambda_1 pz + \lambda_2 p)W' - 2\lambda_1 q W = 0.$$

This equation is of Painlevé type, provided $q = 2p$ and $r = 3p^2/2$. Indeed, integrating once results in

$$W'' + p^2 W^3/2 + (\lambda_2 p - 4\lambda_1 pz)W = \lambda_3,$$

where λ_3 is a constant of integration. Clearly, this equation is nothing but a slightly modified second Painlevé equation.

3. The *spherical Boussinesq equation*, see Bassom, Clarkson and Hicks [2],

$$u_{tt} + \frac{2}{t}u_t + \frac{\gamma}{2t}u_{xx} - 3(u^2)_{xx} - u_{xxxx} = 0 \tag{49.24}$$

may be reduced by

$$u(x,t) = -\frac{1}{4t}\left(\frac{dw}{dz} + w^2 + 2zw\right), \quad z = x/(2\sqrt{t}),$$

to the fourth Painlevé equation (P_4) with respect to $w(z)$, where $\alpha = 1 - \frac{1}{2}\gamma$.

4. We shortly remark the *shallow water wave equation*

$$u_{xxxt} + \alpha u_x u_{xt} + \beta u_t u_{xx} - u_{xt} - u_{xx} = 0, \tag{49.25}$$

see Clarkson and Mansfield [1], which admits the symmetry reduction

$$u(x,t) = f(t)w(z) + x/\alpha + t/\beta, \quad z = xf(t)$$

to

$$zw^{(4)} + 4w''' + (\alpha+\beta)zw'w'' + \beta ww'' + 2\alpha(w')^2 = 0.$$

This equation is of Painlevé type if and only if $\alpha = \beta$ or if $\alpha = 2\beta$. These two special cases are solvable in terms of the solutions of (P_3) and (P_5).

5. The *modified Boussinesq equation*

$$\frac{1}{3}q_{tt} - q_t q_{xx} - \frac{3}{2}q_x^2 q_{xx} + q_{xxxx} = 0 \tag{mB}$$

has been treated by Quispel, Nijhoff and Capel [1]. Again, a connection to (P_2) and (P_4) may be easily found. Applying first the transformation

$$q(x,t) = x(K+at)/3 + \theta(\tau), \quad V(\tau) = -\theta'(\tau)/2, \quad \tau = x - Kt - \frac{1}{2}at^2,$$

we reduce (mB) to the equation (P_2) in the form

$$V'' = 2V^3 + a\tau V/3 - K^2V/6 + \mu - 1/2,$$

with parameters a, K, μ. On the other hand, if we put

$$q = \gamma \ln t + Q(\tau), \quad \tau = xt^{-1/2},$$

then we obtain the equation

$$Q^{(4)} - \frac{3}{2}Q''(Q')^2 + \frac{1}{2}\tau Q''Q' + \frac{1}{12}\tau^2 Q'' - \gamma Q'' + \frac{1}{4}\tau Q' - \frac{1}{3}\gamma = 0$$

for $Q(\tau)$. Integrating this equation, we further substitute

$$Q'(\tau) = \frac{2}{K}w(\tau) - \frac{1}{3}\tau, \quad zl = \tau, \quad K = -\frac{1}{4}l^3, \quad l^4 = 16.$$

The function $w(z)$ now satisfies the fourth Painlevé equation (P_4) where $\alpha = -\frac{1}{4}l^2\gamma$ and β is an arbitrary constant of integration.

6. We finally mention here the pair

$$u_t + v_x + uu_x = 0, \quad v_t + u_{xxx} + (uv)_x = 0, \tag{49.26}$$

of differential equations equivalent with the Boussinesq equation. This pair of equations may be solved by

$$u(x,t) = t + 2w(z), \quad v(x,t) = -z - 2w^2(z),$$

where $z = x - t^2/2 - K$, K being an arbitrary constant, and $w(z)$ being a solution of the second Painlevé equation. Similarly as to several occasions above, also here the fourth Painlevé equation may be obtained. Indeed, (49.26) also admits a solution

$$u(x,t) = \lambda\big(w(z) + 2z\big)(2t)^{-1/2}, \quad v(x,t) = -\big(w^2(z) + 2zw(z) - 2\alpha\big)(2t)^{-1},$$

where $2\mu z = x(2t)^{-1/2}$, $2\mu^2 = 1$, $\lambda\mu = 1$ such that $w(z)$ is a solution of (P_4) with parameters α and β.

C. Sine–Gordon type equations

The sine–Gordon equations were originally discovered in differential geometric considerations related to the theory of surfaces of constant negative curvature. Moreover, the idea of Bäcklund transformations was also discovered in relation to these considerations. Although sine–Gordon equations come out from a purely mathematical background, it has later appeared that they admit numerous physical applications. These include applications in nonlinear optics, solid state physics, ferromagnetism and elementary particle physics.

1. In some sense, the mathematical prototype for the sine–Gordon equations is the *Liouville equation*

$$u_{xt} = \exp u. \tag{49.27}$$

This equation appears in a variety of physical problems and may be applied in differential geometry and in the theory of automorphic functions, see Vekua [1]. In particular, as has been shown by Barbashov, Nesterenko and Chervyakov [1], several nonlinear models, e.g., in the general relativity with the constant curvature, the Born–Infeld scalar field and the relativistic string, can be described in terms of the equation (49.27).

The general solution of (49.27) is known. In fact, if $w(z)$ is a solution of the equation

$$w'' = \frac{(w')^2}{w} - \frac{1}{z}w' + \frac{1}{z}w^2, \tag{49.28}$$

then the function u determined by the transformation $u = \ln w(z)$, $z = xt$ is a solution of (49.27). Clearly, (49.28) is exactly the third Painlevé equation in the case of parameters $\alpha = 1$, $\beta = \gamma = \delta = 0$. We recall that in this case, (P_3) can be integrated in terms of elementary functions, see §29.

2. The sine–Gordon equation

$$u_{XX} - \frac{1}{c^2}u_{TT} = \sin(u) \tag{SG}$$

may be expressed, in terms of the transformation

$$x = \frac{1}{2}(X - cT), \quad t = \frac{1}{2}(X + cT),$$

in the form

$$u_{xt} = \sin u. \tag{49.29}$$

Clearly, the sine–Gordon equation is invariant under $u \mapsto u + 2n\pi$. Applying the transformation

$$\phi(\tau, \eta) = u(x, t), \quad 2x = \tau + \eta, \quad 2t = \tau - \eta,$$

we observe that (49.29) may be also represented in the from

$$\phi_{\tau\tau} - \phi_{\eta\eta} = \sin\phi. \tag{49.30}$$

The transition from (49.29) to (49.30) may be understood as the transition from the "light cone" (x, t) to the laboratory coordinate system (τ, η).

Looking now at the third Painlevé equation

$$w'' = \frac{(w')^2}{w} - \frac{1}{z}w' + \frac{1}{2z}(w^2 - 1)$$

with the equation parameters $\alpha = -\beta = 1/2$, $\gamma = \delta = 0$, it is an elementary computation to see that whenever $w(z)$ solves this equation, then the function $u(x,t) = -i \log w(z)$, $z = xt$, $i^2 = -1$ is a solution of (49.29), see Ablowitz and Segur [1].

3. Similarly as to the sine–Gordon equation (49.29), the hyperbolic sine-Gordon equation

$$u_{xt} = \sinh u \tag{49.31}$$

also relates to the third Painlevé equation (P_3). In fact, if $w(z)$ is a solution of (P_3) with the same parameter values as in the case of the sine-Gordon equation above, the function $u(x,t) = \ln w(z)$, $z = xt$ is a solution of (49.31).

4. The sine–Gordon equation may also be considered with an additional term in the form

$$u_{xt} = \nu \sin u + \mu x t \sin 2u, \tag{49.32}$$

where ν, μ are arbitrary parameters, see Gromak [6]. Of course, the sine–Gordon equation (49.29) follows from (49.32) when $\nu = 1$, $\mu = 0$. If $\nu = 0$, $\mu \neq 0$ in (49.32), then (49.32) may be reduced to the sine–Gordon equation (49.29) by the transformation

$$u_1 = 2u, \quad x_1 = x^2, \quad t_1 = t^2(\mu/2).$$

By a linear transformation of the independent variables, we may assume that $\mu = 2$ in (49.32) without loss of generality, provided $\mu \neq 0$. Denoting $z := xt$, let $u = u(z)$ be an arbitrary solution. Then the function $w(z) = \exp(-iu(z))$ is a solution of the equation (P_3) with the parameter values $2\alpha = -2\beta = \nu$, $\gamma = -\delta = 1$.

5. An even more generalized variant of the sine–Gordon equation is the equation

$$u_{xt} = \frac{1}{xt}\sec^2\left(\frac{1}{2}u\right)\left(a\cot^3\left(\frac{1}{2}u\right) + b\tan\left(\frac{1}{2}u\right)\right) + c\sin u + d\,xt\sin 2u, \tag{49.33}$$

where a, b, c, d are arbitrary parameters. Indeed, if $a = b = d = 0$, $c = 1$, then we obtain the sine–Gordon equation, while if $a = b = 0$, then (49.33) reduces back to (49.32). The case $b = d = 0$ when $xt = z$ has been considered by Jimbo [1] in relation to nonlinear field theory and Salihoğlu [1] considered the connection of (49.33) with nonlinear field theory, provided $d = 0, c = -1, xt = z$. The general case of (49.33) has been treated by Gromak and Tsegelnik [1]. They showed that whenever $w(z)$ is a solution of (P_5) with parameters α, β, γ, δ, then the function $u(x,t) := 2\,\mathrm{arccot}\sqrt{-w(z)}$, $z = xt$ is a solution of the equation (49.33) with parameters $a = -\alpha, b = -\beta, 2c = -\gamma, 4d = -\delta$. Using this transformation, special classes of exact self-similar solutions for the equation (49.33) may be easily constructed.

6. We are closing this part of §49 by looking at the *two-dimensional sine–Gordon equation* in the form

$$u_{tt} - u_{xx} - u_{yy} + \nu \sin u + \mu f(x, y, t) \sin 2u = 0, \tag{49.34}$$

where ν, μ are arbitrary parameters and $f(x, y, t) \neq 0$ is an additional function to be fixed to obtain self-similar solutions, i.e. solutions depending on the variable $z = xt$. Of course, the three-dimensional sine–Gordon equation follows from (49.34) by fixing $\nu = 1$, $\mu = 0$. In this case, exact self-similar reductions to (P_3) were obtained by Lakshmanan and Kaliappan [1]. The form of the additional function $f(x, y, t)$ for which the equation (49.34) produces self-similar solutions generated by solutions of (P_3) has been offered in Gromak [7]. In particular, if

$$h = C^2 - A^2 - B^2 \neq 0, \quad \varsigma_1 = t^2 - x^2 - y^2,$$
$$\varsigma_2 = Ct - Bx - Ay, \quad f(x, y, t) = \varsigma, \quad \varsigma = \varsigma_1 - \varsigma_2^2/h,$$

then $u(\varsigma) = i \ln w(\varsigma)$ is solution of (49.34), provided $w(\varsigma)$ is a solution of (P_3) with parameter values $\alpha = -\beta = -\nu/8$, $\gamma = -\delta = -\mu/8$. See also Clarkson, Mansfield and Milne [1] for the exact self-similar reductions of two-dimensional system of sine–Gordon equations to ordinary differential equations of Painlevé type.

D. Nonlinear equations of Schrödinger type

1. The basic form of the nonlinear Schrödinger equation is

$$iu_t + u_{xx} + \gamma u^2 \bar{u} = 0, \tag{NS}$$

where $\bar{u}$ stands for the complex conjugate of u, and i is the imaginary unit. The nonlinear Schrödinger equation (NS) is one of the fundamental equations of the nonlinear mathematical physics, being applied to describe the evolution of the hydrodynamic waves in deep water, optical waves in the nonlinear crystals and fiberglass, plasma waves and thermal waves in solids. In particular, the emergence of the laser technology in 1960's even increased the fundamental role of the equation (NS) and of its solitary solutions in nonlinear optics.

To describe the connection of the equation (NS) with Painlevé equations, we may seek, following Boiti and Pempinelli [1], solutions of (NS) in the form

$$u(x, t) = (2t)^{-1/2} v(\tau) \exp(i Q(\tau)), \quad \tau = xt^{-1/2}/2$$

with real functions $v(\tau)$ and $Q(\tau)$, to obtain the equations

$$Q''v + 2v'Q' - 2\tau v' - 2v = 0, \quad v'' - (Q')^2 v + 2\tau Q'v + 2\gamma v^3 = 0.$$

Substituting now

$$Q'(\tau) = \tau + g/g_\tau, \quad v^2 = 2g_\tau$$

and integrating, we get for g the equation

$$(g'')^2 + 4(\tau g' - g)^2 + 8\gamma(g')^3 + 2\gamma\mu^2 g' = 0, \tag{49.35}$$

where μ is an arbitrary parameter of integration. Fixing now $\gamma = -1$ in (49.35), as we may do without loss of generality, and substituting

$$g = w(w - \tau)^2/2 + (w_\tau^2 - 2w_\tau + 1 - \mu^2)/(8w),$$

the nonlinear Schrödinger equation (NS) is finally transformed into a Painlevé equation (P_4) of the form

$$ww'' = \frac{1}{2}(w')^2 - 6w^4 + 8\tau w^3 - 2\tau^2 w^2 - (\mu - 1)^2/2.$$

2. We next consider the nonlinear Schrödinger equation with an additional term in the form

$$iu_t + u_{xx} + \varepsilon i(u^2\bar{u})_x + \mu/(t^2 u\bar{u}^2) = 0, \tag{49.36}$$

where $\varepsilon^2 = 1$ and μ is an arbitrary parameter. If $\mu = 0$, then $u = t^{-1/4} f(\eta)$, $f = \sigma e^{i\theta}$, $w(\eta) = \sigma^2$, $\eta = xt^{-1/2}$ satisfies (P_4), see Ablowitz, Ramani and Segur [1]. Moreover, if $\mu \neq 0$, then $w(z) = \sigma^2/\lambda$, $z = \eta/k$, $\lambda^4 = -1$, $k = -2\varepsilon\lambda$ satisfies (P_4) with parameter values $\alpha = 2\gamma\varepsilon\lambda^2$, $\beta = 8(\gamma^2 - \mu)$, while $\theta(\eta)$ satisfies the equation $\sigma^2\theta' = \sigma^2\eta/4 - 3\varepsilon\sigma^4/4 + \gamma$, see Gromak [10].

3. The equation of the form (Kawamoto [1])

$$iu_t + u_{xx} + (-2\alpha x + 2|u|^2)u = 0 \tag{49.37}$$

is connected with (NS). The equation (49.37) may be reduced by the transformation $u(x,t) = f(x)\exp(iKt)$ to a slightly modified form of (P_2): $f'' - (2\alpha x + K)f + 2|f|^2 f = 0$.

E. Equations of Einstein type

There are a number of cases where equations of Einstein type are known to reduce to some of the Painlevé equations, see Wils [1] and Schief [1]. We are restricted ourselves to presenting an example of the complex Ernst equation

$$\nabla^2 u = 2\bar{u}(\nabla u)^2/(u\bar{u} - 1), \tag{49.38}$$

where

$$u = u(\rho, z), \quad \nabla^2 u = \frac{\partial^2 u}{\partial^2 \rho} + \frac{1}{\rho}\left(\frac{\partial u}{\partial \rho}\right) + \frac{\partial^2 u}{\partial^2 z}, \quad (\nabla u)^2 = \left(\frac{\partial u}{\partial \rho}\right)^2 + \left(\frac{\partial u}{\partial z}\right)^2.$$

This equation appears when investigating axially symmetric vacuum solutions of the equations of the general theory of relativity. If now $w(\rho)$ is a solution of the equation

(P_5), where $\alpha = -\beta = -\frac{1}{2}b^2$, $\gamma = 0$, $\delta = -2a^2$, and a, b are real numbers, such that $w > 0$ on some set of values $\tau \in \mathbb{E} \subset \mathbb{R}$, then the function $u = \exp(iaz+is)w(\rho)^{1/2}$, where s is a real solution of the equation $ds/d\rho = b(1-w)^2/(2\rho w)$, is a solution of the equation (49.38), see Léauté and Marcilhacy [1].

F. Equations of Toda type

The Toda lattice is a model of the particles on the line interacting with their closest neighbors only and with unit mass. The equation of their motion takes form $Q_n'' = \exp(Q_{n-1} - Q_n) - \exp(Q_n - Q_{n+1})$, where $Q_n(t)$ is the distance from the point of equilibrium to the nth particle. If we now define $Q_n - Q_{n-1} = -\ln p_n$, then p_n satisfies the equation

$$\frac{d^2(\log p_n)}{dt^2} = p_{n+1} - 2p_n + p_{n-1}, \tag{49.39}$$

which may be written, equivalently, in the form

$$q_n' = p_{n-1} - p_n, \; p_n' = p_n(q_n - q_{n+1}), \tag{49.40}$$

where $q_n = q_n(t)$, $p_n = p_n(t)$, $n \in \mathbb{Z}$. Both of the equations (49.39) and (49.40) are called as the Toda equation.

Let now

$$H = H_2(q, p, k) = \frac{1}{2}p^2 - \left(q^2 + \frac{t}{2}\right)p - kq$$

be the Hamiltonian for the second Painlevé equation $q'' = 2q^3 + tq + k - 1/2$, where k is a complex parameter, see §43. Writing the variables as $q(k)$ and $p(k)$ and the Hamiltonian as $H(k)$, the Bäcklund transformation and the corresponding change of the Hamiltonian can be presented as follows:

$$\begin{aligned} q(k-1) &= -q(k) + \frac{k-1}{p(k) - 2q(k)^2 - t}, \\ p(k-1) &= -p(k) + 2q(k)^2 + t, \\ H(k-1) &= H(k) + q(k). \end{aligned}$$

The τ-function of the second Painlevé equation will now be defined as

$$\frac{d}{dt}\log \tau(k) = H(k),$$

up to a multiplicative constant. It is known that the τ-function satisfies the Toda equation

$$\frac{d^2}{dt^2}\log \tau(k) = \frac{\tau(k-1)\tau(k+1)}{\tau(k)^2},$$

see Okamoto [3].

If we eliminate the function p_n from the system

$$\begin{aligned} p_n' &= p_n(p_n + 2q_n + 3t) + 3n - 1, \\ q_n' &= -q_n(2p_n + q_n + 3t) - 3n - 1, \end{aligned} \tag{49.41}$$

then the function $y_n(x) := (2/3)^{1/2} q_n(\sqrt{2/3}x)$ satisfies the Painlevé equation (P_4) with $\alpha = n$, $\beta = -2(3n+1)^2/9$. The pair (49.41) may be rewritten in the form

$$\begin{aligned} p_n' &= p_n(q_n - q_{n-1}), \\ q_n' &= q_n(p_{n+1} - p_n). \end{aligned} \tag{49.42}$$

Substituting $r_n = p_n q_n$, $s_n = -p_n - q_{n-1}$, $u_n = p_{n+1} q_n$, $v_n = -p_n - q_n$ into (49.42), we see that the pairs (r_n, s_n), (u_n, v_n) are solutions of the Toda equation $s_n' = r_{n-1} - r_n$, $r_n' = r_n(s_n - s_{n+1})$.

§50 Discrete Painlevé equations

This final section is devoted to offering the reader a short overview of discrete Painlevé equations as the second aspect of applications. However, by lack of space, we have no possibility of striving to completeness here either. For a more detailed exposition, we propose an interested reader to consult the excellent survey article due to Grammaticos, Nijhoff and Ramani [1]. Within of our limited scope, we first proceed to explain how discrete Painlevé equations may be constructed, and how their continuous counterparts, Painlevé differential equations, may be obtained from the discrete ones by a suitable limiting process. From the several possible methods applicable into this direction, we select the singularity confinement method, being perhaps the most natural from the complex analytic point of view. Having found examples of discrete Painlevé, we are going to look at their properties as complex difference equations. Unfortunately, this field has called little attention recently, although two excellent books covering the basic theory may be found, written by Nörlund [1] and Meschkowski [1]. Indeed, since their publication, very few studies exist devoted to meromorphic solutions of difference equations.

We now proceed to describe, in rough terms, how the singularity confinement method, see Grammaticos, Ramani and Papageorgiou [1], applies to construct complex difference equations corresponding, in some sense, to Painlevé differential equations. The idea here, corresponding to the Painlevé property that all singularities of solutions outside of the fixed singularities are poles only, is that a singularity of solution of a discrete equation does not propagate, i.e. it disappears after finitely many iterations.

As the first concrete example, let us consider the following complex difference equation

$$f(z+1) + f(z) + f(z-1) = a(z) + b(z)/f(z), \tag{50.1}$$

leading by some computations to a difference equation corresponding to (P_1). Observe that the argumentation of these computations is by no means strict in mathematical sense, being rather of heuristic nature. Therefore, clearly, the process to construct discrete counterparts of Painlevé differential equations is far from being unique. For convenience, we now assume that the coefficients $a(z), b(z)$ are polynomials. The idea, as presented in Grammaticos, Nijhoff and Ramani [1], is to assume that $f(z-1)$ takes a finite, non-zero value, while $f(z) = \varepsilon$, and then to select the coefficients to meet the requirement that after finitely many iterations, the dependence of $f(z+n)$ does not contain negative powers of ε. To this end, let us assume that $b(z) \neq 0$. A direct computation now results in

$$f(z+1) = \frac{b(z)}{\varepsilon} + a(z) - f(z-1) - \varepsilon, \tag{50.2}$$

$$f(z+2) = -\frac{b(z)}{\varepsilon} + a(z+1) - a(z) + f(z-1) + \frac{b(z+1)}{b(z)}\varepsilon + O(\varepsilon^2) \tag{50.3}$$

and

$$f(z+3) = a(z+2) - a(z+1) + \frac{b(z) - b(z+1) - b(z+2)}{b(z)}\varepsilon + O(\varepsilon^2). \tag{50.4}$$

If we now continue from (50.4) to the next iteration step

$$\begin{aligned} f(z+4) = {} & \frac{b(z)}{\varepsilon} + a(z) - a(z+2) + a(z+3) - f(z-1) \\ & + \frac{b(z+3)}{a(z+2) - a(z+1)} + O(\varepsilon), \end{aligned} \tag{50.5}$$

diverging as $\varepsilon \to 0$. To avoid this, we assume that $a(z+2) - a(z+1) = 0$, which means that $a(z)$ has to be a constant, say a. Under this assumption, reapplication of (50.4) now results in

$$f(z+4) = \frac{b(z)\big(b(z) - b(z+1) - b(z+2) + b(z+3)\big)}{b(z) - b(z+1) - b(z+2)}\frac{1}{\varepsilon} + O(1). \tag{50.6}$$

To avoid divergence of (50.6), we now require that

$$b(z) - b(z+1) - b(z+2) + b(z+3) = 0. \tag{50.7}$$

Denoting $c(z) := b(z) - b(z+1)$, we conclude from (50.7) that the polynomial $c(z)$ has to periodic, $c(z) - c(z+2) = 0$, hence a constant, say $c(z) = -\alpha$. Therefore, $b(z+1) = b(z) + \alpha$, and so $b'(z+1) = b'(z)$ implies, by periodicity again, that b' is constant, hence $b(z) = \alpha z + \beta$. So, we obtain the following candidate

$$f(z+1) + f(z) + f(z-1) = a + \frac{\alpha z + \beta}{f(z)} \tag{dP_1}$$

as for a discrete counterpart of the first Painlevé differential equation (P_1).

To show now that (dP_1) may be understood as a discrete counterpart of (P_1), we apply a transformation of type $z = \varepsilon t$, striving to the continuous limit through $\varepsilon \to 0$. More precisely, we define $f(z) = -\frac{1}{2} + \varepsilon^2 u(t)$, $a := -3$, $\alpha t + \beta = -\frac{1}{4}(3 + 2\varepsilon^4 t)$, expressing $f(z+1)$, $f(z-1)$ in (dP_1) as Taylor series around z in terms of ε. Letting now $\varepsilon \to 0$, one may immediately see that $u(t)$ satisfies the first Painlevé differential equation (P_1), see e.g. Fokas, Grammaticos and Ramani [1].

To continue, a rather similar reasoning may be applied to show that for a suitable choice of polynomial coefficients, the equation

$$f(z+1) + f(z-1) = \frac{a(z) + b(z)f(z)}{1 - f(z)^2} \tag{50.8}$$

appears to be a discrete counterpart of (P_2). Indeed, assuming that $f(z-1)$ does not imply a singularity, and that $f(z) = \sigma + \varepsilon$ with $\sigma = \pm 1$, the first steps of iteration result in

$$f(z+1) = -\frac{b(z) + \sigma a(z)}{2\varepsilon} + \frac{a(z) - \sigma b(z)}{4} - f(z-1) + O(\varepsilon) \tag{50.9}$$

and

$$f(z+2) = -\sigma + \frac{2b(z+1) - b(z) - \sigma a(z)}{b(z) + \sigma a(z)}\varepsilon + O(\varepsilon^2). \tag{50.10}$$

To obtain (50.9) and (50.10), one may expand the expressions for $f(z+1)$, $f(z+2)$ in Taylor series of ε around $\varepsilon = 0$. While continuing, we obtain

$$b(z) - 2b(z+1) + b(z+2) + \sigma\big(a(z) - a(z+2)\big) = 0. \tag{50.11}$$

Therefore, $a(z) - a(z+2) = 0$, hence $a(z) = \delta$ has to be a constant by periodicity. Moreover, $b(z) - 2b(z+1) + b(z+2) = 0$, and writing this as $b(z) - b(z+1) = b(z+1) - b(z+2)$, we easily conclude that $b(z) = \alpha z + \beta$ has to be a linear polynomial. Therefore, (50.8) reduces to

$$f(z+1) + f(z-1) = \frac{\delta + (\alpha z + \beta) f(z)}{1 - f(z)^2}. \tag{dP_2}$$

Again, the simple transformation $f(z) = \varepsilon u(t)$, $\alpha z + \beta = 2 + \varepsilon^2 t$, $\delta = \varepsilon^3 \mu$ results in the second Painlevé equation (P_2) with parameter μ.

Similarly as presented in §42 for Painlevé differential equations, their discrete counterparts permit a similar coalescence property, which we now demonstrate by showing that (dP_1) may be obtained from (dP_2) by a simple limiting process. Indeed, substituting in (dP_2) $f(z) = 1 + \nu g(z)$, $\alpha = -2A\nu^2$, $\beta = -4 - 2C\nu$, $\delta = 4 + 2C\nu - 2B\nu^2$, a straightforward computation results in

$$\begin{aligned} &B + Az + Cg(z) + A\nu z g(z) - g(z)g(z-1) - g(z)^2 \\ &\quad - \frac{1}{2}\nu g(z)^2 g(z-1) - g(z)g(z+1) - \frac{1}{2}\nu g(z)^2 g(z+1) = 0. \end{aligned}$$

Letting now $\nu \to 0$, we immediately get (dP_1) in the form

$$g(z+1) + g(z) + g(z-1) = C + \frac{B + Az}{g(z)}.$$

We are not going to continuing further in detail, restricting ourselves just to list here the basic discrete counterparts of the remaining Painlevé equations (P_3) to (P_6), see Ramani, Grammaticos and Hietarinta [1] and Grammaticos and Ramani [1] for their construction and downward coalescence:

$$f(z+1)f(z-1) = \frac{(a + f(z))(b + f(z))}{(cq^z f(z) + 1)(dq^z + 1)}, \tag{dP_3}$$

$$\big(f(z+1) + f(z)\big)\big(f(z) + f(z-1)\big) = \frac{(f(z)^2 - a^2)(f(z) - b^2)}{(f(z) - (\alpha z + \beta))^2 - c^2}, \tag{dP_4}$$

$$\begin{aligned}&\big(f(z+1)f(z) - 1\big)\big(f(z)f(z-1) - 1\big)\\ &\quad = (\alpha z + \beta)\gamma q^z \frac{(f(z) + a)(f(z) + 1/a)(f(z) + b)(f(z) + 1/b)}{(f(z) + \gamma q^z)(f(z) + \delta q^z)},\end{aligned} \tag{dP_5}$$

$$\begin{aligned}&\frac{\big(f(z+1)f(z) - \gamma^2 q^z q^{z+1}\big)\big(f(z)f(z-1) - \gamma^2 q^z q^{z-1}\big)}{\big(f(z+1)f(z) - 1\big)\big(f(z)f(z-1) - 1\big)}\\ &\quad = \frac{(f(z) - \gamma q^z a)(f(z) - \gamma q^z/a)(f(z) - \gamma q^z b)(f(z) - \gamma q^z/b)}{(f(z) - c)(f(z) - 1/c)(f(z) - d)(f(z) - 1/d)},\end{aligned} \tag{dP_6}$$

where $a, b, c, q, \alpha, \beta, \gamma, \delta$ are constants.

From the general point of view of this book it would be interesting to find out purely complex analysis arguments for constructing the discrete Painlevé equations. By Chapter 2, the first natural idea would be to insist that the difference equations in question, at least in the case of (dP_1), (dP_2) and (dP_4) admit at least some solutions in the complex plane of finite order of growth. Observe that for complex difference equations, in a very general sense, replacing for a solution $f(z)$, the argument z by a suitable periodic entire function, one may always construct another solution of infinite order of growth. Therefore, in this situation, one has to look at the minimal growth of solutions, or perhaps the minimal growth for certain classes of solutions. This kind of investigations have started very recently, see Ablowitz, Halburd and Herbst [1], Heittokangas et al. [1] and Grammaticos et al. [1]. However, the case is that this topic is in its very beginning.

By the preceding list of basic discrete Painlevé equations, we immediately see that all of them are special cases of the following general type of complex difference equation: Let $c_1, \dots, c_n \in \mathbb{C}$ be distinct complex numbers, and J be a collection of non-empty subsets of $\{1, 2, \dots, n\}$. We then consider a difference equation of type

$$\sum_J \alpha_J(z)\Big(\prod_{j\in J} f(z + c_j)\Big) = R(z, f) := \frac{a_0(z) + a_1(z)f + \cdots + a_p(z)f^p}{b_0(z) + b_1(z)f + \cdots + b_q(z)f^q}, \tag{50.12}$$

where the coefficients α_J, $a_j(z)$, $b_k(z)$ are meromorphic and of small growth $S(r, f)$ in the Nevanlinna theory sense, see Appendix B. We also assume that a_p, b_q and at least one of α_J are non-vanishing functions, such that the corresponding product $\prod_{j\in J}$ contains at least one $f(z+c_j)$ with $c_j \neq 0$. Moreover, we assume that the right hand side is irreducible with respect to f over the field of meromorphic functions small in the sense of $S(r, f)$, and we denote $d := \max(p, q)$. We then obtain

Theorem 50.1. *Let f be a meromorphic solution of (50.12). If $d > n$, then the lower order $\mu(f) = \infty$.*

Proof. Take $\varepsilon > 0$ small enough to satisfy $\gamma := \frac{d}{n(1+\varepsilon)} > 1$. Combining Lemma B.14, Theorem B.16, Theorem B.17 and recalling Lemma B.10, we immediately obtain from the equation (50.12),

$$\begin{aligned} d < dT(r, f) &= T(r, R) + S(r, f) \\ &= T\Big(r, \sum_J \alpha_J(z) \prod_{j\in J} f(z+c_j)\Big) + S(r, f) \\ &\leq \sum_{j=1}^{n} T\big(r, f(z+c_j)\big) + S(r, f) \\ &\leq n(1+\varepsilon)T(r+C, f) + M + S(r, f) \\ &\leq n(1+\varepsilon)T(\sigma r + C, f), \end{aligned} \tag{50.13}$$

for all $r \geq r_0 = r_0(\sigma) \geq 1/\varepsilon$, where $M = M(\varepsilon)$ and C are some positive numbers, and σ is an arbitrary number such that $\sigma > 1$. Inductively, for any $k \in \mathbb{N}$, we obtain from (50.13),

$$T\left(\sigma^k r + \frac{\sigma^k - 1}{\sigma - 1} C, f\right) \geq \gamma^k T(r, f) \tag{50.14}$$

for all $r \geq r_0$. Take now $r \in [r_0, \sigma r_0 + C)$ and denote

$$s = s(k, r) = \sigma^k r + \frac{\sigma^k - 1}{\sigma - 1} C.$$

Observe that $s = s(k, r)$ covers each value in $[\sigma r_0 + C, \infty]$, provided (k, r) covers $\mathbb{N} \times [r_0, \sigma r_0 + C)$. But then

$$\begin{aligned} k &= \frac{1}{\log\sigma} \log\left(\frac{s(\sigma-1)+C}{r(\sigma-1)+C}\right) \\ &\geq \frac{1}{\log\sigma} \log\left(\frac{s(\sigma-1)+C}{(\sigma r_0 + C)(\sigma-1)+C}\right) \\ &\geq \frac{1}{\log\sigma}\left(\log s - \log\left(\sigma r_0 + \frac{\sigma}{\sigma-1}C\right)\right). \end{aligned}$$

Hence, we may use (50.14) to see that for all $s \geq \sigma r_0 + C$,

$$\log T(s, f) \geq \frac{\log \gamma}{\log \sigma} \left(\log s - \log \left(\sigma r_0 + \frac{\sigma}{\sigma - 1} C \right) \right) + \log T(r_0, f).$$

Dividing now through by $\log s$ and letting $s \to \infty$, we observe that

$$\mu(f) \geq \frac{\log \gamma}{\log \sigma}.$$

Letting now $\sigma \to 1+$, we are done. □

As shown in Grammaticos et al. [1], the requirement that (transcendental) meromorphic solutions exist for a complex difference equation, may perhaps be used, by a suitable reformulation of the equation in question, to obtain other restrictions than what has been indicated in Theorem 50.1 above. Unfortunately, there does not exist presently, according to our knowledge, purely complex analytic requirements which would force a difference equation to reduce into the discrete Painlevé equations listed above, or into some other existing versions of the counterparts to Painlevé differential equations. A potential idea, not yet explored, might be to try some requirements related to the value distribution of the solutions, see Heittokangas et al. [1], Theorem 3.1.

Appendix A

Local existence and uniqueness of solutions of complex differential equations

In this appendix, we have collected, for the convenience of the reader, some basic results concerning the local existence and uniqueness of analytic solutions of complex differential equations. We are restricting ourselves to such local existence and uniqueness results only, which are explicitly needed in the actual chapters of this book. Essentially, two types of results have been included: We first offer some very basic local existence and uniqueness results usually proved by the Picard successive approximations or by the Cauchy majorant method. Although basically familiar, there are not so many standard references actually proving these results in the complex plane case. Secondly, we include a couple of important results for the local existence and uniqueness of solutions for complex differential equations in a neighborhood of a singularity of regular type.

We omit most proofs in this appendix. However, for the convenience of the reader again, we have included explicit references for complete proofs in standard books.

To start with, we first fix some notations in the n-dimensional complex space $\mathbb{C}^n$, equipped with the standard euclidean topology obtained by identifying $\mathbb{C}^n$ with $\mathbb{R}^{2n}$, as well as in an open set $\Omega \subset \mathbb{C}^n$. For a point $(z_1, \ldots, z_n) \in \Omega$, we denote by $z_j = x_{j,1} + i x_{j,2}$, $j = 1, \ldots, n$, the corresponding real coordinates. Looking at a function $f : \Omega \to \mathbb{C}$, continuously differentiable with respect to all corresponding real coordinates, we may define the complex partial derivatives of f by

$$\frac{\partial f}{\partial z_j} := \frac{1}{2}\left(\frac{\partial f}{\partial x_{j,1}} - i\frac{\partial f}{\partial x_{j,2}}\right), \quad j = 1, \ldots, n,$$

$$\frac{\partial f}{\partial \bar{z}_j} := \frac{1}{2}\left(\frac{\partial f}{\partial x_{j,1}} + i\frac{\partial f}{\partial x_{j,2}}\right), \quad j = 1, \ldots, n.$$

Then we may define analyticity in several complex variables by

Definition A.1. A continuously differentiable function $f : \Omega \to \mathbb{C}$ is analytic (holomorphic) in Ω, if

$$\overline{\partial f} = \sum_{j=1}^{n} \frac{\partial f}{\partial \bar{z}_j}(dx_{j,1} - i dx_{j,2}) = 0$$

in Ω.

It is now well-known that an analytic function $f : \Omega \to \mathbb{C}$ may be uniquely represented, locally, in a power series

$$f(z_1, \ldots, z_n) = \sum_{k_1,\ldots,k_n=0}^{\infty} a_{k_1,\ldots,k_n}(z_1 - \beta_1)^{k_1} \ldots (z_n - \beta_n)^{k_n}$$

converging uniformly in compact subsets of a polydisc $B(\beta, r) = \prod_{j=1}^{n} B(\beta_j, r_j)$. Moreover, if $|f(z_1, \ldots, z_n)| \leq M$ in $B(\beta, \rho)$ for every ρ such that $0 < \rho_j < r_j$, $j = 1, \ldots, n$, then the Cauchy estimates

$$|a_{k_1,\ldots,k_n}| \leq M/(r_1^{k_1} \ldots r_n^{k_n})$$

follow. For the practical applications of the following basic existence and uniqueness results, it is important to recall the classical theorem, due to Hartogs, that whenever $f : \Omega \to \mathbb{C}$ is analytic in each of its variables $z_1, \ldots, z_n$ separately, while all other variables have arbitrarily given fixed values, then f is analytic in the sense of Definition A.1.

Starting with the most elementary first order differential equation

$$w' = f(z, w),$$

we may recall the following local existence and uniqueness theorem. For a complete proof (by the Cauchy majorant method), see e.g. Herold [1], Satz 2.1, or Hille [2], Theorem 2.4.1.

Theorem A.2. *Suppose $f : \Omega \to \mathbb{C}$ is analytic in a domain $\Omega \subset \mathbb{C}^2$ and $(z_0, w_0) \in \Omega$. Also assume that*

$$\overline{B}(z_0, w_0; r) := \overline{B}(z_0, r) \times \overline{B}(w_0, r) \subset \Omega$$

and that $|f(z, w)| \leq M$ for all $(z, w) \in \overline{B}(z_0, w_0; r)$. Then the initial value problem

$$w' = f(z, w), \quad w(z_0) = w_0$$

admits a unique analytic solution $w(z)$ in a neighborhood U of z_0 such that

$$B\big(z_0, r(1 - \exp(-1/(2M)))\big) \subset U.$$

By a slight modification of the proof referred to above, Theorem A.2 may be extended to corresponding systems of first order differential equations in the complex plane, see Herold [1] or Hille [2] again. Actually, the following theorem is the key result to proving that all solutions of the Painlevé differential equations (P_1)–(P_5), suitably modified for (P_3) and (P_5), are meromorphic functions, see Chapter 1.

Theorem A.3. *Suppose that $f_j : \Omega \to \mathbb{C}$, $j = 1, \ldots, n$ in a domain $\Omega \subset \mathbb{C}^{n+1}$ are analytic and that $(z_0, w_{1,0}, \ldots, w_{n,0}) \in \Omega$. Also assume that for some $r > 0$,*

$$\overline{B}(z_0, w_{1,0}, \ldots, w_{n,0}; r) = \overline{B}(z_0, r) \times \prod_{j=1}^{n} \overline{B}(w_{j,0}, r) \subset \Omega$$

and that

$$|f_j(z, w_1, \ldots, w_n)| \leq M, \quad j = 1, \ldots, n,$$

for all $(z, w_1, \ldots, w_n) \in \overline{B}(z_0, w_{1,0}, \ldots, w_{n,0}; r)$. Then the initial value problem

$$w_j'(z) = f_j(z, w_1(z), \ldots, w_n(z)), \quad w_j(z_0) = w_{j,0}, \quad j = 1, \ldots, n,$$

admits a unique analytic solution $(w_1(z), \ldots, w_n(z))$ in a neighborhood U of z_0 such that

$$B\left(z_0, r\left(1 - \exp\left(\frac{-1}{(n+1)M}\right)\right)\right) \subset U.$$

Corollary A.4. *Consider the system of differential equations*

$$w_j' = f_j(z, w_1, \ldots, w_n), \quad j = 1, \ldots, n, \tag{A.1}$$

where f_j are functions satisfying the same suppositions in $\bar{B}(z_0, w_{1,0}, \ldots, w_{n,0}; r)$ as in Theorem A.3. Let γ be a curve terminating in $z = z_0$ and let $\{a_k\}_{k=1}^{\infty} \subset \gamma$ be a sequence such that $a_k \to z_0$ as $k \to \infty$. Suppose that a solution $(w_1(z), \ldots, w_n(z))$ of (A.1) is meromorphic along $\gamma \setminus \{z_0\}$, and that $w_j(a_k) \to w_{j,0}$, $j = 1, \ldots, n$, as $k \to \infty$. Then, the solution is analytic at $z = z_0$ as well and satisfies $w_j(z_0) = w_{j,0}$.

Proof. By Theorem A.3, if a_k satisfies $|a_k - z_0| + |w_j(a_k) - w_{j,0}| < r/2$, then $w_j(z)$, $j = 1, \ldots, n$ are analytic in $B\big(a_k, (r/2)\big(1 - \exp\big(\frac{-1}{(n+1)M}\big)\big)\big)$. From this fact the corollary immediately follows. □

Another slight modification of the Cauchy majorant proof, see e.g. Herold [1], Satz 3.1, results in the following basic existence and uniqueness result for linear differential equations:

Theorem A.5. *Let $b_{j,k}(z), a_j(z)$, $j, k = 1, \ldots, n$, be analytic functions in a disc $B(z_0, r)$. Then the initial value problem*

$$w_j' = \sum_{k=1}^{n} b_{j,k}(z) w_k + a_j(z), \quad j = 1, \ldots, n,$$

$$w_j(z_0) = w_{j,0}, \quad j = 1, \ldots, n,$$

admits a unique solution $(w_1(z), \ldots, w_n(z))$ analytic in $B(z_0, r)$.

Writing an nth order linear differential equation as a system of first order linear differential equations, we obtain an immediate

Corollary A.6. *Let $p_1(z), \ldots, p_n(z), a(z)$ be analytic in a disc $B(z_0, r)$. Then the initial value problem*

$$w^{(n)} + p_1(z)w^{(n-1)} + \cdots + p_n(z)w = a(z),$$
$$w^{(j)}(z) = w_{j,0}, \quad j = 0, \ldots, n-1,$$

admits a unique solution $w(z)$ analytic in $B(z_0, r)$.

Corollary A.7. *If all the coefficients in Corollary* A.5. *are entire functions, then the solution $w(z)$ is entire as well.*

In several occasions of our actual chapters, Painlevé differential equations appear to be closely connected with Riccati differential equations. Therefore, we close the first part of this appendix by

Theorem A.8. *Suppose the coefficients $a_0(z), a_1(z), a_2(z)$ of a Riccati differential equation*

$$w' = a_0(z) + a_1(z)w + a_2(z)w^2 \tag{A.2}$$

are analytic in a disc $B(z_0, r)$. Then all local solutions of (A.2) *in a domain $\Omega \in B(z_0, r)$ admit a meromorphic continuation over the whole disc $B(z_0, r)$.*

For a proof of Theorem A.8, see e.g. Laine [1], Theorem 9.1.3, and Jank and Volkmann [1], Satz 20.2.

Corollary A.9. *If all the coefficients of* (A.2) *are entire functions, then all local solutions of* (A.2) *admit a meromorphic continuation over the whole complex plane $\mathbb{C}$.*

We now proceed to the second part of this appendix, the local behavior of solutions in a neighborhood of a singularity of regular type of the differential equation, resp. a system of differential equations, in question. To this end, we consider a system of differential equations with a singular point of regular type at $z = 0$, i.e. a system of the form

$$zy_j' = f_j(z, y_1, \ldots, y_n), \quad j = 1, \ldots, n, \tag{A.3}$$

where the right hand sides $f_j(z, y_1, \ldots, y_n)$ are analytic at $z = 0, y_1 = 0, \ldots, y_n = 0$ and satisfy the initial conditions $f_j(0, \ldots, 0) = 0$ for $j = 1, \ldots, n$. Such a system is called a Briot–Bouquet system. The equation (A.3) can be written in the vectorial form:

$$zY' = F(z, Y), \tag{A.4}$$

where $Y := {}^t(y_1, \ldots, y_n) \in \mathbb{C}^n$ is the unknown vector and

$$F(z, Y) = {}^t(f_1(z, Y), \ldots, f_n(z, Y))$$

is a vector-valued function. $F(z, Y)$ may be expanded into a power series

$$F(z, Y) = \sum_{j+|K|\geq 1} A_{jK} z^j Y^K \tag{A.5}$$

convergent, hence bounded, in some domain around the origin:

$$|z| < r_0, \quad \|Y\| < r_1. \tag{A.6}$$

Here $K := (k(1), \dots, k(n)) \in (\mathbb{N} \cup \{0\})^n$, $|K| := k(1) + \cdots + k(n)$, the coefficient $A_{jK} := {}^t(a_{jK}^{(1)}, \dots, a_{jK}^{(n)}) \in \mathbb{C}^n$, $\|Y\| = \max_j |y_j|$ and $Y^K = y_1^{k(1)} \dots y_n^{k(n)}$.

Theorem A.10. *If none of the eigenvalues of the matrix*

$$A = \begin{pmatrix} \frac{\partial f_1}{\partial y_1}(0, \mathbf{0}) & \cdots & \frac{\partial f_1}{\partial y_n}(0, \mathbf{0}) \\ \vdots & & \vdots \\ \frac{\partial f_n}{\partial y_1}(0, \mathbf{0}) & \cdots & \frac{\partial f_n}{\partial y_n}(0, \mathbf{0}) \end{pmatrix} \tag{A.7}$$

is a positive integer, then the equation (A.4) *admits a solution of the form*

$$Y(z) = \sum_{j=1}^{\infty} C_j z^j, \quad C_j \in \mathbb{C}^n, \tag{A.8}$$

convergent at $z = 0$. Furthermore, there is a unique solution analytic around $z = 0$ and satisfying $Y(0) = \mathbf{0}$.

Proof. See Iwasaki, Kimura, Shimomura and Yosida [1], p. 259–263. □

Corollary A.11. *If $\lambda \in \mathbb{C} \backslash \mathbb{N}$, then the differential equation*

$$zy' = \lambda y + p_{10} z + \sum_{j+k\geq 2} p_{jk} z^j y^k$$

with constant coefficients admits a unique solution analytic around $z = 0$ such that $y(0) = 0$.

Proof. For a proof by using the majorant method, see Herold [1], p. 55–56. □

To compare two multi-indices $K = (k(1), \dots, k(n))$ and $L = (l(1), \dots, l(n))$, we define a relation $K \succ L$, if either (a) or (b) holds:

(a) $k(1) + \cdots + k(n) > l(1) + \cdots + l(n)$;

(b) $k(1) + \cdots + k(n) = l(1) + \cdots + l(n)$,

and, for some j satisfying $1 \le j \le n$,

$$k(i) = l(i) \text{ for all } i < j \text{ and } k(j) > l(j).$$

We also write $K \succeq L$ if $K \succ L$ or $K = L$, and $K \prec L$, if $L \succ K$.

If it happens that the assumption concerning the eigenvalues of the matrix (A.7) in Theorem A.10 fails, we can anyway prove the convergence of a formal solution as follows:

Theorem A.12. *Suppose that the system* (A.4) *with* (A.5) *and* (A.6) *admits a formal solution of the form* (A.8). *Then* (A.8) *converges around* $z = 0$.

Proof. The proof may be reduced to Theorem A.10. Indeed, for the eigenvalues $\lambda_1, \dots, \lambda_n$ of the matrix A, take a positive integer j_0 such that $j_0 \ge \max(\lambda_1, \dots, \lambda_n) + 1$. By the change of variables $Y = W + \sum_{j=1}^{j_0} C_j z^j$, the system (A.4) transforms into

$$zW' = H(z, W) \tag{A.9}$$

with $H(z, W) = \sum_{j+|K|\ge 1} H_{jK} z^j W^K$ satisfying $H_{1,\mathbf{0}} = \cdots = H_{j_0,\mathbf{0}} = \mathbf{0}$ and $\frac{\partial H(0,\mathbf{0})}{\partial W} = A$. Moreover,

$$H_{jK} = {}^t\big(h_{jK}^{(1)}, \dots, h_{jK}^{(n)}\big)$$

is a column vector in $\mathbb{C}^n$. Then the system (A.9) admits a formal solution $W(z) = \sum_{j=j_0+1}^{\infty} C_j z^j$. The further change of variables $W = z^{j_0} V$ reduces (A.9) into the system of the form (A.4) with the Jacobian matrix $(\partial F/\partial V)(0, \mathbf{0}) = A - j_0 I$, which admits a formal solution $V = \sum_{k=1}^{\infty} C_{j_0+k} z^k$. If j_0 is sufficiently large, then all eigenvalues of $A - j_0 I$ are negative numbers. From this fact the conclusion follows. □

Let us finally consider a more general case of the matrix A, assuming that the eigenvalues $\lambda_1, \dots, \lambda_n$ of the matrix A satisfy the following two conditions:
(A1) For every $(j, L) \succeq (1, \mathbf{0})$ where $L := (l(1), \dots, l(n))$, and for $\nu = 1, \dots, n$,

$$j + \lambda_1 l(1) + \cdots + \lambda_n l(n) - \lambda_\nu \ne 0.$$

(A2) (Poincaré condition) The convex hull of the points $1, \lambda_1, \dots, \lambda_n$ in the complex plane does not contain the origin, i.e. we can draw a line l through the origin so that $1, \lambda_1, \dots, \lambda_n$ are situated on one side of l. Equivalently, we may say that there exists a complex number t such that $e^t, e^{\lambda_1 t}, \dots, e^{\lambda_n t}$ are all small simultaneously.

Theorem A.13. *Under the assumptions* (A1) *and* (A2), *the equation* (A.4) *admits a solution of the form*

$$Y = \Phi(z, z^\Lambda C)$$

with the following properties:

(1) *The series*

$$\Phi(z, Z) := \sum_{j+|L|\geq 1} P_{jL} z^j Z^L,$$

where $P_{jL} \in \mathbb{C}^n$, $Z = {}^t(z_1, \ldots, z_n)$, $P_{0E[\nu]} = e_\nu = {}^t(0, \ldots, 0, 1, 0, \ldots, 0)$, *with* 1 *on the* ν^{th} *place, converges for* $|z| < r$, $\|Z\| < r$, *where* r *is a sufficiently small positive constant.*

(2) $Z = z^\Lambda C$ *is a solution of the equation*

$$zZ' = \Lambda Z,$$

where $C = {}^t(c_1, \ldots, c_n) \in \mathbb{C}^n$.

Proof. See Iwasaki, Kimura, Shimomura and Yosida [1], p. 278–289. □

Appendix B
Basic notations and facts in the Nevanlinna theory

This section is devoted to reviewing those basic facts in the Nevanlinna theory, which are necessary in the actual chapters devoted to Painlevé equations. For more material, for proofs omitted here and for most of the notations, the reader may consult Hayman [1], Jank and Volkmann [1] and Laine [1]. In addition, we frequently apply the following notations: Given $\Phi : [r_0, +\infty) \to \mathbb{R}_+$, $\Psi : [r_0, +\infty) \to \mathbb{R}_+$, we denote $\Phi(r) \ll \Psi(r)$, resp. $\Psi(r) \gg \Phi(r)$, provided for some $r_0 \geq 0$, $\Phi(r) = O(\Psi(r))$ as $r \to +\infty$, and $\Phi(r) \asymp \Psi(r)$, if $\Phi(r) \ll \Psi(r)$ and $\Psi(r) \ll \Phi(r)$ hold simultaneously.

Let $f(z)$ be an arbitrary meromorphic function in $\mathbb{C}$. For $r > 0$, denote by $n(r, f)$ the cardinal number of the poles of $f(z)$ in the disk $|z| \leq r$, each pole being counted according to its multiplicity. Then the *counting function* of $f(z)$ is defined by

$$N(r, f) := \int_0^r \frac{1}{t}\big(n(t, f) - n(0, f)\big)\, dt + n(0, f) \log r, \tag{B.1}$$

which measures the average frequency of poles in the disk $|z| < r$. For $x > 0$, we write $\log^+ x = \max\{\log x, 0\}$. The *proximity function* of $f(z)$ is defined by

$$m(r, f) := \frac{1}{2\pi} \int_0^{2\pi} \log^+ |f(re^{i\theta})| d\theta, \tag{B.2}$$

which measures the average magnitude on the circle $|z| = r$. Then we put

$$T(r, f) := m(r, f) + N(r, f), \tag{B.3}$$

which is called the *characteristic function* of $f(z)$. It is easy to see that, for meromorphic functions $f(z)$, $g(z)$, formulas such as

$$\begin{aligned} m(r, \alpha f + \beta g) &\leq m(r, f) + m(r, g) + O(1), \\ m(r, fg) &\leq m(r, f) + m(r, g), \\ T(r, \alpha f + \beta g) &\leq T(r, f) + T(r, g) + O(1), \\ T(r, fg) &\leq T(r, f) + T(r, g) \end{aligned}$$

$(\alpha, \beta \in \mathbb{C})$ are valid. Let (a_j) and (b_k), respectively, be the zeros and the poles of $g(z)$ in the disk $|z| < r$, each repeated according to its multiplicity. By the Poisson–Jensen formula, for every z satisfying $|z| < r$,

$$\begin{aligned} \log |g(z)| = {} & \frac{1}{2\pi} \int_0^{2\pi} \log |g(re^{i\theta})| \cdot \frac{r^2 - |z|^2}{|re^{i\theta} - z|^2} d\theta \\ & + \sum_{|a_j|<r} \log \left| \frac{r(z - a_j)}{r^2 - \bar{a}_j z} \right| - \sum_{|b_k|<r} \log \left| \frac{r(z - b_k)}{r^2 - \bar{b}_k z} \right|. \end{aligned} \tag{B.4}$$

Substituting $g(z) = z^{-p}f(z) = c_p + O(z)$ ($c_p \neq 0$, $p \in \mathbb{Z}$) into (B.4) with $z = 0$, we have

$$m(r, f) + N(r, f) = m(r, 1/f) + N(r, 1/f) + \log|c_p|. \tag{B.5}$$

Replacing $f(z)$ by $f(z) - a$, $a \in \mathbb{C}$, we obtain the first main theorem:

Theorem B.1. *For an arbitrary meromorphic function $f(z)$ and an arbitrary $a \in \mathbb{C}$,*

$$T(r, 1/(f - a)) = T(r, f) + O(1).$$

The characteristic function $T(r, f)$ measures the complexity of the behavior, namely the transcendency of $f(z)$. For example, $T(r, e^z) \asymp r$, $T(r, \exp(z^2)) \asymp r^2$.

Proposition B.2. *A meromorphic function $f(z)$ satisfies $T(r, f) = O(\log r)$, if and only if $f(z)$ is a rational function.*

The *order* of $f(z)$ is defined by

$$\rho(f) := \limsup_{r\to\infty} \frac{\log T(r, f)}{\log r}. \tag{B.6}$$

For example, $\rho(e^z) = 1$, $\rho(\exp(z^2)) = 2$, and $\rho(q) = 0$ for any rational function $q(z)$. From an identity due to H. Cartan,

$$T(r, f) = \frac{1}{2\pi}\int_0^{2\pi} N(r, 1/(f - e^{i\theta}))d\theta + \log^+|f(0)|,$$

we obtain

$$\frac{dT(r, f)}{d\log r} = \frac{1}{2\pi}\int_0^{2\pi} n(r, 1/(f - e^{i\theta}))d\theta,$$

provided that $|f(0)| \neq \infty$. Combining this with the relation $T(r, f) = T(r, 1/f) + \log|c_p|$ in the remaining case $f(0) = \infty$, we derive

Proposition B.3. *$T(r, f)$ is increasing and convex with respect to* $\log r$.

Furthermore, we have

Proposition B.4. *A meromorphic function $f(z)$ is transcendental if and only if $\log r/T(r, f) = o(1)$ as $r \to \infty$.*

Proof. Suppose that $f(z)$ is transcendental. We regard $T(r, f) = T_*(r)$ as a function of $\log r$. If $dT_*(r)/d\log r \leq B$ for some $B > 0$, then $T_*(r) \leq B\log r - O(1) \ll \log r$, which contradicts the supposition. Since $dT_*(r)/d\log r$ is increasing with respect to r by Proposition B.3, $dT_*(r)/d\log r \to \infty$ as $r \to \infty$. For any $\varepsilon > 0$, there exists $r_\varepsilon > 0$ such that $dT_*(r)/d\log r \geq 1/\varepsilon$, $r \geq r_\varepsilon$ and hence $T_*(r) \geq \varepsilon^{-1}\log r + O(1)$, $r \geq r_\varepsilon$, which implies that $\log r/T(r, f) \to 0$ as $r \to \infty$. □

Define, for an arbitrary $a \in \mathbb{C}$,

$$\delta(a, f) := \liminf_{r \to \infty} \frac{m(r, 1/(f-a))}{T(r, f)}, \tag{B.7}$$

and

$$\delta(\infty, f) := \liminf_{r \to \infty} \frac{m(r, f)}{T(r, f)}. \tag{B.8}$$

These quantities are called the *deficiency* of a, and that of ∞, respectively. For a transcendental meromorphic function $f(z)$, by Theorem B.1, if $\delta(a, f) < 1$ then $f(z)$ admits infinitely many a-points. If $f(z)$ admits a finite number of a-points in $\mathbb{C}$ only, then $\delta(a, f) = 1$.

To count the multiple poles, we put $n_1(r, f) = \sum_{|\tau| \le r} (\mu(\tau) - 1)$, where $\mu(\tau)$ denotes the multiplicity of a pole τ and $\sum_{|\tau| \le r}$ denotes the summation over all poles in the disk $|z| \le r$. Then

$$N_1(r, f) := \int_0^r \frac{1}{t} \big(n_1(t, f) - n_1(0, f)\big)\, dt + n_1(0, f) \log r \tag{B.9}$$

measures the frequency of the multiple poles. We then define, for $a \in \mathbb{C}$,

$$\vartheta(a, f) := \liminf_{r \to \infty} \frac{N_1(r, 1/(f-a))}{T(r, f)}, \tag{B.10}$$

and

$$\vartheta(\infty, f) := \liminf_{r \to \infty} \frac{N_1(r, f)}{T(r, f)}, \tag{B.11}$$

which are called the *ramification index* of a, and that of ∞, respectively. If all a-points are simple, then $\vartheta(a, f) = 0$, and if they are double, then $\vartheta(a, f) \le 1/2$.

Let $\phi(r)$ be a function defined on an interval $[r_0, +\infty)$, where $r_0 > 1$. We write

$$\phi(r) = S(r, f)$$

if $\phi(r) = o(T(r, f))$ as $r \to +\infty$ outside of a possible exceptional set of finite linear measure. Applying the differential operator $(\partial/\partial x - i\partial/\partial y)$ to (B.4) (with $g = f$), we obtain

$$\begin{aligned} \frac{f'(z)}{f(z)} &= \frac{1}{\pi} \int_0^{2\pi} \log |f(re^{i\theta})| \cdot \frac{re^{i\theta}}{(re^{i\theta} - z)^2} d\theta \\ &\quad + \sum_{|a_j| < r} \left(\frac{1}{z - a_j} + \frac{\bar{a}_j}{r^2 - \bar{a}_j z} \right) - \sum_{|b_k| < r} \left(\frac{1}{z - b_k} + \frac{\bar{b}_k}{r^2 - \bar{b}_k z} \right) \end{aligned}$$

for $|z| < r$. By this inequality, we get the basic estimates for logarithmic derivatives:

Proposition B.5. *For an arbitrary non-constant meromorphic function $f(z)$, and for an arbitrary $k \in \mathbb{N}$, $m(r, f^{(k)}/f) = S(r, f)$. In particular, if $\rho(f) < +\infty$, then $m(r, f^{(k)}/f) = O(\log r)$.*

If z_0 is a pole of $f(z)$ with multiplicity μ_0, then z_0 is a pole of $f^{(k)}(z)$ with multiplicity $\mu_0 + k$. Hence, $N(r, f^{(k)}) \leq (k+1)N(r, f)$. By Proposition B.5, $m(r, f^{(k)}) \leq m(r, f) + m(r, f^{(k)}/f) = m(r, f) + S(r, f)$. Hence $T(r, f^{(k)}) \leq (k+1)T(r, f) + S(r, f)$. Thus we have

Corollary B.6. *For every $k \in \mathbb{N}$, $T(r, f^{(k)}) \leq (k+1)T(r, f) + S(r, f)$.*

Observing that

$$m(r, f^{(q)}/f^{(p)}) \leq \sum_{j=p}^{q-1} m(r, f^{(j+1)}/f^{(j)}), \quad q > p, \quad p, q \in \mathbb{N},$$

and that $m(r, f^{(j+1)}/f^{(j)}) = S(r, f^{(j)}) = S(r, f)$, see Corollary B.6, we obtain

Corollary B.7. *For arbitrary positive integers p and q such that $p < q$, we have $m(r, f^{(q)}/f^{(p)}) = S(r, f)$.*

Remark. In Proposition B.5, Corollaries B.6 and B.7, $S(r, f)$ can be replaced by $O\big(\log(rT(r, f))\big)$ as $r \to +\infty$ outside of a possible exceptional set of finite linear measure.

The second main theorem in the Nevanlinna theory, which is another basic result, is stated as follows:

Theorem B.8. *For an arbitrary non-constant meromorphic function $f(z)$ and for an arbitrary number of distinct points $a_1, \dots, a_q \in \mathbb{C}$, $q \in \mathbb{N}$, we have*

$$m(r, f) + \sum_{j=1}^{q} m(r, 1/(f - a_j)) + N(r, 1/f') + N_1(r, f) \leq 2T(r, f) + S(r, f).$$

From this theorem, we immediately obtain

$$\delta(\infty, f) + \vartheta(\infty, w) + \sum_{j=1}^{q} \big(\delta(a_j, f) + \vartheta(a_j, f)\big) \leq 2.$$

This implies that, for each $n \in \mathbb{N}$, the number of the points $a \in \mathbb{C}$ such that $\delta(a, f) \geq 1/n$ (resp. $\vartheta(a, f) \geq 1/n$) does not exceed $2n$, and hence all points with a non-zero deficiency (resp. a non-zero ramification index) constitute a countable set. Thus we have

Corollary B.9. *For an arbitrary non-constant meromorphic function $f(z)$,*

$$\sum_{a \in \widehat{\mathbb{C}}} \big(\delta(a, f) + \vartheta(a, f)\big) \leq 2, \quad \widehat{\mathbb{C}} = \mathbb{C} \cup \{\infty\},$$

where the summation ranges over all points in $\widehat{\mathbb{C}}$ such that $\delta(a, f) > 0$ or $\vartheta(a, f) > 0$.

To eliminate the exceptional set, the following elementary lemma sometimes appears useful:

Lemma B.10. *Let $\phi(r)$ and $\psi(r)$ be monotone increasing functions on $(0, +\infty)$ satisfying $\phi(r) \leq \psi(r)$ outside of an exceptional set of finite linear measure. Given $\alpha > 1$, there exists a number $r_0 > 0$ such that $\phi(r) \leq \psi(\alpha r)$ on $(r_0, +\infty)$.*

The following lemma is due to Clunie [1], see also Laine [1]:

Lemma B.11. *Let f be a transcendental meromorphic function such that $f^{p+1} = Q(z, f)$, $p \in \mathbb{N}$, where $Q(z, u)$ is a polynomial in u and its derivatives with meromorphic coefficients $a_\mu(z)$, $\mu \in M$. Suppose that the total degree of $Q(z, u)$ as a polynomial in u and its derivatives does not exceed p. Then*

$$m(r, f) = O\Big(\sum_{\mu \in M} m(r, a_\mu)\Big) + S(r, f). \tag{B.12}$$

Moreover, if $\rho(f) < +\infty$, the error term $S(r, f)$ can be replaced by $O(\log r)$ with no exceptional set.

Proof. For $r > 0$, consider the subinterval defined by

$$I(r) = \big\{\theta \in [0, 2\pi] \mid |f(re^{i\theta})| > 1\big\}.$$

It is easy to see that

$$m(r, f) = \frac{1}{2\pi} \int_{\theta \in I(r)} \log^+ |f(re^{i\theta})| d\theta. \tag{B.13}$$

We write $Q(z, f)$ in the form

$$Q(z, f) = \sum_{\mu \in M} a_\mu(z) f^{i_0(\mu)} (f')^{i_1(\mu)} \dots (f^{(l)})^{i_l(\mu)},$$
$$l, i_j(\mu) \in \mathbb{N} \cup \{0\} \ (j = 0, 1, \dots, l), \quad \iota(\mu) = i_0(\mu) + i_1(\mu) + \dots + i_l(\mu) \leq p.$$

Then,

$$f = \sum_{\mu \in M} a_\mu(z) f^{\iota(\mu) - p} (f'/f)^{i_1(\mu)} \dots (f^{(l)}/f)^{i_l(\mu)}.$$

Hence, for $r > 0$ and for $\theta \in I(r)$,

$$\log^+ |f(re^{i\theta})|$$
$$\leq \sum_{\mu \in M} \Big(\log^+ |a_\mu(re^{i\theta})| + p \sum_{j=1}^{l} \log^+ \big|f^{(j)}(re^{i\theta})/f(re^{i\theta})\big|\Big) + O(1).$$

Substituting this into (B.13), we get

$$m(r, f) \ll \sum_{\mu \in M} m(r, a_\mu) + \sum_{j=1}^{l} m(r, f^{(j)}/f). \tag{B.14}$$

By Corollary B.7, we obtain (B.12), which completes the proof. □

The following lemma, due to Mohon'ko and Mohon'ko [1], see also Laine [1], gives an estimate for the proximity function of the reciprocal of $f(z) - c$:

Lemma B.12. *Let $F(z, u)$ be a polynomial in u and its derivatives with meromorphic coefficients $b_\kappa(z)$, $\kappa \in K$. Suppose that f is a transcendental meromorphic function satisfying $F(z, f) = 0$, and that c is a complex number. If $F(z, c) \not\equiv 0$, then*

$$m(r, 1/(f - c)) = O\Big(\sum_{\kappa \in K} T(r, b_\kappa)\Big) + S(r, f). \tag{B.15}$$

Moreover, if $\rho(f) < +\infty$, the error term $S(r, f)$ can be replaced by $O(\log r)$ with no exceptional set.

Proof. Put $g = f - c$. Then $F(z, g + c) = 0$, namely

$$G(z, g) \equiv F(z, g + c) - F(z, c) = -F(z, c). \tag{B.16}$$

Since $G(z, 0) \equiv 0$, we may write (B.16) in the form

$$G(z, g) = \sum_{\lambda \in \Lambda} B_\lambda(z) g^{i_0(\lambda)} (g')^{i_1(\lambda)} \dots (g^{(l)})^{i_l(\lambda)} = -F(z, c) \not\equiv 0, \tag{B.17}$$

$$l, i_j(\lambda) \in \mathbb{N} \cup \{0\}\ (j = 0, 1, \dots, l), \quad i_0(\lambda) + i_1(\lambda) + \dots + i_l(\lambda) \geq 1, \tag{B.18}$$

where $B_\lambda(z)$ ($\lambda \in \Lambda$) and $-F(z, c)$ are linear combinations of $b_\kappa(z)$, $\kappa \in K$, with complex coefficients. For $r > 0$, consider the subinterval

$$J(r) = \big\{\theta \in [0, 2\pi] \mid |g(re^{i\theta})| < 1\big\}.$$

From (B.16) and (B.18), we derive that, for $\theta \in J(r)$,

$$\begin{aligned} \log^+ |1/g(re^{i\theta})| &\leq \log^+ |1/F(re^{i\theta}, c)| \\ &+ \sum_{\lambda \in \Lambda} \Big(\log^+ |B_\lambda(re^{i\theta})| + \sum_{j=1}^{l} i_j(\lambda) \log^+ \big|g^{(j)}(re^{i\theta})/g(re^{i\theta})\big| \Big) + O(1). \end{aligned} \tag{B.19}$$

Observe that

$$m(r, 1/g) = \frac{1}{2\pi} \int_{\theta \in J(r)} \log^+ |1/g(re^{i\theta})| d\theta,$$

and that

$$m(r, 1/F(z,c)) \leq T(r, F(z,c)) \ll \sum_{\kappa \in K} T(r, b_\kappa).$$

Then, from (B.19), it follows that

$$m(r, 1/g) \ll \sum_{\kappa \in K} T(r, b_\kappa) + \sum_{j=1}^{l} m(r, g^{(j)}/g). \tag{B.20}$$

Combining (B.20) with Proposition B.5, we obtain (B.15). □

Remark. In Lemma B.11 or B.12, $S(r, f)$ can be replaced by $O\big(\log(rT(r, f))\big)$ as $r \to +\infty$ outside of a possible exceptional set of finite linear measure.

We also give here another version of the error term in Lemma B.11 or Lemma B.12 for a special class of meromorphic functions.

Lemma B.13. *Suppose that f is a transcendental meromorphic function such that $\log T(r, f) = O(r)$. Then $m(r, f^{(k)}/f) = O(r)$ and $\log T(r, f^{(k)}) = O(r)$ with no exceptional set for every $k \in \mathbb{N}$. Furthermore, the error term $S(r, f)$ in Lemma* B.11 *or Lemma* B.12 *can be replaced by $O(r)$ with no exceptional set.*

Proof. It is not difficult to see that $m(r, f'/f) \ll \log T(2r, f) + \log r = O(r)$ with no exceptional set. By the same argument as in the proof of Proposition B.5, we can inductively show that $m(r, f^{(k)}/f) = O(r)$ with $\log T(r, f^{(k)}) = O(r)$ for every $k \in \mathbb{N}$. From this estimate the conclusion immediately follows. □

We close this appendix by giving some results from Nevanlinna theory needed in our considerations related to discrete Painlevé equations in Chapter 10. Basically, these results are elementary. However, their proofs usually don't appear in standard references of value distribution theory.

Lemma B.14. *Given $\varepsilon > 0$, $c \in \mathbb{C}$ and a meromorphic function f, then*

$$T(r, f(z+c)) \leq (1+\varepsilon)T(r+|c|, f(z)) + M$$

for all $r \geq r_c$, for some constant $M = M(c, \varepsilon)$.

Proof. This follows by a slight modification of the proof of Lemma 1 in Ablowitz, Halburd and Herbst [1]. □

Proposition B.15. *Given three distinct meromorphic functions f, g, h,*

$$T(r, fg + gh + hf) \leq T(r, f) + T(r, g) + T(r, h) + O(1).$$

Proof. Looking at pole multiplicities, summing over $|z| \leq r$ and integrating logarithmically, we immediately get

$$N(r, fg + gh + hf) \leq N(r, f) + N(r, g) + N(r, h).$$

To estimate $\log^+ |fg + gh + hf|$, we first observe that

$$\log^+ |fg + gh + hf| \leq \log^+ 3 = \log^+ |f| + \log^+ |g| + \log^+ |h| + \log 3$$

whenever $|f| \leq 1, |g| \leq 1, |h| \leq 1$. If, on the other hand, all of f, g, h are > 1 by modulus, then

$$\begin{aligned}\log^+ |fg + gh + hf| &= \log^+ \left(|fgh| \left(\frac{1}{|f|} + \frac{1}{|g|} + \frac{1}{|h|} \right) \right) \\ &\leq \log^+ |f| + \log^+ |g| + \log^+ |h| + \log 3.\end{aligned}$$

If one of f, g, h only is ≤ 1 by modulus, say f, then we have

$$\begin{aligned}\log^+ |fg + gh + hf| &\leq \log^+ (|g| + |gh| + |h|) \leq \log^+ (3|gh|) \\ &\leq \log^+ |f| + \log^+ |g| + \log^+ |h| + \log 3.\end{aligned}$$

Finally, if exactly two of f, g, h are ≤ 1 by modulus, say f and g, then

$$\begin{aligned}\log^+ |fg + gh + hf| &\leq \log^+ (1 + 2|h|) \leq \log^+ (3|h|) \\ &\leq \log^+ |f| + \log^+ |g| + \log^+ |h| + \log 3.\end{aligned}$$

Therefore, by integration,

$$m(r, fg + gh + hf) \leq m(r, f) + m(r, g) + m(r, h) + \log 3,$$

and the assertion follows. □

A slight modification of the proof above immediately results in

Theorem B.16. *Given distinct meromorphic functions $f_1, \ldots f_n$, let $\{J\}$ denote the collection of all non-empty subsets of $\{1, \ldots, n\}$, and suppose that $\alpha_J \in \mathbb{C}$ for each $J \in \{J\}$. Then*

$$T\Big(r, \sum_J \alpha_J \Big(\prod_{j \in J} f_j \Big)\Big) \leq \sum_{k=1}^{n} T(r, f_k) + O(1).$$

Remark. It is a straightforward modification to obtain the corresponding estimate as in the preceding theorem, if the constant coefficients α_J are replaced by small meromorphic functions $\alpha_J(z)$.

Theorem B.17. *Let f be a meromorphic function and*

$$R(z, f) = \frac{a_0(z) + a_1(z)f + \cdots + a_p(z)f^p}{b_0(z) + b_1(z)f + \cdots + b_q(z)f^q}$$

be irreducible rational in f with meromorphic coefficients $a_j(z)$, $b_k(z)$ of small growth $S(r, f)$. Then

$$T\big(r, R(z, f(z))\big) = \max(p, q)T(r, f(z)) + S(r, f).$$

Proof. This is usually known as the Valiron–Mohon'ko theorem. For a proof, see Laine [1], Theorem 2.2.5. □

Bibliography

Ablowitz, M. J. and Clarkson, P. A.

[1] Solitons, nonlinear evolution equations and inverse scattering, London Math. Soc. Lecture Note Ser. 149, Cambridge University Press, Cambridge, 1991.

Ablowitz, M. J., Halburd, R. and Herbst, B.

[1] On the extension of the Painlevé property to difference equations, Nonlinearity 13 (2000), 889–905.

Ablowitz, M. J., Ramani, A. and Segur, H.

[1] Nonlinear evolution equations and ordinary differential equations of Painlevé type, Lett. Nuovo Cimento (2) 23 (1978), 333–338.

[2] A connection between nonlinear evolution equations and ordinary differential equations of P-type, I, J. Math. Phys. 21 (1980), 715–721.

[3] A connection between nonlinear evolution equations and ordinary differential equations of P-type, II, J. Math. Phys. 21 (1980), 1006–1015.

Ablowitz, M. J. and Segur, H.

[1] Exact linearization of a Painlevé transcendent, Phys. Rev. Lett. 38 (1977), 1103–1106.

[2] Solitons and the inverse scattering transform, SIAM Stud. Appl. Math. 4, Philadelphia 1981.

Abramowitz, M. and Stegun, I. A.

[1] Handbook of mathematical functions, Dover, New York, 1965.

Adler, V. E.

[1] Nonlinear chains and Painlevé equations, Phys. D 73 (1994), 335–351.

Ahlfors, L.

[1] Complex analysis. Second edition, Mc Graw-Hill Book Co., New York–Toronto–London, 1966.

Airault, H.

[1] Rational solutions of Painlevé equations, Stud. Appl. Math. 61 (1979), 31–53.

Barbašov, B. M., Nesterenko, V. V. and Červjakov, A. M.

[1] Solitons in geometric field theories, Teoret. Mat. Fiz. 40 (1979), 15–27 (in Russian).

Bassom, A. P. and Clarkson, P. A.

[1] New exact solutions of the discrete fourth Painlevé equation, Phys. Lett. A 194 (1994), 358–370.

Bassom, A. P., Clarkson, P. A. and Hicks, A. C.

[1] Bäcklund transformations and solution hierarchies for the fourth Painlevé equation, Stud. Appl. Math. 95 (1995), 1–71.

[2] On the application of solutions of the fourth Painlevé equation to various physically motivated nonlinear partial differential equations, Adv. Differential Equations 1 (1996), 175–198.

Bassom, A. P., Clarkson, P. A., Hicks, A. C. and McLeod, J. B.

[1] Integral equations and exact solutions for the fourth Painlevé equation, Proc. Roy. Soc. London Ser. A 437 (1992), 1899, 1–24.

Bieberbach, L.

[1] Theorie der gewöhnlichen Differentialgleichungen auf funktionentheoretischer Grundlage dargestellt, 2. Auflage, Springer-Verlag, Berlin–New York, 1965.

Boiti, M. and Pempinelli, F.

[1] Nonlinear Schrödinger equation, Bäcklund transformations and Painlevé transcendents, Nuovo Cimento Soc. Ital. Fis. B (11) 59 (1980), 40–58.

Boutroux, P.

[1] Sur quelques propriétés des fonctions entières, Acta Math. 28 (1904), 97–224.

[2] Recherches sur les transcendents de M. Painlevé et l'étude asymptotique des équations différentielles du second order, Ann. Sci. École Norm. Sup. 30 (1913), 255–375, and 31 (1914), 99–159.

Bureau, F. J.

[1] Differential equations with fixed critical points, II, Ann. Mat. Pura Appl. (4) 66 (1964), 1–116.

Calogero, F. and Degasperis, A.

[1] Spectral transform and solitons, Vol I. Tools to solve and investigate nonlinear evolution equations, North-Holland Publishing Co., Amsterdam–New York, 1982.

Calvert, G. and Woodhouse, N. M. J.

[1] Painlevé transcendants and Einstein's equation, Classical Quantum Gravity 13 (1996), L33–L39.

Clarkson, P. A., Joshi, N. and Pickering, A.

[1] Bäcklund transformations for the second Painlevé hierarchy: a modified truncation approach, Inverse Problems 15 (1999), 175–187.

Clarkson, P. A. and Kruskal, M. D.

[1] New similarity reductions of the Boussinesq equation, J. Math. Phys. 30 (1989), 2201–2213.

Clarkson, P. A. and Mansfield, E. L.

[1] On a shallow water wave equation, Nonlinearity 7 (1994), 975–1000.

Clarkson, P. A., Mansfield, E. L. and Milne, A. E.

[1] Symmetries and exact solutions of a (2 + 1)-dimensional sine-Gordon system, Philos. Trans. Roy. Soc. London Ser. A 354 (1996), 1807–1835.

Clarkson, P. A. and Winternitz, P.

[1] Symmetry reduction and exact solutions of nonlinear partial differential equations, in: The Painlevé property, CRM Ser. Math. Phys., Springer-Verlag, New York, 1999, 591–660.

Clunie, J.

[1] On integral and meromorphic functions, J. London Math. Soc. 37 (1962), 17–27.

Conte, R. and Musette, M.

[1] First-degree birational transformations of the Painlevé equations and their contiguity relations, J. Phys. A 34 (2001), 10507–10522.

Cosgrove, C. M.

[1] All binomial-type Painlevé equations of the second order and degree three or higher, Stud. Appl. Math. 90 (1993), 119–187.

Davis, H., Scott, W., Springer, G. and Resch, D.

[1] Studies in differential equations, Northwestern University Press, Evanston, 1956.

Debnath, L.

[1] Nonlinear partial differential equations for scientists and ehgineers, Birkhäuser, Boston–Basel–Berlin, 1997.

Dubrovin, B. and Mazzocco, M.

[1] Monodromy of certain Painlevé IV transcendents and reflection groups, Invent. Math. 141 (2000), 55–147.

Erëmenko, A. È.

[1] Meromorphic solutions of algebraic differential equations, Uspekhi Mat. Nauk 37 (4) (1982), 53–82 (in Russian).

Erugin, N. P.

[1] On the theory of the first Painlevé equation, Dokl. Acad. Nauk. BSSR 2 (1958), 3–6 (in Russian).

Flaschka, H. and Newell, A. C.

[1] Monodromy- and spectrum-preserving deformations, I, Comm. Math. Phys. 76 (1980), 65–116.

Fokas, A. S. and Ablowitz, M. J.

[1] Linearization of the Korteweg-de Vries and Painlevé II equations, Phys. Rev. Lett. 47 (1981), 1096–1100.

[2] On a unified approach to transformations and elementary solutions of Painlevé equations, J. Math. Phys. 23 (1982), 2033–2042.

Fokas, A. S., Grammaticos B. and Ramani A.

[1] From continuous to discrete Painlevé equations, J. Math. Anal. Appl. 180 (1993), 342–360.

Fokas, A. S. and Its, A. R.

[1] The isomonodromy method and the Painlevé equations, in: Important developments in soliton theory, Springer Ser. Nonlinear Dynamics, Springer-Verlag, Berlin, 1993, 99–122.

Fokas, A. S., Mugan, U. and Ablowitz, M. J.

[1] A method of linearization for Painlevé equations: Painlevé IV, V, Phys. D 30 (1988), 247–283.

Fokas, A. S., Yorsos, Y. C.

[1] The transformation properties of the sixth Painlevé equation and one-parameter families of solutions, Lett. Nuovo Cimento 30 (1981), 539–544.

Fuchs, R.

[1] Über lineare homogene Differentialgleichungen zweiter Ordnung mit drei im Endlichen gelegenen wesentlich singulären Stellen, Math. Ann. 63 (1907), 301–321.

Fukutani, S., Okamoto, K. and Umemura, H.

[1] Special polynomials and the Hirota bilinear relations of the second and the fourth Painlevé equations, Nagoya Math. J. 159 (2000), 179–200.

Gambier, B.

[1] Sur les équations différentielles de second ordre et du premier degré dont l'intégrale générale est à points critiques fixes, Acta Math. 33 (1910), 1–55.

Gardner, C. S., Greene, J. M., Kruskal, M. D. and Miura, M. R.

[1] Method for solving the Korteweg-de Vries equation, Phys. Rev. Lett. 19 (1967), 1095–1097.

Garnier, R.

[1] Sur des équations différentielles du troisième ordre dont l'intégrale générale est uniforme et sur une classe d'équations nouvelles d'ordre supérieur dont l'intégrale générale a ses point critiques fixes, Ann. Sci. École Norm. Sup. 29 (1912), 1–126.

[2] Étude de l'intégrale générale de l'équation (VI) de M. Painlevé dans le voisinage de ses singularités tanscendents, Ann. Sci. École Norm. Sup. 34 (1917), 239–253.

[3] Solution du problème de Riemann pour les systèmes différentielles linéaires du second ordre, Ann. Sci. École Norm. Sup. 43 (1926), 177–307.

[4] Contribution à l'étude des solutions de l'équation (V) de Painlevé, J. Math. Pures Appl. 46 (1967), 353–412.

Gibbon, J. D., Newell, A. C., Tabor, M. and Zeng, Y. B.

[1] Lax pairs, Bäcklund transformations and special solutions for ordinary differential equations, Nonlinearity 1 (1988), 481–490.

Golubew, W.

[1] To the theory of the Painlevé equations, Acta Math. 28 (1912), 323–249 (in Russian).

[2] Lectures on the analytic theory of differential equations, Gosudarstv. Izdat. Tehn.-Teor. Lit., Moscow-Leningrad, 1950 (in Russian).

Grammaticos, B., Nijhoff, F.W. and Ramani, A.

[1] Discrete Painleve Equations, in: The Painlevé property, CRM Ser. Math. Phys., Springer-Verlag, New York, 1999, 413–516.

Grammaticos, B. and Ramani, A.

[1] On a novel q-discrete analogue of the Painlevé VI equation, Phys. Lett. A 257 (1999), 288–292.

Grammaticos, B., Ramani, A. and Papageorgiou, V.

[1] Do integrable mappings have the Painlevé property?, Phys. Rev. Letters 67 (1991), 1825–1828.

Grammaticos, B., Tamizhmani, T., Ramani, A. and Tamizhmani, K.

[1] Growth and integrability in discrete systems, J. Phys. A 34 (2001), 3811–3821.

Gromak, V. I.

[1] The solutions of the third Painlevé equation, Differ. Uravn. 9 (1973), 2082–2083; Differential Equations 9 (1973), 1599–1600.

[2] On the theory of Painlevé's equations, Differ. Uravn. 11 (1975), 373–376; Differential Equations 11 (1975), 285–287.

[3] The solutions of the Painlevé's fifth equation, Differ. Uravn. 12 (1976), 740–742; Differential Equations 12 (1976), 519–521.

[4] Single-parameter families of solutions of Painlevé equations, Differ. Uravn. 14 (1978), 2131–2135; Differential Equations 14 (1978), 1510–1513.

[5] Algebraic solutions of the third Painlevé equation, Dokl. Akad. Nauk BSSR 23 (1979), 499–502 (in Russian).

[6] The sine-Gordon equation and the third Painlevé equation, Differ. Uravn. 18 (1982), 1984–1986 (in Russian).

[7] The two-dimensional sine-Gordon equation and the third Painlevé equation, Dokl. Akad. Nauk BSSR 27 (1983), 872–874.

[8] Nonlinear evolution equations and equations of P-type, Differ. Uravn. 20 (1984), 2042–2048; Differential Equations 20 (1984), 1435–1439.

[9] Reducibility of the Painlevé equations, Differ. Uravn. 20 (1984), 1674–1683; Differential Equations 20 (1984), 1191–1198.

[10] On the theory of the fourth Painlevé equation, Differ. Uravn. 23 (1987), 760–768; Differential Equations 23 (1987), 506–513.

[11] Autotransformations of the Painlevé equations, Dokl. Akad. Nauk BSSR 32 (1988), 395–398 (in Russian).

[12] On the transcendence of the Painlevé equations, Differ. Uravn. 32 (1996), 154–160; Differential Equations 32 (1996), 156–162.

[13] On a general representation of solutions of the Painlevé equations, Differ. Uravn. 33 (1997), 614–617; Differential Equations 33 (1997), 617–620.

[14] Bäcklund transformations of Painlevé equations and their applications, in: The Painlevé property, CRM Ser. Math. Phys., Springer-Verlag, New York, 1999, 687–734.

[15] Discrete Painlevé equations of the higher order, Trudy Inst. Mat. NAN Belarus 6 (2000), 65–74 (in Russian).

Gromak, V. I. and Lukashevich, N. A.

[1] Analytic properties of solutions of Painlevé equations, Universitetskoe, Minsk, 1990 (in Russian).

Gromak, V. I. and Prokasheva, V. A.

[1] Special classes of solutions of the sixth Painlevé equation, Differ. Uravn. 20 (1984), 1438–1439 (in Russian).

Gromak, V. I. and Tsegel'nik, V. V.

[1] Nonlinear two-dimensional field theory models and Painlevé equations, Teoret. Mat. Fiz. 55 (1983), 189–196; Theoret. and Math. Phys. 55 (1983), 440–445.

[2] A system of 3-wave resonant interaction and equations of P-type, Teoret. Mat. Fiz. 78 (1989), 22–34; Theoret. and Math. Phys. 78 (1989), 14–23.

[3] On solutions of the fourth Painlevé equation and its applications, Differ. Uravn. 27 (1991), 1303–1312; Differential Equations 27 (1991), 910–918.

[4] On the properties of Bäcklund transformations of Painlevé equations, Differ. Uravn. 32 (1996), 1018–1023; Differential Equations 32 (1996), 1024–1029.

Gundersen, G.

[1] On the real zeros of solutions of $f'' + A(z)f = 0$ where $A(z)$ is entire, Ann. Acad. Sci. Fenn. Ser. A I Math. 11 (1986), 275–294.

Gundersen, G. and Steinbart, E.

[1] A generalization of the Airy integral for $f'' + z^n f = 0$, Trans. Amer. Math. Soc. 337 (1993), 737–755.

Hahn, L. and Epstein, B.

[1] Classical complex analysis, Jones and Bartlett Publishers, Boston–London–Singapore, 1996.

Hayman, W.

[1] Meromorphic functions, Clarendon Press, Oxford, 1964.

Heittokangas, J., Korhonen, R., Laine, I., Rieppo J. and Tohge, K.

[1] Complex difference equations of Malmquist type, Comput. Math. Function Theory 1 (2001), 27–39.

Herold, H.

[1] Differentialgleichungen im Komplexen, Vandenhoeck & Ruprecht, Göttingen, 1975.

Hille, E.

[1] Lectures on ordinary differential equations, Addison-Wesley, Reading–London–Don Mills, 1969.

[2] Ordinary differential equations in the complex domain, Wiley-Interscience [John Wiley & Sons], New York–London–Sydney, 1976.

Hinkkanen, A. and Laine, I.

[1] Solutions of the first and second Painlevé equations are meromorphic, J. Anal. Math. 79 (1999), 345–377.

[2] Solutions of a modified third Painlevé equation are meromorphic, J. Anal. Math. 85 (2001), 323–337.

[3] Solutions of a modified fifth Painlevé equation are meromorphic, Report Univ. Jyväskylä 83 (2001), 133–146.

[4] The meromorphic nature of the sixth Painlevé transcendents, to appear.

Horn, J. and Wittich, H.

[1] Gewöhnliche Differentialgleichungen, Walter de Gruyter, Berlin, 1960.

Hotzel, R.

[1] Algebraische Differentialgleichungen mit zulässigen Lösungen: ein algebraischer Ansatz, M.Sc.-thesis, Aachen 1994.

Hukuhara. M.

[1] Meromorfeo de la solvo de (F) $y'' = 6y^2 + x$, unpublished lecture notes.

[2] La ekvacio (VI) de Painlevé-Gambier, unpublished lecture notes.

Ince, E. L.

[1] Ordinary differential equations, Dover Publications, New York, 1944.

Its, A. R., Fokas, A. S. and Kapaev, A. A.

[1] On the asymptotic analysis of the Painlevé equations via the isomonodromy method, Nonlinearity 7 (1994), 1291–1325.

Its, A. R. and Novokshenov, V. Y.

[1] The isomonodromic deformation method in the theory of Painlevé equations, Lecture Notes in Math. 1191, Springer-Verlag, Berlin, 1986.

Iwasaki, K., Kimura, H., Shimomura, S. and Yoshida, M.

[1] From Gauss to Painlevé. A modern theory of special functions, Aspects of Mathematics E16, Friedr. Vieweg & Sohn, Braunschweig, 1991.

Jank, G. and Volkmann, L.

[1] Einführung in die Theorie der ganzen und meromorphen Funktionen mit Anwendungen auf Differentialgleichungen, Birkhäuser, Basel–Boston, 1985.

Jimbo, M.

[1] An inverse scattering problem and the fifth Painlevé transcendent, Prog. Theoret. Phys. 61 (1979), 359–360.

Jimbo, M. and Miwa, T.

[1] Monodromy preserving deformation of linear ordinary differential equations with rational coefficients, II, Phys. D 2 (1981), 407–448.

[2] Monodromy preserving deformation of linear ordinary differential equations with rational coefficients, III, Phys. D 4 (1981/82), 26–46.

Jimbo, M., Miwa, T. and Ueno, K.

[1] Monodromy preserving deformation of linear ordinary differential equations with rational coefficients, I. General theory and τ-function, Phys. D 2 (1981), 306–352.

Joshi, N. and Kitaev, A. V.

[1] On Boutroux's tritronquée solutions of the first Painlevé equation, Stud. Appl. Math. 107 (2001), 253–291.

Joshi, N. and Kruskal, M. D.

[1] A direct proof that solutions of the six Painlevé equations have no movable singularities except poles, Stud. Appl. Math. 93 (1994), 187–207.

Kametaka, Y.

[1] On poles of the rational solution of the Toda equation of Painlevé-II type, Proc. Japan. Acad. Ser. A Math. Sci. 59 (1983), 358–360.

[2] On poles of the rational solution of the Toda equation of Painlevé-IV type. Proc. Japan. Acad. Ser. A Math. Sci. 59 (1983), 453-455.

Kawamoto, S.

[1] The nonlinear Schrödinger-type equations reducible to the second Painlevé equation, J. Phys. Soc. Japan 52 (1983), 2613–2616.

Kießling, H.

[1] Zur Werteverteilung der Lösungen algebraischer Differentialgleichungen, Ph.D.-thesis, Berlin, 1996.

Kimura, H.

[1] The construction of a general solution of a Hamiltonian system with regular type singularity and its application to Painlevé equations, Ann. Mat. Pur. Appl. (4) 134 (1983), 363–392.

Kimura, H. and Okamoto, K.

[1] On the polynomial Hamiltonian structure of the Garnier systems, J. Math. Pures Appl. (9) 63 (1984), 129–146.

Kitaev, A. V.

[1] Self-similar solutions of a modified Schrödinger equation, Teoret. Mat. Fiz. 64 (1985), 347–369 (in Russian).

[2] On similarity reduction of the three-wave system to the Painlevé equations, J. Phys. A 23 (1990), 3543–3553.

[3] Elliptic asymptotics of the first and second Painlevé transcendents, Russian Math. Surveys 49 (1994), 81–150.

Kitaev, A. V. and Korotkin, D. A.

[1] On solutions of the Schlesinger equations in terms of Θ-functions, Internat. Math. Res. Notices 97 (1998), 877–905.

Kitaev, A. V., Law, C. K. and McLeod, J. B.

[1] Rational solutions of the fifth Painlevé equation, Differential Integral Equations 7 (1994), 967–1000.

Korhonen, R.

[1] Meromorphic solutions of algebraic differential and difference equations with deficient slowly moving targets, Ann. Acad. Sci. Fenn. Math. Diss. 129 (2002).

Kowalevski, S.

[1] Sur le problème de la rotation d'un corps solide autour d'un point fixe, Acta Math. 12 (1889), 177–232.

[2] Sur une propriété du système d'équations différentielles qui définit la rotation d'un corps solide auteur d'un point fixe, Acta Math. 14 (1890), 81–93.

Kruskal, M. D. and Clarkson, P. A.

[1] The Painlevé-Kowalevski and poly-Painlevé tests for integrability, Stud. Appl. Math. 86 (1992), 87–165.

Kudryashov, N. A.

[1] The first and second Painlevé equations of higher order and some relations between them, Phys. Lett. 224A (1997), 353–360.

Laine, I.

[1] Nevanlinna theory and complex differential equations. Walter de Gruyter, Berlin–New York, 1993.

Lakshmanan, M. and Kaliappan, P.

[1] Lie transformations, nonlinear evolution equations, and Painlevé forms, J. Math. Phys. 24 (1983), 795–806.

Lamb, G. L.

[1] Bäcklund transformations at the turn of the century, in: Bäcklund transformations (R. Miura, ed.), Lecture Notes in Math. 515 (1976), Springer-Verlag, Berlin–Heidelberg–New York, 69–79.

Langley, J.

[1] Private communication.

Lax, P. D.

[1] Integrals of nonlinear equations of evolution and solitary waves, Comm. Pure Appl. Math. 21 (1968), 467–490.

[2] Almost periodic solutions of the KdV equation, SIAM Rev. 18 (1976), 351–375.

Léauté, B. and Marcilhacy, G.

[1] Separable solutions of a nonlinear complex field Lorenz-invariant equation, Nuovo Cimento Soc. Ital. Fis. B 68 (1982), 147–160.

[2] A new transcendent solution of Einstein's equations, Phys. Lett. A 87 (1981/1982), 159–161.

Levi, D. and Winternitz, P., Eds

[1] Painlevé transcendents. Their asymptotics and physical applications, Plenum Press, New York, 1992.

Li, Y. and He, Y.

[1] On analytic properties of higher analogs of the second Painlevé equation, J. Math. Phys. 43 (2002), 1106–1115.

Ludlow, D. K. and Clarkson, P. A.

[1] Symmetry reductions and exact solutions for a generalised Boussinesq equation, in: Applications of analytic and geometric methods to nonlinear differential equations (Exeter, 1992), Kluwer Acad. Publ., Dortrecht, 1993, 415–430.

Lukashevich, N. A.

[1] Theory of the fourth Painlevé equation, Differ. Uravn. 3 (1967), 771–780; Differential Equations 3 (1967), 395–399.

[2] On the theory of the third Painlevé equation, Differ. Uravn. 3 (1967), 1913–1923; Differential Equations 3 (1967), 994–999.

[3] The second Painlevé equation, Differ. Uravn. 7 (1971), 1124-1125; Differential Equations 7 (1971), 853–854.

[4] On the theory of Painlevé's sixth equation, Differ. Uravn. 8 (1972), 1404–1408 (in Russian).

Lukashevich, N. A. and Yablonski, A. I.

[1] On a class of solutions of the sixth Painlevé equations, Differ. Uravn. 3, (1967), 520–523; Differential Equations 3 (1967), 264–267.

Malmquist J.

[1] Sur les equations differentielles du second ordre, don't l'integral generale a ses points critiques fixes, Ark. för Mat., Astron. och Fys. 17, No. 8 (1923).

Manin, Yu. I.

[1] Sixth Painlevé equation, universal elliptic curve, and mirror of P^2, in: Geometry of differential equations, Amer. Math. Soc. Transl. Ser. 2 186, Amer. Math. Soc., Providence, RI, 1998, 131–151.

Mazzocco, M.

[1] Rational solutions of the Painlevé VI equation, J. Phys. A 34 (2001), 2281–2294.

McLeod, J. B. and Olver, P. J.

[1] The connection between partial differential equations soluble by inverse scattering and ordinary differential equations of Painlevé type, SIAM J. Math. Anal. 14 (1983), 488–506.

Meschkowski, H.

[1] Differenzengleichungen, Vandenhoeck & Ruprecht, Göttingen, 1959.

Milne, A. E., Clarkson P. A., Bassom A. P.

[1] Bäcklund transformations and solution hierarchics for the third Painlevé Equation, Stud. Appl. Math. 98 (1997), 139–194.

Mohon'ko, A. and Mohon'ko, V.

[1] Estimates for the Nevanlinna characteristics of some classes of meromorphic functions and their applications to differential equations, Sibirsk Mat. Zh. 15 (1974), 1305–1322 (in Russian).

Mues, E. and Redheffer, R.

[1] On the growth of logarithmic derivatives, J. London Math. Soc. 8 (1974), 412–425.

Murata, Y.

[1] Rational solutions of the second and the fourth Painlevé equations, Funkcial. Ekvac. 28 (1985), 1–32.

[2] Classical solutions of the third Painlevé equations, Nagoya Math. J. 139 (1995), 37–65.

Musette, M.

[1] Painlevé analysis for Nonlinear Partial Differential Equations, in: The Painlevé property, CRM Ser. Math. Phys., Springer-Verlag, New York, 1999, 517–572.

Nishioka, K.

[1] A note on the transcendency of Painlevé's first transcendent, Nagoya Math. J. 109 (1988), 63–67.

[2] Irreducibility of the linear differential equation attached to Painlevé's first equation, Tokyo J. Math. 16 (1993), 171–177.

[3] The first equation of Painlevé–differential algebraic aspects, in: Proc. of the 5th International Colloq. on Differential Equations, Vol. 2 (Plovdiv, 1994), Sci. Cult. Technol. Publ., Singapore, 1995, 123–129.

[4] Linear differential equation attached to Painlevé's first equation, Funkcial. Ekvac. 38 (1995), 277–282.

Nishitani, T. and Tajiri, M.

[1] On similary solutions of the Boussinesq equation, Phys. Lett. A 89 (1982), 379–380.

Nörlund, N.

[1] Vorlesungen über Differenzenrechnung, Springer-Verlag, Berlin, 1924.

Noumi, M., Okada, M., Okamoto, K. and Umemura, H.

[1] Special polynomials associated with the Painlevé equations, II, in: Integrable systems and algebraic geometry (Kobe/Kyoto, 1997), World Sci. Publishing, River Edge, NJ, 1998, 349–372.

Noumi, M. and Okamoto, K.

[1] Irreducibility of the second and the fourth Painlevé equations, Funkcial Ekvac. 40 (1997), 139–163.

Noumi, M. and Yamada, Y.

[1] Higher order Painlevé equations of type $A_l^{(1)}$, Funkcial. Ekvac. 41 (1998), 483–503.

[2] Affine Weyl groups, discrete dynamical systems and Painlevé equations, Comm. Math. Phys. 199 (1998), 281–295.

[3] Umemura polynomials for the Painlevé V equation, Phys. Lett. A 247 (1998), 65–69.

[4] Symmetries in the fourth Painlevé equation and Okamoto polynomials, Nagoya Math. J. 153 (1999), 53–86.

Okamoto, K.

[1] Polynomial Hamiltonians associated with Painlevé equations, I, Proc. Japan Acad. Ser. A Math. Sci. 56 (1980), 264–268.

[2] Polynomial Hamiltonians associated with Painlevé equations, II, Proc. Japan Acad. Ser. A Math. Sci. 56 (1980), 367–371.

[3] On the τ-function of the Painlevé equations, Phys. D 2 (1981), 525–535.

[4] Studies on the Painlevé equations, III. Second and fourth Painlevé equations, P_{II} and P_{IV}, Math. Ann. 275 (1986), 221–255.

[5] Isomonodromic deformation and Painlevé equations, and the Garnier system, J. Fac. Sci. Univ. Tokyo Sect. IA, Math 33 (1986), 575–618.

[6] Studies on the Painlevé equations, I. Sixth Painlevé equation P_{VI}, Ann. Mat. Pura Appl. (4) 146 (1987), 337–381.

[7] Studies on the Painlevé equations, II. Fifth Painlevé equation P_{V}, Japan J. Math. (N.S.) 13 (1987), 47–76.

[8] Studies on the Painlevé equations, IV. Third Painlevé equation P_{III}, Funkcial. Ekvac. 30 (1987), 305–332.

[9] Algebraic relations among six adjacent τ-functions related to the fourth Painlevé system, Kyushu J. Math. 50 (1996), 513–532.

[10] The Hamiltonian associated the Painlevé Equations, in: The Painlevé property, CRM Ser. Math. Phys., Springer-Verlag, New York, 1999, 735–787.

Okamoto, K. and Takano, K.

[1] The proof of the Painlevé property by Masao Hukuhara (Esperanto), Funkcial. Ekvac. 44 (2001), 201–217.

Painlevé, P.

[1] Leçons sur la théorie analytique des équations différentielles, profesées à Stockholm, 1895, Hermann, Paris, 1897.

[2] Mémoire sur les équations différentielles dont l'intégrale générale est uniforme, Bull. Soc. Math. France 28 (1900), 201–261.

[3] Sur les équations différentielles du second ordre et d'ordre supérieur dont l'intégrale générale est uniforme, Acta Math. 25 (1902), 1–85.

[4] Oeuvres de Paul Painlevé, Tome I, II, III, Éditions du Centre National de la Recherche Scientifique, Paris, 1973–1975.

Picard, E.

[1] Mémoire sur la théorie des fonctions algébriques de deux variables, J. de Math. (4)5 (1889), 135–319.

Quispel, G. R. W., Nijhoff, F. W. and Capel, H. W.

[1] Linearization of the Boussinesq equation and the modified Boussinesq equation, Phys. Lett. A 91 (1982), 143–145.

Rainville, E.

[1] Special Functions, Chelsea Publ. Co., Bronx, New York, 1971.

Ramani, A., Grammaticos, B. and Hietarinta, J.

[1] Discrete versions of the Painlevé equations, Phys. Rev. Letters 67 (1991), 1829–1832.

Ramani, A., Grammaticos, B., Tamizhmani, T. and Tamizhmani, K.

[1] Special function solutions of the discrete Painlevé equations, Comput. Math. Appl. 42 (2001), 603–614.

Rogers, C. and Shadwick, W. F.

[1] Bäcklund transformation and their applications, Academic Press, New York–London, 1982.

Saks, S. and Zygmund, A.

[1] Analytic functions. Third edition, Elsevier Publ. Co., Amsterdam–London–New York; Polish Scientific Publ., Warsaw, 1971.

Salihoğlu, S.

[1] Two-dimensinal O(N) nonlinear σ-model and the fifth Painlevé transcendent, Phys. Lett. B 89 (1980), 367–368.

Schief, W. K.

[1] The Painlevé III, V and VI transcendents as solutions of the Einstein-Weyl equations, Phys. Lett. A 267 (2000), 265–275.

Schubart, H.

[1] Zur Wertverteilung der Painlevéschen Transzendenten, Arch. Math. (Basel) 7 (1956), 284–290.

Schubart, H. and Wittich, H.

[1] Über die Lösungen der ersten beiden Painlevéschen Differentialgleichungen, Math. Z. 66 (1957), 364–370.

Shimomura, S.

[1] On solutions of the fifth Painlevé equation on the positive real axis II, Funkcial. Ekvac. 30 (1987), 203–224.

[2] Value distribution of Painlevé transcendents of the third kind, Complex Variables Theory Appl. 40 (1999), 51–62.

[3] Value distribution of Painlevé transcendents of the fifth kind, Results Math. 38 (2000) 348–361.

[4] Value distribution of Painlevé trancendents of the first and the second kind, J. Anal. Math. 82 (2000) 333–346.

[5] Painlevé property of a degenerate Garnier system of (9/2)-type and of a certain fourth order non-linear ordinary differential equation, Ann. Scuola Norm. Sup. Pisa Cl. Sci. (4) 29 (2000) 1–17.

[6] The first, the second and the fourth Painlevé trancendents are of finite order, Proc. Japan Acad. Ser. A 77 (2001) 42–45.

[7] On deficiencies of small functions for Painlevé trancendents of the fourth kind, Ann. Acad. Sci. Fenn. Ser. A I Math. 27 (2002), 109–120.

[8] Growth of the first, the second and the fourth Painlevé transcendents, to appear in Math. Proc. Cambridge Philos. Soc.

[9] Growth of modified Painlevé transcendents of the fifth and the third kind, to appear in Forum Math.

[10] Lower estimates for the growth of Painlevé transcendents, to appear in Funkcial. Ekvac.

[11] Proofs of the Painlevé property for all Painlevé equations, to appear.

Steinmetz, N.

[1] On the primeness of Painlevé transcendents, Lecture Notes in Pure and Appl. Math. 78 (1982), 119–128.

[2] Zur Wertverteilung der Lösungen der vierten Painlevéschen Differentialgleichung, Math. Z. 181 (1982), 553–561.

[3] Über eine Klasse von Painlevéschen Differentialgleichung, Arch. Math. (Basel) 41 (1983), 261–266.

[4] On Painlevé's equations I, II and IV, J. Anal. Math. 82 (2000), 363–377.

[5] Value distribution of the Painlevé transcendents, Israel J. Math. 128 (2002), 29–52.

Tajiri, M. and Kawamoto, S.

[1] Reduction of the KdV and cylindrical KdV equations to Painlevé equation, J. Phys. Soc. Japan 51 (1982), 1678–1681.

Takano, K.

[1] A 2-parameter family of solutions of Painlevé equation (V) near the point of infinity, Funkcial. Ekvac. 26 (1983), 79–113.

[2] Reduction for Painlevé equations at the fixed singular points of the first kind, Funkcial. Ekvac. 29 (1986), 99–119.

Takasaki, K.

[1] Painlevé-Calogero correspondence revisited, J. Math. Phys. 42 (2001), 1443–1473.

Umemura, H.

[1] On the irreducibility of the first differential equation by Painlevé, in: Algebraic geometry and commutative algebra, Vol. II, Kinokuniya, Tokyo, 1988, 771–789.

[2] Second proof of irreducibility of the first differential equation by Painlevé, Nagoya Math. J. 117 (1990), 125–171.

[3] On the transformation group of the second Painlevé equation, Nagoya Math. J. 157 (2000), 15–46.

Umemura, H. and Watanabe, H.

[1] Solutions of the second and fourth Painlevé equations, I, Nagoya Math. J. 148 (1997), 151–198.

[2] Solutions of the third Painlevé equation, I, Nagoya Math. J. 151 (1998), 1–24.

Vekua, I. N.

[1] Remarks on properties of solution of the equation $\Delta u = -2Ke^u$, Sibirsk Mat. Zh. 1 (1960), 331–342 (in Russian).

Vorob'ev, A. P.

[1] On the rational solutions of the second Painlevé equation, Differ. Uravn. 1 (1965), 79–81; Differential Equations 1 (1965), 58–60.

Walter, W.

[1] Differential and integral inequalities, Springer-Verlag, Berlin–Heidelberg–New York, 1970.

Wasow, W. R.

[1] Asymptotic expansions for ordinary differential equations, John Wiley & Sons, New York–London–Sydney, 1965.

Watanabe, H.

[1] Solutions of the fifth Painlevé equation, I, Hokkaido Math. J. 24 (1995), 231–267.

[2] Birational canonical transformations and classical solutions of the sixth Painlevé equation, Ann. Scuola Norm. Sup. Pisa Cl. Sci. 27 (1998), 379–425.

[3] On the defining variety and birational canonical transformations of the fifth Painlevé equation, Analysis (Munich) 18 (1998), 351–357.

[4] On the defining variety and birational canonical transformations of the fourth Painlevé equation, Funkcial. Ekvac. 42 (1999), 1–7.

Weiss, J.

[1] The Painlevé property for partial differential equations. II. Bäcklund transformations, Lax pairs, and the Schwarzian derivative, J. Math. Phys. 24 (1983), 1405–1413.

[2] On classes of integrable systems and the Painlevé property, J. Math. Phys. 25 (1984), 13–24.

Weiss, J., Tabor, M. and Carnevale, G.

[1] The Painlevé property for partial differential equations, J. Math. Phys. 24 (1983), 522–526.

Wils, P.

[1] Painlevé solutions in general relativity, Classical and Quantum Gravity 6 (1989), 1231- 1241.

Yablonskiĭ, A.I.

[1] On the rational solutions of the second Painlevé equation, Izv. Akad. Nauk. BSSR, 3 (1959), 30–35 (in Russian).

[2] Residues of poles of solutions of Painlevé's second equation, Dokl. Akad. Nauk BSSR 4 (1960), 47–50 (in Russian).

Yoshida, S.

[1] 2-parameter family of solutions for Painlevé equations (I)-(V) at an irregular singular point, Funkcial. Ekvac. 28 (1985), 233–248.

Yuan, W. and Li, Y.

[1] Rational solutions of Painlevé equations, Canad. J. Math. 54 (2002), 648–670.

Zaharov, V., Manakov, S., Novikov, S. and Pitaevsky, L.

[1] Theory of solitons. The method of the inverse problem, Nauka, Moscow, 1980 (in Russian).

Index